3 (Issue 2)

Corus UK Limited
Swinden Technology Centre

Library and Information Services
Moorgate, Rotherham   S60 3AR

NOTES TO BORROWERS

1. This publication should be returned or the loan renewed within 21 days, i.e. by the last date shown below.

2. As this Library is primarily a reference tool for the Technology Centre, borrowed publications should be readily accessible on demand.

3. It is a condition of this loan that, in case of damage or loss, this publication shall be replaced at the cost of the borrower

Name	Address	Date Due	Initial Before Returning
G. Watts		2/9/07	G Jw 26/3/09
Thomas Axe -		13-8-09	

REGRESSION ANALYSIS AND ITS APPLICATION

STATISTICS: Textbooks and Monographs

A SERIES EDITED BY

D. B. OWEN, Coordinating Editor
Department of Statistics
Southern Methodist University
Dallas, Texas

Volume 1: The Generalized Jackknife Statistic, *H. L. Gray and W. R. Schucany*

Volume 2: Multivariate Analysis, *Anant M. Kshirsagar*

Volume 3: Statistics and Society, *Walter T. Federer*

Volume 4: Multivariate Analysis: A Selected and Abstracted Bibliography, 1957-1972, *Kocherlakota Subrahmaniam and Kathleen Subrahmaniam* (out of print)

Volume 5: Design of Experiments: A Realistic Approach, *Virgil L. Anderson and Robert A. McLean*

Volume 6: Statistical and Mathematical Aspects of Pollution Problems, *John W. Pratt*

Volume 7: Introduction to Probability and Statistics (in two parts)
Part I: Probability; Part II: Statistics, *Narayan C. Giri*

Volume 8: Statistical Theory of the Analysis of Experimental Designs, *J. Ogawa*

Volume 9: Statistical Techniques in Simulation (in two parts), *Jack P. C. Kleijnen*

Volume 10: Data Quality Control and Editing, *Joseph I. Naus*

Volume 11: Cost of Living Index Numbers: Practice, Precision, and Theory, *Kali S. Banerjee*

Volume 12: Weighing Designs: For Chemistry, Medicine, Economics, Operations Research, Statistics, *Kali S. Banerjee*

Volume 13: The Search for Oil: Some Statistical Methods and Techniques, *edited by D. B. Owen*

Volume 14: Sample Size Choice: Charts for Experiments with Linear Models, *Robert E. Odeh and Martin Fox*

Volume 15: Statistical Methods for Engineers and Scientists, *Robert M. Bethea, Benjamin S. Duran, and Thomas L. Boullion*

Volume 16: Statistical Quality Control Methods, *Irving W. Burr*

Volume 17: On the History of Statistics and Probability, *edited by D. B. Owen*

Volume 18: Econometrics, *Peter Schmidt*

Volume 19: Sufficient Statistics: Selected Contributions, *Vasant S. Huzurbazar (edited by Anant M. Kshirsagar)*

Volume 20: Handbook of Statistical Distributions, *Jagdish K. Patel, C. H. Kapadia, and D. B. Owen*

Volume 21: Case Studies in Sample Design, *A. C. Rosander*

Volume 22: Pocket Book of Statistical Tables, *compiled by R. E. Odeh, D. B. Owen, Z. W. Birnbaum, and L. Fisher*

Volume 23: The Information in Contingency Tables, *D. V. Gokhale and Solomon Kullback*

Volume 24: Statistical Analysis of Reliability and Life-Testing Models: Theory and Methods, *Lee J. Bain*

Volume 25: Elementary Statistical Quality Control, *Irving W. Burr*

Volume 26: An Introduction to Probability and Statistics Using BASIC, *Richard A. Groeneveld*

Volume 27: Basic Applied Statistics, *B. L. Raktoe and J. J. Hubert*

Volume 28: A Primer in Probability, *Kathleen Subrahmaniam*

Volume 29: Random Processes: A First Look, *R. Syski*

Volume 30: Regression Methods: A Tool for Data Analysis, *Rudolf J. Freund and Paul D. Minton*

Volume 31: Randomization Tests, *Eugene S. Edgington*

Volume 32: Tables for Normal Tolerance Limits, Sampling Plans, and Screening, *Robert E. Odeh and D. B. Owen*

Volume 33: Statistical Computing, *William J. Kennedy, Jr. and James E. Gentle*

Volume 34: Regression Analysis and Its Application: A Data-Oriented Approach, *Richard F. Gunst and Robert L. Mason*

OTHER VOLUMES IN PREPARATION

REGRESSION ANALYSIS AND ITS APPLICATION

A Data-Oriented Approach

RICHARD F. GUNST
Department of Statistics
Southern Methodist University
Dallas, Texas

ROBERT L. MASON
Automotive Research Division
Southwest Research Institute
San Antonio, Texas

MARCEL DEKKER, INC.   New York and Basel

Library of Congress Cataloging in Publication Data

Gunst, Richard F
   Regression analysis and its application.

   (Statistics, textbooks and monographs ; 34)
   Bibliography : p
   Includes index.
   1. Regression analysis. I. Mason, Robert Lee, joint author
   II. Title.
QA278.2.G85   519.5'36   80-18101
ISBN 0-8247-6993-7

MARCEL DEKKER, INC.

270 Madison Avenue, New York, New York 10016

Current printing (last digit):

10 9 8 7 6 5 4 3 2 1

PRINTED IN THE UNITED STATES OF AMERICA

To Ann and Carmen

PREFACE

Regression analysis is considered indispensable as a data analysis technique in a variety of disciplines. In recent years many research articles dealing with new techniques for regression analysis have appeared in the statistical and applied literature but few of these recent advances have appeared in regression textbooks written for data analysts. **Regression Analysis and Its Application: A Data-Oriented Approach** bridges the gap between a purely theoretical coverage of regression analysis and the needs of the data analyst who requires a working knowledge of regression techniques. Data analysts, consultants, graduate and upper-level undergraduate students, faculty members, research scientists, and employees of governmental data-collection agencies comprise only a few of the groups whose research activities can benefit from the material in this book.

The main prerequisites for reading this book are a first course in statistical methods and some college-level mathematics. A first course in statistical methods is required so that summation notation, basic probability distributions (normal, t, chi-square, F), hypothesis testing, and confidence interval estimation are already familiar to the reader. Some mathematical knowledge of algebra, functional relationships [$f(x) = x^2$, x^{-1}, $\ln(x)$], and solving equations is also important to an appreciation of the material covered in the text. Although manipulation of vectors and matrices are essential to the understanding of the topics covered in the last two-thirds of this book, we do not presume that readers have had a course in vector algebra. Rather, we include an introduction to the properties and uses of vectors and matrices in Chapter 4.

Regression Analysis and Its Application: A Data-Oriented Approach contains two features which set it apart from other books on the same subject. As the title indicates, regression analysis is viewed in the general context of data analysis rather than strictly in the classical parametric model formulation. Understanding the implications of how data are collected, restrictions on the data, and the consequences of ignoring these characteristics of regression data are stressed prior to discussions of how to define regression models and they are restressed throughout the remainder of the book. The data base, model specification, estimation of parameters, and inferential procedures are thus seen to be inseparable components of a thorough regression analysis.

A second distinguishing feature of this book is its emphasis on the analysis of real data sets. Throughout every chapter of the text data sets are used to illustrate and justify the procedures being discussed. No contrived examples are used. In addition to several small data sets presented in the text, ten larger data sets are contained in Appendices A and B. All the data sets were selected because they illustrate either actual uses of regression analysis or concepts involved in interpreting the results of such analyses, and because no special background is needed to understand the purposes of the analyses. The data sets are not specifically oriented toward any discipline and are small enough to be implemented for classroom or private use. They are sufficiently complex, however, that they exhibit many of the analytic difficulties likely to be faced in practice.

Turning to the subject-matter content, this text is not written to be a "cookbook" just as it is not written as a theoretical treatise. Some theoretical derivations are necessary for a complete understanding of the proper utilization of regression techniques. Our goal, however, is not to comprehensively cover theoretical derivations and properties of regression estimators but only to employ them when they provide insight to the topics being discussed. For this reason, most chapters contain technical appendices which can be skipped without losing essential understanding of the topical coverage. This blend of a minimum of theoretical derivation in the body of the text and supplemental theory in the chapter appendices enables readers with different mathematical and statistical backgrounds to gain an appreciation of regression analysis that is suited to their backgrounds and needs. On the other hand, the absence of any theoretical derivations too often tends to relegate regression analysis on the stature of a "black box": mystical and mechanical.

The topical coverage of regression analysis is structured into ten chapters, of which Chapters 3 to 8 would be similar to a standard one-semester regression course. Chapters 1 and 2 stress the importance of data-collection and initial model specification prior to discussions of formal regression analysis. Chapter 3 is a comprehensive treatment of single-variable regression, including intercept and no-intercept models, assumptions underlying the use of single-variable models, and estimation of models. Chapter 4 is an introduction to vector and matrix algebra which is intended to be a preparation for the discussion of multiple-variable regression models and estimation in Chapter 5. Inferential techniques for single- and multiple-variable regression models is the subject of Chapter 6. Following a discussion of model assumptions, both tests of hypothesis and interval estimation techniques for regression models are presented in this chapter.

Beginning with Chapter 7, topics of a more specialized nature are covered. Chapter 7 treats residual analysis with a special emphasis on the assessment of model assumptions. Variable selection techniques, including forward selection and backward elimination, is the subject of Chapter 8. The effects of multicollinearities on estimation of model parameters and inference procedures is detailed in Chapter 9. Finally, three of the more popular biased regression estimators are briefly introduced in Chapter 10. Much of the content of these last four chapters is taken from recent statistical literature and reflects popular current trends in regression analysis.

A book of this nature must invariably fail to include topics that are of interest to readers. Our emphasis is on presenting classical regression analysis, updated with many of the newer analysis techniques, and with major attention devoted to the problems and possible solutions associated with multicollinear predictor variables. Nonliner regression, path analysis, simultaneous equation systems, two- and three-stage least squares, random coefficient models, and Bayesian modelling are some of the topics that are beyond the intended scope of this text. To have included treatment of these other topics would have required elimination of some of the classical material or a cutback in the data analysis, both of which we were reluctant to do.

Many individuals have contributed to the successful completion of this effort. Professionally we are deeply indebted to Professor J. T.Webster who kindled our interest in regression analysis and continues to provide advice and stimulating critiques of our joint efforts. A special thanks is also due Dr. I. N. Shimi and the U. S. Air Force Office of Scientific Research for research support over the past several years that led to the development of some of the techniques presented in this book. Similarly, Mr. George Lawrason, Southwest Research Institute, provided staff assistance and personal encouragement without which our efforts would have been exceedingly more difficult.

Early drafts of this manuscript were greatly improved by the suggestions offered by Dr. Alan Agresti, University of Florida, and Dr. Dallas Johnson, Kansas State University. The final version of this text is substantially altered from the first one thanks to the incisive critiques of both these reviewers. Numerous corrections in the text, especially the numerical examples, were made after careful checking by Messrs. Michael Conerly and Tsu-Shung Hua, graduate students at Southern Methodist University.

Typing of this manuscript was painstakingly accomplished by Mrs. Marilyn Reeves and Mrs. Dee Patterson. All authors who have published manuscripts containing a large amount of technical typing know that this is a laborious and often thankless task. Mr. Michael Scofield also contributed substantially to the quality of the manuscript by making initial drafts of all figures.

The final copy of this manuscript was designed and typeset on a word processor by Ms. Billie Jean Ford, an editorial coordinator at Southwest Research Institute. Her many extra hours of labor and effort in setting the text and her congenial spirit throughout our text corrections are most appreciated.

We wish to express our appreciation to the Biometrika Trustees for permission to reproduce Tables C.4, C.5, and C.6. Other acknowledgements for tables and data sets appear in the text.

Finally, we could not conclude these acknowledgements without expressing our deepest appreciation to our families. To both our parents who provided us with a love for learning (and many other gifts) and our wives and children who sacrificed many hours over the last few years without us, we can only offer our heartfelt thanks.

Robert L. Mason
San Antonio, Texas

Richard F. Gunst
Dallas, Texas

CONTENTS

1. **INTRODUCTION**   1
  1.1   DATA COLLECTION   2
    1.1.1   Data Base Limitations   3
    1.1.2   Data-Conditioned Inferences   5
  1.2   REGRESSION ANALYSIS   6
    1.2.1   Linear Regression Models   6
    1.2.2   Regression vs. Correlation   8
  1.3   USES OF REGRESSION ANALYSIS   9
    1.3.1   Prediction   9
    1.3.2   Model Specification   10
    1.3.3   Parameter Estimation   11
  1.4   ABUSES OF REGRESSION ANALYSIS   12
    1.4.1   Extrapolation   12
    1.4.2   Generalization   15
    1.4.3   Causation   17

2. **INITIAL DATA EXPLORATION**   19
  2.1   PRELIMINARY DATA EDITING   23
    2.1.1   Obvious Anomalies   23
    2.1.2   Descriptive Statistics   26
    2.1.3   Graphical Aids   28
  2.2   INITIAL MODEL SPECIFICATION   33
    2.2.1   Categorical Predictor Variables   33
    2.2.2   Interactions   37
    2.2.3   Smoothing   39

2.3   REEXPRESSING VARIABLES   42
     2.3.1   Special Functions   42
     2.3.2   Predictor Variable Transformations   45
     2.3.3   Linearization of the Response   47
     2.3.4   Prediction Versus Model Building   48
EXERCISES   50

3.   SINGLE-VARIABLE LEAST SQUARES   52
 3.1   APPROPRIATENESS   56
     3.1.1   Theoretical Validity   57
     3.1.2   Approximations in Restricted Regions   61
     3.1.3   No-Intercept Models   63
 3.2   LEAST SQUARES PARAMETER ESTIMATION   66
     3.2.1   The Principle of Least Squares   66
     3.2.2   Model Assumptions   71
     3.2.3   Standardization   72
 3.3   ADDITIONAL MEASURES OF FIT   77
     3.3.1   Partitioning Variability   78
     3.3.2   Coefficient of Determinatio   82
     3.3.3   Residual Variability   83
APPENDIX   85
     3.A   Derivation of Least Squares Estimators   85
     3.B   Coefficient Estimates for Standardized Variables   87
     3.C   Partitioning the Total Sum of Squares   88
     3.D   Calculation of the Coefficient of Determination   89
EXERCISES   90

4.   MULTIPLE-VARIABLE PRELIMINARIES   92
 4.1   REVIEW OF MATRIX ALGEBRA   94
     4.1.1   Notation   94
     4.1.2   Vector and Matrix Operations   96
     4.1.3   Model Definition   103
     4.1.4   Latent Roots and Vectors   104
     4.1.5   Rank of a Matrix   106
 4.2   CARE IN MODEL BUILDING   108
     4.2.1   Misspecification Bias   108
     4.2.2   Overspecification Redundancy   110
 4.3   STANDARDIZATION   111
     4.3.1   Benefits   112
     4.3.2   Correlation Form of $X'X$ 114

4.4   MULTICOLLINEARITY   115
  4.4.1   Definition   115
  4.4.2   Detection   118
  4.4.3   Sample vs. Population Multicollinearities   120
APPENDIX   122
  4.A   Determinants and Inverse Matrices   122
  4.B   Determinants and Inverses Using Latent Roots and
        Latent Vectors   124
EXERCISES   126

5. MULTIPLE-VARIABLE LEAST SQUARES   128
  5.1   PARAMETER ESTIMATION   131
    5.1.1   Matrix Algebra Formulation   132
    5.1.2   Fitting by Stages   135
    5.1.3   Orthogonal Predictor Variables   140
    5.1.4   Standardization   142
    5.1.5   No-Intercept Models   148
  5.2   INTERPRETATION OF FITTED MODELS   149
    5.2.1   General Interpretation   150
    5.2.2   Adjustment Computations   152
    5.2.3   Special Case: Indicator Variables   153
  5.3   INITIAL ASSESSMENT OF FIT   154
    5.3.1   Analysis of Variance Table   154
    5.3.2   Error Variance   158
  APPENDIX   160
    5.A   Derivation of Least Squares Estimators   160
    5.B   Algebraic Derivation of Fitting by Stages   161
    5.C   Calculating SSR   164
    5.D   Equivalence of Residual Expressions   165
  EXERCISES   166

6. INFERENCE   167
  6.1   MODEL DEFINITION   169
    6.1.1   Four Key Assumptions   169
    6.1.2   Alternative Assumptions   172
  6.2   ESTIMATOR PROPERTIES   174
    6.2.1   Geometrical Representation   174
    6.2.2   Expectation   178
    6.2.3   Variances and Covariances   183
    6.2.4   Probability Distributions   187

6.3   TESTS OF HYPOTHESIS   189
  6.3.1   Tests on Individual Parameters   190
  6.3.2   Analysis of Variance Tests   195
  6.3.3   Repeated Predictor Variable Values (Lack of Fit Test)   198
6.4   INTERVAL ESTIMATION   202
  6.4.1   Confidence Intervals and Regions   202
  6.4.2   Response Intervals   206
APPENDIX   210
  6.A   Expectation of Least Squares Estimators   210
  6.B   Variances and Covariances of Least Squares Estimators   213
  6.C   Distribution of Least Squares Estimators   216
EXERCISES   218

7. RESIDUAL ANALYSIS   220
7.1   TYPES OF RESIDUALS   223
  7.1.1   Distinction Between Residuals and Errors   224
  7.1.2   Raw and Scaled Residuals   225
  7.1.3   Deleted Residuals   229
7.2   VERIFICATION OF ERROR ASSUMPTIONS   231
  7.2.1   Checks for Random Errors   232
  7.2.2   Test for Serial Correlation   234
  7.2.3   Detecting Heteroscedasticity   236
  7.2.4   Normal Probability Plots   239
7.3   MODEL SPECIFICATION   241
  7.3.1   Plots of Residuals Versus Predictor Variables   242
  7.3.2   Partial Residual Plots   247
7.4   OUTLIER DETECTION   252
  7.4.1   Plotting Techniques   252
  7.4.2   Statistical Measures   255
APPENDIX   258
  7.A   Derivation of Deleted Residual   258
EXERCISES   259

8. VARIABLE SELECTION TECHNIQUES   262
8.1   BASIC CONSIDERATIONS   264
  8.1.1   Problem Formulation   265
  8.1.2   Additional Selection Criteria   267
8.2   SUBSET SELECTION METHODS   268
  8.2.1   All Possible Regressions   269
  8.2.2   Best Subset Regression   270
  8.2.3 C_k Plots   272
  8.2.4   Additional Search Routines   275

8.3   STEPWISE SELECTION METHODS   278
  8.3.1   Forward Selection Method   278
  8.3.2   Backward Elimination Method   282
  8.3.3   Stepwise Procedure   284
APPENDIX   286
  8.A   Derivation of Equation (8.3.4)   286
EXERCISES   288

9.   MULTICOLLINEARITY EFFECTS   290
9.1   COEFFICIENT ESTIMATORS   293
  9.1.1   Variances and Covariances   293
  9.1.2   Estimator Effects   299
9.2   INFERENCE PROCEDURES   302
  9.2.1 t Statistics   302
  9.2.2   Other Selection Criteria   304
  9.2.3   Effects on Subset Selection Methods   306
9.3   POPULATION-INHERENT MULTICOLLINEARITIES   308
  9.3.1   Variable Selection   308
  9.3.2   Prediction   310
EXERCISES   313

10.   BIASES REGRESSION ESTIMATORS   315
10.1   PRINCIPAL COMPONENT REGRESSION   317
  10.1.1   Motivation   318
  10.1.2   Analysis of Variance   322
  10.1.3   Inference Techniques   326
10.2   LATENT ROOT REGRESSION ANALYSIS   329
  10.2.1   Motivation   330
  10.2.2   Predictive Multicollinearities   332
  10.2.3   Analysis of Variance   335
  10.2.4   Inference Techniques   338
10.3   RIDGE REGRESSION   340
  10.3.1   Motivation   341
  10.3.2   Selecting Ridge Parameters   343
  10.3.3   Analysis of Variance   346
10.4   FINAL REMARKS   348
EXERCISES   350

APPENDIX A.   DATA SETS ANALYZED IN THIS TEXT   352
  A.1   Selected Demographic Characteristics of Countries of the
     World   358
  A.2   Final High School Average Grades and First Year of
     College Average Grades   359

A.3   Homicides in Detroit, 1961-1973   360
A.4   Educational Status Attainment   361
A.5   Nitrous Oxide Emissions Modelling   362
A.6   Anthropometric and Physical Fitness Measurements on
       Police Department Applicants   363
A.7   Housing Rent Study   365
A.8   Body Measurements on Police Department Applicants   367
APPENDIX B.   DATA SETS FOR FURTHER STUDY   368
B.1   Mortality and Pollution Study   370
B.2   Solid Waste Data   372
APPENDIX C.   STATISTICAL TABLES   373
C.1   Cumulative Standard Normal Distribution   376
C.2   Cumulative Student t Distribution   377
C.3   Cumulative Chi-Square Distribution   378
C.4   Cumulative F Distribution   380
C.5   Critical Values for Runs Test   386
C.6   Critical Values for Durbin-Watson Test Statistic   388

BIBLIOGRAPHY   389

INDEX   398

REGRESSION ANALYSIS AND ITS APPLICATION

CHAPTER 1

INTRODUCTION

Data analysis of any kind, including a regression analysis, has the potential for far-reaching consequences. Conclusions drawn from small laboratory experiments or extensive sample surveys might only influence one's colleagues and associates or they could form the basis for policy decisions by governmental agencies which could conceivably affect millions of people. Data analysts must, therefore, have an adequate knowledge of and a healthy respect for the procedures they utilize.

Consider as an illustration of the potential for far-reaching effects of a data analysis one of the most massive research projects ever undertaken, the Salk polio vaccine trials (Meier, 1972). The conclusions drawn from the results of this study ultimately culminated in a nationwide polio immunization program and virtual elimination of this tragic disease in the United States. The foresight and competence of the principal investigators of the study prevented ambiguity of the results and possible criticism of the conclusions. The handling of this experiment provides valuable lessons in the overall role of data analysis and the care with which it must be approached.

Polio in the early 1950's was a mysterious disease. No one could predict where or when it would strike. It did not affect a large segment of any community but those it did strike, mostly children, were often left paralyzed. Its crippling effect on young children and the sporadic nature of its occurrence led to demands for a major effort in eradicating the disease. Salk's vaccine was one of the most promising ones available, but it had not been sufficiently tested.

Since the occurrence of polio in any specific community could not be predicted and only a small portion of the population actually contracted the disease in any year, a large-scale experiment including many communities was necessitated. In the end over one million children participated in the study, some receiving the vaccine and others just a placebo.

In allowing their children to participate, many parents insisted on knowing whether their child received the vaccine or the placebo. These children constituted the "observed-placebo" group (Meier, 1972). The planners of the experiment, realizing potential difficulties in the interpretation of the results, insisted that there be a large number of communities for which neither child, parent, nor diagnosing physician knew whether the child received the vaccine or the placebo. This group of children made up the "placebo-control" group.

For both groups of children the incidence of polio was lower for those vaccinated than for those who were not vaccinated. The conclusion was unequivocal: the Salk vaccine proved effective in preventing polio. This conclusion would have been compromised, however, had the planners of the study not insisted that the placebo-control group be included. Doubts that the observed-placebo group could reliably indicate the effectiveness of the vaccine were raised both before and after the experiment. The indicators of polio are so similar to those of some other diseases that the diagnosing physician might tend to diagnose polio if he knew the child had not been vaccinated and diagnose one of the other diseases if he knew the child had been vaccinated. After the experiment was conducted, analysis of the data for the observed-control group indicated that the vaccine was effective but the differences were not large enough to prevent charges of (unintentional) physician bias. Differences in the incidence of polio between vaccinated and nonvaccinated children in the placebo-control group were larger than those in the observed-control group and the analysis of this data provided the definitive conclusion. Thus due to the careful planning and execution of this study, including the data collection and analysis, the immunization program that was later implemented has resulted in almost complete eradication of polio in the United States.

1.1 DATA COLLECTION

Data can be compiled in a variety of ways. For specific types of information, the U. S. Bureau of the Census can rely on nearly complete enumerations of the U. S. population or on data collected using sophisticated sample survey designs. The Bureau of the Census can insure that all segments of the population are represented in most of the analyses that they desire to perform. Many research endeavors, however, are conducted on a relatively smaller scale and are limited by time, manpower, or economics. Characteristic of these studies is a data base that is restricted by the data-collection techniques.

So important is the data base to a regression analysis that we begin our development of multiple linear regression with the data-collection phase. The emphasis of this section is on an understanding of the benefits associated with a good data collection effort and the influence on the interpretation of fitted models when the data base is restricted. While it may not always be possible to build a data base as large or as representative as one might desire, knowledge of the limitations of a data base can prevent many incorrect applications of regression methodology.

1.1.1 Data-Base Limitations

Regression analysis provides information on relationships between a response variable and one or more predictor variables but only to the degree that such information is contained in the data base. Whether the data are compiled from a complete enumeration of a population, an appropriate sample survey, a haphazard tabulation, or by simply inventing data, regression coefficients can be estimated and conclusions can be drawn from the fitted model. The quality of the fit and accuracy of conclusions, however, depend on the data used: data that are not representative or not properly compiled can result in poor fits and erroneous conclusions.

One of many studies that illustrates the problems that arise when one is forced to draw inferences from a potentially nonrepresentative sample is found in Crane (1965). In her attempt to assess the influence of graduate school prestige and current academic affiliation on productivity and peer recognition of university professors, she surveyed faculty members in three disciplines from three universities on the east coast of the United States. The responses were voluntary and presumably not all professors in these disciplines participated in the study. Although Crane's study did not call for a regression analysis, the interpretation problems that occur as a result of her data-collection effort are applicable regardless of the type of analysis performed.

Questions naturally arise concerning any conclusions that would be drawn from a study with the data-base limitations of this one. Do these three disciplines truly represent all academic disciplines? Can these three universities be said to be typical of all universities in the United States? If some professors chose not to participate in the study, are the responses thereby biased? These questions cannot be answered from Crane's data. Only if additional studies provide results similar to hers for other disciplines and other schools can global conclusions be drawn concerning the influence of graduate school and current academic affiliation on recognition and productivity of university professors. No amount of statistical analysis can compensate for these data-base limitations.

Criticisms of limited data bases and disagreements with conclusions drawn from the analysis of them are common. Nevertheless, the choice is often between conducting no investigations at all or analyzing restricted sets of data. We do not advocate the former position; however, it is the obligation of the data analyst to investigate the data-collection process, discover any limitations in the data collected, and restrict conclusions accordingly. Another example will stress these points and the consequences of underrating their importance.

A well-publicized study on male sexuality (Kinsey et al., 1948) evoked widespread criticism both because of its controversial subject matter and because of its data-collection procedures. Responses were solicited from males belonging to a large number of groups in order to make the sampling more feasible. About 5,300 males were interviewed in prisons, mental institutions, rooming houses, etc. By interviewing volunteers from groups such as this, a large sample of responses could be obtained without exhaustive effort and expense. The convenience of selecting responses in this fashion is the primary factor contributing to the debate over the results of the study.

Among the criticisms raised about the Kinsey report, most centered on the data-collection process. Some groups (such as college men) were overrepresented while others (such as Catholics) were underrepresented and still others (such as Blacks) were completely excluded. The subjects were all volunteers and this fact led to further charges of unrepresentativeness. Additional criticisms centered on the interview technique which relied solely on an individual's ability to recall events in his past.

The statistical methodology used in the Kinsey report was highly praised although it was descriptive and relatively simple (Cochran, Mosteller, and Tukey, 1954). In response to the criticisms of the Kinsey report, moreover, the investigators argued that this study was just a pilot study for a much larger sexual attitude survey. Nevertheless, in numerous instances the conclusions drawn from the study went beyond bounds that could be substantiated by the data. Actually, the conclusions are quite limited in generality. The two examples just discussed demonstrate the problems that can arise from the absence of an adequately representative data base. Regardless of the sophistication of statistical analyses of the data, deficiencies in the data base can preclude valid conclusions. In particular, interpreting fitted regression models and comparing estimated model parameters in a regression analysis can lead to erroneous inferences if problems with the data go undetected or are ignored.

1.1.2 Data-Conditioned Inferences

Of particular relevance to a discussion of data-collection problems is the nature of the inferences that can be drawn once the data are collected. Data bases are generally compiled to be representative of a wide range of conditions but they can fail to be as representative as intended even when good data-collection techniques are employed. One can be led to believe that broad generalizations from the data are possible because of a good data-collection effort when a closer inspection of the data might reveal that deficiencies exist in the data base.

Equality of Educational Opportunity, also known as the Coleman report (Coleman et al., 1966), suffers from problems associated with the data actually collected rather than the data-collection plan. It is generally agreed that the plan for the data collection was adequate. Under severe time constraints mandated by the Civil Rights Act of 1964 that authorized the study, however, the data had to be analyzed without sufficient time to correct imperfections. Only 59% of the high schools surveyed responded and these were mainly in suburban areas of the country. The survey was further criticized for poor choices of measures of school resources and facilities. Measures of social background were also lacking in the final analysis (see Mosteller and Moynihan, 1972).

Each of these particular problems of the Coleman report were noted during or after the data-analysis phase of the study and are not specifically attributable to poor planning of the data collection. Given additional time, the nonresponse rate certainly could have been lessened and perhaps some of the other problems just mentioned could have been rectified. The inability to correct these data inadequacies cancels much of the beneficial effect of the planning phase. Unknown biases incurred by the large nonresponse rate and the imprecise measures of school resources, facilities, and socioeconomic factors cast doubt on the conclusions just as forcefully as poor planning of the data collection would have.

Unlike the Coleman report wherein the nature of possible nonrepresentativeness is unknown (again, due to unknown biases from the nonrespondents and the absence of the other measures mentioned above), known characteristics of a set of data can allow modified inferences to be drawn when the nature of the nonrepresentativeness can be identified. For example, it is known that Crane only surveyed professors from three disciplines at three universities on the east coast. This information is acknowledged in her article and can form the basis for inferences for a population of professors from these three disciplines who teach at universities similar in nature to the three she surveyed. Thus these inferences are conditioned on known characteristics of her data base. In this conditional framework her data and conclusions can be of great value, provided that either the nonresponse rate is small enough to be ignored or that it can be ascertained that the nonrespondents would reply similarly to those who did respond (this latter point holds true also for Coleman's data but the large nonresponse rate cannot be ignored).

Deficiencies in the data base that can be identified, therefore, may enable conditional inferences or conclusions to be drawn. Poor data-collection procedures that result in suspected data deficiencies of an unknown nature can render any attempt at analysis of the data fruitless.

1.2   REGRESSION ANALYSIS

Regression analysis consists of graphic and analytic methods for exploring relationships between one variable, referred to as a response variable, and one or more other variables, called predictor variables. Regression analysis is distinguished from other types of statistical analyses in that the goal is to express the response variable as a function of the predictor variables. Once such an expression is obtained the relationship can be utilized to predict values of the response variable, identify which variables most affect the response, or verify hypothesized causal models of the response.

This section is devoted to formal definition of the types of regression models that are to be discussed in this book. Following an example of a current application of regression analysis, the next subsection presents an algebraic definition of a regression model. The subsequent one briefly distinguishes regression analysis from correlation analysis.

1.2.1   Linear Regression Models

A recent application of regression analysis to a contemporary problem of considerable interest illustrates one of the uses of a regression analysis. Lave and Seskin (1979) studied the relationship between mortality (deaths per 100,000 population) and air pollution. The complexity of the task is apparent when one realizes that mortality can be influenced by many factors, only one of which might be air pollution. Among the groups of variables mentioned by the authors as potentially influential on mortality are:

Physical:	Age, Sex, Race
Socioeconomic:	Income, Occupation, Housing Density, Migration
Personal:	Smoking Habits, Medical Care, Exercise Habits, Nutrition, Genetics
Environmental:	Air Pollution Levels, Radiation Levels, Climate, Domestic Factors.

In this study mortality would be regarded as the response variable and the above characteristics as possible predictor variables.

A host of difficulties must be addressed before the technical details of a regression analysis can be performed to examine the relationship between mortality and these predictor variables. Among the problems addressed by Lave and Seskin are the lack of adequate information on many of the variables, ambiguity in the definition of others, errors in measurement, and the controversy over causal assumptions; e.g., if air pollution levels are found to be beneficial as predictor variables does this imply that air pollution causes increases in mortality? Putting these questions aside for the moment, the authors obtained prediction equations ("fitted" or estimated regression models) for mortality using air pollution and socioeconomic variables. One of the equations, using 1960 data from the 117 largest Standard Metropolitan Statistical Areas (SMSA's), is

Mortality Rate = 301.205 + 0.631·(Minimum Sulfate Level) + 0.452·(Mean Particulate Level) + 0.089·(Population per square mile) + other terms.

Using this fitted regression model, mortality rates for SMSA's can be estimated by inserting values for minimum sulfate level, etc., and performing the multiplications and additions indicated in the prediction equation. The value of each predictor variable can be assessed through statistical tests on the estimated coefficients (multipliers) of the predictor variables. These procedures and other evaluations of the prediction equation will be detailed in later chapters. We now turn to a formal algebraic definition of a regression model.

All applications of linear regression methodology involve the specification of a linear relationship between the response and predictor variables. Denoting the response variable by Y and the p predictor variables by $X_1, X_2, ..., X_p$, the linear relationship takes the form

$$Y = \alpha + \beta_1 X_1 + \beta_2 X_2 + ... + \beta_p X_p + \varepsilon. \qquad (1.2.1)$$

In expression (1.2.1), α, β_1, ..., β_p are unknown model parameters called regression coefficients. The last term, ε, of this relationship is inserted to reflect the fact that observed responses are subject to variability and cannot be expressed exactly as a linear combination of the predictor variables. The error, ε, can be due to only random fluctuation of the responses, to predictor variables that have erroneously been left out of the relationship, to incorrect specification of the relationship [e.g., if X_1 should actually appear in eqn. (1.2.1) as X_1^2], or to some combination of these.

In a study of student achievement, for instance, one might postulate that test scores on a national achievement exam (Y) are a linear function of high school grade point average (X_1) and a student's I.Q. score (X_2). One could then model the test scores as

$$Y = \alpha + \beta_1 X_1 + \beta_2 X_2 + \varepsilon$$

where ε measures the error in determining Y from only X_1 and X_2. Numerical procedures discussed in later chapters allow α, β_1, and β_2 to be estimated as well as the probable size of the error.

The term "linear" is used to distinguish the type of regression models that are analyzed in this book: the unknown parameters in eqn. (1.2.1) occur as simple multipliers of the predictor variables or, in the case of α, as additive constants. If one assigns numerical values to any $(p-1)$ of the predictor variables, the relationship between the response variable and the remaining predictor variable is, apart from the error term, a straight line. For example, in the model

$$Y = 10 + 5X_1 + 3X_2 - 4X_3 + \varepsilon$$

assigning values $X_1 = 2$ and $X_3 = 1$ results in a straight line relationship between Y and X_2 (again, apart from the error term):

$$Y = 16 + 3X_2 + \varepsilon.$$

Models in which functions of the predictor variables are incorporated can also be linear regression models. This is true for models such as the following:

$$Y = \alpha + \beta_1 X_1 + \beta_2 X_1^2 + \beta_3 X_2 + \beta_4 X_1 X_2 + \varepsilon.$$

It is important to realize that in this model functions of predictor variables such as X_1^2 or $X_1 X_2$ are not themselves functions of unknown parameters, only of other predictor variables.

1.2.2 Regression vs. Correlation

Regression analysis has been loosely described as a study of the relationships between one variable, the response variable, and one or more other variables, the predictor variables. Implied in these discussions is the notion that regression analysis specifically addresses how the predictor variables influence, describe, or control the response. The relationship is intended to be directional in that one ignores whether the response variable affects the predictors.

In some situations, notably laboratory experiments, the researcher can choose values for the predictor variables and then observe the magnitude of the response for the preselected value of the predictors. Here the relationships are clearly unidirectional. In many studies, however, the researcher can only observe simultaneously the response and predictor variables. Regression analysis of the data may still be called for if the goal of the study prescribes that one variable be examined in terms of its responsiveness to one or more other variables.

Correlation analysis is called for if one desires to assess the simultaneous variability of a collection of variables. Relationships among variables in a correlation analysis are generally not directional. One usually does not desire to study how some variables respond to others, but rather how they are mutually associated. A simple example of the need for a correlation analysis is if one desires to study the simultaneous changes (with age) of height and weight for a population. By sampling all age groups in the population one could obtain data on these two variables. The goal here is not to describe how weight influences height or vice-versa but to assess jointly height and weight changes in the population from infants to adults. On the other hand, if one did desire to predict height from weight a regression analysis would be appropriate.

1.3   USES OF REGRESSION ANALYSIS

Among the many techniques available for fitting regression models, precisely which ones are most appropriate for a particular study depend on the goals of the study. There are, for example, many schemes available for estimating the parameters of the regression model. Some are more beneficial than others if one needs to insure that the individual parameters are estimated with as much accuracy as possible. Other estimators might be more appropriate if the study demands good prediction of the response variable but not necessarily highly accurate parameter estimates.

This section outlines some of the major uses of regression methodology. Most uses can be placed in three broad categories: prediction, model specification, and parameter estimation. We now describe each of these uses.

1.3.1   Prediction

Consider the need for a municipality to be able to predict energy consumption. This is vitally important to any community since the local officials must be able to budget projected energy costs and arrange for the delivery of sufficient quantities to meet their needs. The primary focus in establishing a prediction equation for energy consumption is that the resulting equation is able to predict accurately. It is not of paramount importance that any specific set of predictor variables be included in the model nor that the regression coefficients are estimated precisely. All that is required is accuracy of prediction.

A prediction equation for energy consumption would have a form similar to eqn. (1.2.1), namely

$$\hat{Y} = \hat{\alpha} + \hat{\beta}_1 X_1 + \hat{\beta}_2 X_2 + \ldots + \hat{\beta}_p X_p. \tag{1.3.1}$$

In this expression $\hat{Y}$ represents a predicted value for energy consumption based on specified values of the predictor variables $X_1, X_2, \ldots, X_p$. The estimated regression coefficients are denoted $\hat{\alpha}, \hat{\beta}_1, \hat{\beta}_2, \ldots, \hat{\beta}_p$. Of concern in prediction models is the accuracy with which $\hat{Y}$ approximates Y. Criteria for determining the parameter estimates of the model should incorporate a measure of the accuracy of prediction. Note that such a criterion could result in poor parameter estimates so long as prediction was deemed suitably accurate.

To summarize, some prediction equations are designed to do one thing well, predict. Of secondary importance is the correct specification of the model and the accuracy of individual parameter estimates. To be sure, if worthwhile predictor variables are erroneously left out of the initial model specification, good prediction may never be attained. Nevertheless, good prediction can sometimes result even if a model is misspecified.

1.3.2 Model Specification

Model specification is critically important in regression analyses whose objective is the assessment of the relative value of individual predictor variables on the prediction of the response. In order to gauge which variables contribute most to the prediction of the response, one must first insure that all relevant variables are contained in the data base and that the prediction equation is defined with the correct functional form for all the predictor variables. Failure to meet either of these two requirements might lead to incorrect inferences.

Suppose, for example, one conducts an investigation into possible wage inequities. Constructing a prediction equation for annual salaries could include predictor variables such as sex, race, job description, number of years employed, and other variables that are believed to be related to wages. One predictor variable that is difficult to define and measure in many employment situations is the quality of the work performed. Yet this might be one of the most important influences on wage differentials. Particularly if one wishes to study sex or race discrimination, this variable should be included. Without it there could appear to be no substantial differences in wages paid to males and females (and so "sex" would be discarded as a poor predictor of annual salaries), whereas if this variable appears in the prediction equation differentials in pay to males and females could become very apparent ("sex" would then be judged a valuable predictor of annual salary).

Similarly, entering "number of years employed" as a variable in a prediction equation for annual salary should be carefully considered. In some employment categories, salaries increase slowly during the first few years of employment and more rapidly thereafter. If a new employee must be trained before becoming a productive member of a company, wages during the first few years of employment might rise slowly. Later when the employee has become technically proficient, wages might rise rapidly. In these circumstances, the number of years employed should enter the prediction equation in such a way that these wage characteristics can be accounted for—a linear term for wages might not be sufficiently accurate.

Regression analyses that require correct model specification are more demanding of the data analyst than ones that only stress good prediction. Attention must be given to both the inclusion of relevant predictor variables and their proper functional expression. Only if this is done can the relative merit of individual predictor variables be correctly assessed.

1.3.3   Parameter Estimation

Regression analyses that are conducted in order to provide good estimates of the model parameters generally are the most difficult to perform. Not only must the model be correctly specified and the prediction be accurate, but the data base must allow for good estimation. Certain characteristics of a data base (such as multicollinearity, defined in Chapter 4) create problems in the estimation of parameters. In the presence of these characteristics, properties of the parameter estimators may dictate that some estimators not be used.

Porter (1974) studied the effect of several variables on occupational attainment. One of his goals was to produce prediction equations for occupational attainment for white and black males in the United States. Predictor variables utilized in his study include socioeconomic status scores, creativity measures, intelligence ratings, the influence of "significant other" persons, school grades, and several other variables. By constructing separate prediction equations for white males and black males, he was able to compare the effect of the predictor variables on occupational attainment for each group by comparing the magnitude of the estimated regression coefficients.

It was necessary for Porter to construct two prediction equations. He wished to examine the different effects of the predictor variables for white and black males. He felt that the predictor variables would influence occupational attainment differently for the two groups so he could not simply construct one equation with an additional variable for race. Thus his comparisons required both appropriate predictor variables and accurate estimates of the parameters in each model.

Limitations of a data base and the inability to measure all predictor variables that are relevant in a study restrict the use of prediction equations. This is especially true when parameter estimation is the primary goal of an investigation. Model specification uses do necessitate reasonable parameter estimates but frequently only crude comparisons of the importance of predictor variables are needed. In the wage study, it might not be necessary to have precise parameter estimates—knowledge of which variables are important predictors may be sufficient. Studies such as Porter's, however, do require precise parameter estimates.

1.4   ABUSES OF REGRESSION ANALYSIS

Just as there are many valuable uses of regression analysis, fitted regression models are often inappropriately used. These abuses are not always apparent to the data analyst. Some misapplications are not the result of the data analysis at all but are due to unsupported or overexaggerated claims made about the fitted models.

Three common misapplications of regression methodology are extrapolation, generalization, and causation. Each of these abuses of regression methodology can be traced to inherent limitations of the data base. We now examine the nature of these limitations.

1.4.1   Extrapolation

Data bases, if they do not comprise a complete enumeration of populations of interest, are intended to be at least representative of those populations. Occasionally it is impossible to collect representative sets of data. One clear example of this is a data base that is to be used to construct a prediction equation for future responses. If either response or predictor variables are affected by time (e.g., economic variables), the data base might not be representative of the future. Prediction of future responses are examples of extrapolations: attempts to infer responses for values of predictor variables that are unrepresentative of the data base.

More specifically, extrapolation usually involves a restriction on the values of predictor variables in a data base when there is not a similar restriction on values for which predicted responses are desired. If one desires to predict suicide rates for white males aged 15 to 80 but the prediction equation is constructed from data for white males aged 20 to 65, attempts to predict suicide rates for ages 15 to 19 and 66 to 80 are extrapolations. In this example the predictor variable age was restricted in the data base to ages 20 to 65 rather than the desired interval 15 to 80.

Population figures are a source of great concern to government, economic, and business leaders. Necessarily, population projections involve extrapolations. It is unknown what the population of the United States will be in the year 2000. Economic changes, social attitudes, religious beliefs, and world events can all cause changes in birth rates and, hence, population totals. Yet plans for future government expenditures, business expansion, etc. require some estimates of future population sizes.

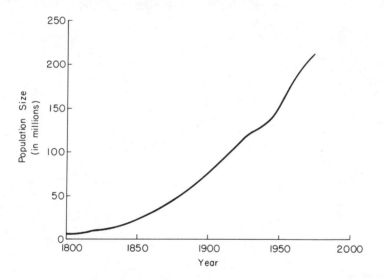

Figure 1.1   Population size of the United States, 1800-1975.
Source: U.S. Bureau of the Census (1961, 1975a, 1976).

Figure 1.1 displays population trends for the United States from 1800 to 1975. Suppose a good fit to the population totals is attainable using some fitted model of the form (1.3.1). From a visual inspection of Figure 1.1, it appears that reasonable prediction equations should predict a rapidly increasing population for the United States. Even though we do not know exactly what the numerical predictions would be, it seems safe to say that unless some drastic changes occur predictions of the population of the United States should increase for many years into the future.

Extrapolations of this kind are extremely dangerous for many reasons. One concern with this type of extrapolation is that the prediction equation does not incorporate information on any other variable that might influence future population totals. A prediction equation using only the data in Figure 1.1 (i.e., only time as a predictor variable) might be able to characterize past population trends well but be far off the mark in predicting future trends.

Demographers consider measures of fertility to be important indicators of future population sizes. The "total fertility rate" is a measure of fertility that can be interpreted as a measure of the total number of children women will bear during their lifetime. It is not a perfect measure of fertility since it uses data on births and women's ages during a single year to estimate how many births these women will have throughout the remainder of their lives. It has, nevertheless, proven valuable in assessing population trends.

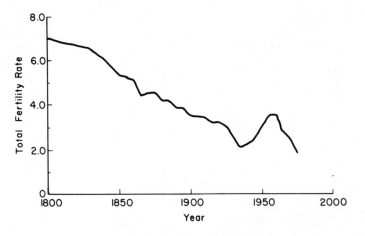

Figure 1.2   Total fertility rate of the United States, 1800-1975.
Source: Coale and Zelnik (1963), National Center for Health
Statistics (1970), and U.S. Bureau of Census (1976).

Figure 1.2 contains estimates of the total fertility rate for U. S. women from 1800 to 1975. The clear downward trend from 1800 was dramatically altered from about 1935 to 1960, the "baby boom" years. Since 1960 the trend is downward again. These trends suggest, contrary to Figure 1.1, that the population size of the United States is not going to continue to rapidly expand but will soon level off and decline. Women are having fewer and fewer children is the indication of Figure 1.2.

This last conclusion is open to as much criticism as was the former one. We are extrapolating the total fertility trend into the future. It should be clear, however, that the total fertility rate should be incorporated in population projections. The U. S. Bureau of the Census uses both population totals and fertility rates in projecting future population totals. Rather than projecting a single population total, the Bureau of the Census calculates three sets of projections, each one based on a different assumption of future fertility rates.

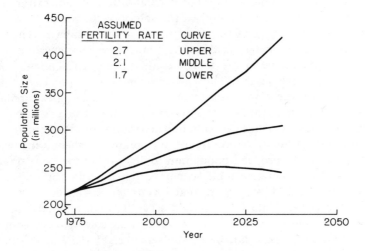

Figure 1.3 Projected population size of the United States, 1975-2035.
Source: U.S. Bureau of the Census (1975b).

Shown in Figure 1.3 are population totals projected from models based on both the data from Figure 1.1 and Figure 1.2. The curves are distinguished by whether one assumes that the total fertility rate will remain at 1.7, 2.1, or 2.7 births per woman. By providing three curves the Bureau of the Census acknowledges the dependence of its extrapolations on information not contained in its data base: future fertility rates. So too, all extrapolations are based on a key assumption: that trends or tendencies in future responses will be similar to those in the data base used to form the prediction equation. All too often this assumption is unrealistic.

1.4.2 Generalization

Generalization from a sample to a population or from one population to another suggests that the results of a regression analysis on one body of the data will be used to make inferences on another body of data. The danger inherent in this process is that the two bodies of data might not possess identical characteristics. Restrictions in the data base used to form a prediction equation may not be present in the population for which inferences are desired and vice-versa. Even if neither the data base nor population contain restrictions on the predictor variables and if all relevant variables are recorded in the data base, the same theoretical regression model might not be valid for both the data base and the population for which inferences are desired.

We distinguish extrapolation from generalization in that extrapolation presumes that predictions are desired for values of the predictor variables that are outside the range of values contained in the data base. Generalization refers more to changes in theoretical models or characteristics of the predictor variables between the data base and the theoretical population for which inferences are desired. Thus, while attempting to predict U.S. population totals for the next century would be regarded as an extrapolation, predicting population totals for European countries using the prediction equation for the United States would be a generalization. Without an analysis of past European population figures it is unknown whether the two sets of data (U.S. and European population totals) exhibit the same trends or whether the same theoretical model can be used to describe both.

Ehrenberg (1968, 1975) argues that investigators who conduct similar studies on different data bases should seek "lawlike relationships" that hold under a wide variety of conditions rather than calculate a separate prediction equation for each data base. Analyzing height and weight data from several studies, Ehrenberg discovered that the relationship (readers not familiar with logarithms should refer to Section 2.3.1)

$$\log (\text{weight in lb.}) = 0.02(\text{height in in.}) + 0.76 \qquad (1.4.1)$$

holds to within ±0.01 units for a wide variety of children. Among the groups whose average weight and height are approximately related by eqn. (1.4.1) are "a variety of Negro, White and Chinese children aged 5-18 in the West Indies" and "children of different ages from 2 to 18 years ... for both sexes (except teenage girls), for black and white races, for different social classes, in Ghana, Rwanda, Burundi, England, France, and Canada, except that no social class differences in the relationship had been found for white children" (Ehrenberg, 1975, p. 240).

Ehrenberg is attempting to combat some of the abuses of both over-generalization and also the feeling that each new data base requires a separate analysis. If a new data base is constructed, he maintains, the data analyst should check to see whether previously fitted models [such as eqn. (1.4.1)] provide adequate fits. If so, the fitted model can be generalized to wider sets of conditions, eventually attaining the status of a "lawlike relationship" similar to clearly established laws in the physical sciences.

Rarely can regression analyses that are performed on a few small data bases cover a wide enough range of conditions on the predictor variables so that generalization to a population can be unequivocal. Only efforts similar in dimension to the Salk polio vaccine trials or carefully selected samples taken by professional sampling experts could evoke such claims. Particularly where much of the data-collection effort consists of observing rather than controlling phenomena, extraneous factors not under the control of the investigator can make generalization very difficult.

1.4.3 Causation

Cause-effect relationships among the predictor and response variables cannot be established solely on the basis of a regression analysis. To assert that predictor variables deemed important in a regression analysis actually *determine* the magnitude of the response variable requires not only that the predictor variables are able to accurately predict the response but that they also control it. Causation demands that changes in the response variable can be induced by changes in the predictor variables and that the predictor variables are the *only* variables that affect the magnitude of the response. These are very stringent requirements.

Regression analysis can aid in the confirmation or refutation of a causal model, but the model itself must be conceived by theoretical considerations. The reasons for the causal influences of the predictor variables must be arguable apart from the data analysis and the choice of specific predictor variables for the regression model should be suggested by the theory. Causal models need not preceed empirical investigations; many are suggested by observable phenomena. The distinction we are making with previous discussions here is that regression analyses cannot prove causality, they can only substantiate or contradict causal assumptions.

The requirements for assessing causality with a regression analysis are akin to those of estimating model parameters. Since causal models must, by assumption, contain all relevant variables in correct functional forms, the need is to estimate the model parameters. Care in constructing the data base and obtaining appropriate parameter estimators is essential. Even so, the data bases are usually samples of populations that are somehow restricted. All the warnings given in this and the previous section should be borne in mind when assessing causal models.

Consider again the height-weight prediction equation (1.4.1). Ehrenberg has shown that this prediction equation is adequate for a wide variety of children. It should be obvious, however, that this is not a causal model. For a given average weight, the average height of a group of children can be predicted but one cannot guarantee that merely by increasing the average weight of a group of children their average heights will be increased as well. Obviously there are underlying nutritional and hereditary factors that influence both height and weight and it is not solely weight that dictates height.

CHAPTER 2

INITIAL DATA EXPLORATION

Underlying virtually all uses of regression models is the (often unstated, perhaps even unrecognized) assumption that the fitted model is capable of adequately predicting the response variable. More critical and unrealistic is the frequent use of the fitted model to describe physical phenomena as though it unerringly defined a functional expression relating the response and predictor variables. Except for the physical sciences, however, there are few theoretical laws that prescribe a functional relationship between response and predictor variables or that even dictate precisely which variables are to be used as predictors. Thus the data analyst often relies on empirical investigations and personal insight to specify prospective models.

Once a particular group of predictor variables has been selected for investigation, data on the response and predictor variables are collected and the researcher embarks on a series of analyses that eventually lead to a final prediction equation. Involved in this process are calculations of measures of the accuracy of potential final models (often differing only in the specific variables included from the original group), statistical tests of the importance of individual variables in predicting the response, and perhaps some analysis of model assumptions such as independence of the error terms.

Although the approach just described might involve sophisticated regression methodology and require the use of advanced computing routines, it still may be a rather superficial analysis. This can be illustrated with an analysis of the data in Table 2.1. Recorded in this table are the average annual salaries of teachers in public elementary and secondary schools in

Table 2.1. Average Annual Salaries of Instructional Staff in Regular Public Elementary and Secondary Schools: United States, 1964-65 ($t = 1$) to 1974-75 ($t = 11$).

Year (t)	Average Annual Salary*
1	$10,606
2	11,249
3	11,186
4	12,035
5	11,911
6	12,123
7	12,480
8	12,713
9	12,835
10	12,496
11	12,070

*converted to 1974-75 dollars

Source: Adapted from U. S. Department of Health, Education, and Welfare (1976), Table 38, p. 90.

the United States from 1964-65 ($t = 1$) to 1974-75 ($t = 11$). The goal is to fit a regression model to this data and predict salaries in future years.

Employing estimation techniques discussed in Chapter 3, a prediction equation for average annual salary (Y) using time (t) as a predictor variable is found to be

$$\hat{Y} = 10,927 + 174t,$$

so that, for example, the predicted average annual salary for instructional staff in 1975-76 ($t = 12$) is

$$\hat{Y} = 10,927 + 174(12) = 13,015.$$

Thus one would estimate the average annual salary to be about $13,015 in 1975-76 and that it would increase by about $174 (the estimated coefficient for time, t) each year. By 1984-85 ($t = 21$) the estimated average wage would be $14,581.

Open to question here is whether these figures are realistic or even accurate. While one cannot gauge the accuracy of future predictions at the same time the prediction equation is being formulated, comparison with the data used to estimate the model is possible. Table 2.2 exhibits the observed and predicted average annual salaries for the 11 years for which data are available. Also displayed are the "residuals," defined to be the difference

Table 2.2. Observed and Predicted Average Annual Salaries of
Instructional Staff: $\hat{Y} = 10,927 + 174t$

Year (t)	Average Salary (Y)	Predicted Salary ($\hat{Y}$)	Residual ($Y - \hat{Y}$)
1	10,606	11,101	-495
2	11,249	11,275	-26
3	11,186	11,449	-263
4	12,035	11,623	412
5	11,911	11,797	114
6	12,123	11,971	152
7	12,480	12,145	335
8	12,713	12,319	394
9	12,835	12,493	342
10	12,496	12,667	-171
11	12,070	12,841	-771

between the observed and predicted salaries for each of the years. Overall the prediction equation appears to perform adequately, although a few points have relatively large residuals, indicating that the predicted responses are not very close to the observed responses. Given the variability inherent in salary data, moreover, many would argue that this model fits the data as well as can be expected. A more critical examination of the data will reveal the dangers of so casual an analysis.

Indispensable in any analysis of regression data is the use of graphs to detect potentially misleading results. With small amounts of data, plots can quickly be sketched by hand; otherwise, most computer libraries contain plotting routines that are easily accessed by users. While graphs are indispensable, it is more an art than a science for the researcher to determine what to graph. Routinely, two-variable plots of response and predictor variables should be performed. Figure 2.1 is such a plot for the data in Table 2.1.

The plotted points in Figure 2.1 suggest that the empirical relationship between these two variables might not be that of a straight line. Particularly the drop in salary over the last couple of years suggests a term indicating curvature should be included with the linear one. The addition of a quadratic term (see Section 2.3.1) results in the following fit (using the procedures of Chapter 5):

$$\hat{Y} = 9,998 + 603t - 36t^2.$$

Comparing predicted values from this equation with the observed salaries in column 2 of Table 2.1 reveals that the largest residual is $326, smaller in

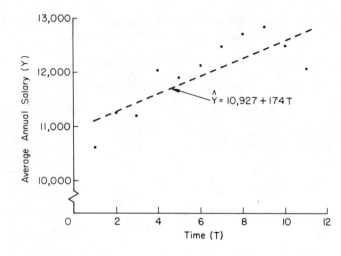

Figure 2.1   Average annual salaries of instructional staff in regular
public elementary and secondary schools: United States, 1964-65
($t = 1$) to 1974-75 ($t = 11$). **Source:** Adapted from U.S. Department of Health, Education and Welfare (1976), Table 38, p. 90.

magnitude than 6 of the 11 residuals in Table 2.2. The fitted values level off
at about $12,500 between 1971-72 and 1972-73 and decline thereafter. If this
fit is accepted as the better of the two, then the interpretation of future salary trends is completely reversed: salaries will decrease from 1972-73
onward so that the estimated average salary in 1984-85 will be $6,785. Thus,
if one believes this second analysis, which was suggested by a plot of the
data, one must conclude that salaries are going to drop over the next several
years—perhaps an unreasonable assertion when compared with current
economic trends.

The resolution of this dilemma—whether to use a relatively poor fitting
but plausible linear prediction equation or a better fitting but possibly
unreasonable quadratic one—is obtained by recalling that graphical techniques are more an art than a science. A quadratic fit is indeed suggested by
the plot of the 11 points in Figure 2.1, but so are many others, including
some that not only fit better than the linear predictor but also allow for
more reasonable future predictions.

Note that there were drops in average annual salary from years 1965-66 to
1966-67 ($t = 2$ to $t = 3$), 1967-68 to 1968-69 ($t = 4$ to $t = 5$), and 1972-73 to
1974-75 ($t = 9$ to $t = 11$). Thus the drop in salary during the last two years
might only be temporary. Perhaps the linear fit is a good one and the inherent year-to-year variability will prevent any better fit from being obtained in
the long run. On the other hand, the apparent drop in wages over the last

two years is not inconsistent with a (perhaps temporary) leveling off of wages.

At this point in our discussion it is not important to select one or the other of these two predictors. The issue raised with this example is that preliminary data exploration can be essential to obtaining an adequate fit to regression data. A more appropriate title to this chapter might be "*Initial Data Exploitation*" since we seek to examine the data *prior* to fitting various models so that the data itself can provide information about alternative model specifications. While an examination of the response and predictor variables requires additional time and effort, the gains are many, including:

(i)   a better appreciation of how the fitted model works (rather than just using it like a "black box"),

(ii)   insight into possible alternative model specifications,

(iii) discovery of obvious errors in data collection or recording, and

(iv) a clearer understanding of the limitations imposed on the fitted model by the data or the functional form of the individual predictors.

2.1 PRELIMINARY DATA EDITING

Screening data prior to formal regression analysis can eliminate costly errors that the formal analysis might not even detect. For example, one or two incorrectly transcribed data values can severely distort estimates of regression coefficients. Yet because the estimated model attempts to fit all the data points, the residuals might all be large whereas elimination of the one or two bad points could lead to a good fit to the remaining ones. The costs to the researcher could involve not only reanalysis if the discrepant points are eventually found, but also the more critical possibility that these bad data values will go undetected and lead to erroneous interpretations of the fitted model and poor predictions of the response variable.

2.1.1 Obvious Anomalies

One of the easiest methods of spotting errors in collecting or recording data is to visually scan the data base. Particularly if the data are to be analyzed on a computer, a listing of the data should be perused if at all possible. The computer output is no better than its input and there are many possible sources of error from the initial data collection to the final keypunching.

Table 2.3 lists observations on 7 demographic variables for the first 20 countries in Data Set A.1, Appendix A. The seven variables are: gross national product (GNP), infant death rate (INFD), number of inhabitants per physician (PHYS), population density (DENS), population density in terms of the amount of land utilized for agriculture (AGDS), population literacy rate (LIT), and a rate of enrollment in higher education (HIED). Note anything suspicious?

Table 2.3. Demographic Characteristics of 20 Countries

Country	GNP	INFD	PHYS	DENS	AGDS	LIT	HIED
Australia	1,316	19.5	860	1	21	98.5	856
Austria	670	37.5	695	84	1,720	98.5	546
Barbados	200	60.4	3,000	548	7,121	91.1	24
Belgium	1,196	35.4	819	301	5,257	967.0	536
British Guiana	235	67.1	3,900	3	192	74.0	27
Bulgaria	365	45.1	740	72	1,380	85.0	456
Canada	1,947	27.3	900	2	257	97.5	645
Chile	379	127.9	1,700	11	1,164	80.1	257
Costa Rica	357	78.9	2,600	24	948	79.4	326
Cyprus	467	29.9	1,400	62	1,042	60.5	78
Czechoslovakia	680	31.0	620	108	1,821	97.5	398
Denmark	1,057	23.7	830	107	1,434	98.5	570
El Salvadore	-219	.763	5,400	127	1,497	39.4	89
Finland	794	21.0	1,600	13	1,512	98.5	529
France	943	27.4	1,014	83	1,288	96.4	667
Guatemala	189	91.9	6,400	36	1,365	29.4	135
Hong Kong	272	41.5	3,300	3,082	98,143	57.5	176
Hungary	490	47.6	650	108	1,370	97.5	258
Iceland	572	22.4	840	2	79	98.5	445
India	73	225.0	5,200	138	2,279	19.3	220

Source: Loether, McTavish, and Voxland (1974). (See Appendix, Data Set A.1.)

At least six values in Table 2.3 appear questionable, three of these are definitely incorrect. Since LIT is expressed as a percentage, the literacy rate for Belgium is in error since no value can be greater than 100%. Comparison with the complete data set in Appendix A confirms that the decimal place is off for the literacy rate in Belgium. The reading should be 96.7.Scanning the gross national product values reveals that the minus sign should be removed from El Salvadore's GNP. An incorrectly typed decimal place is also the problem with the infant death rate for El Salvadore (the correct value is 76.3).

More difficult to deal with are the extremely large values for INFD for India and AGDS and DENS for Hong Kong. That these are the correct observations for these two countries is corroborated by comparison with Data Set A.1. Since they are correctly transcribed, should the data points be deleted because they are clearly extreme (or "outliers")? At this point we say no. We are merely concerned with correcting erroneously recorded observations, which these last three are not.

To appreciate the importance of this initial examination of the data, consider the effect of not doing so. Suppose one desires to regress GNP on LIT. Using the data as listed in Table 2.3 the resulting prediction equation is

$$\hat{Y} = 489.61 + 0.89X,$$

where Y represents GNP and X represents LIT. Now if the first two distorted points mentioned above are corrected, the prediction equation becomes

$$\hat{Y} = -296.28 + 11.51X,$$

a dramatic change! There are other problems that need to be addressed with this data in subsequent sections but note how a quick inspection of the data has eliminated a major obstacle to correctly fitting a model to GNP.

This example can also be used to illustrate an important fact in the correct utilization of regression methodology. As mentioned in the description of Data Set A.1 in Appendix A, the original compilation contains information on several more variables and many additional countries. Only 49 countries were selected from the original data base since only these 49 had recorded values for all seven variables of interest. The incidence of missing data in empirical research is common.* Often the only avenue left open to the researcher is to eliminate observations that do not have complete information on all variables of interest. With this data set we are left with only one-third of the original countries when those that do not have values for all seven variables are eliminated. Will this affect the conclusions to be drawn from subsequent analyses? In general, the answer is yes.

Figure 2.2 plots GNP versus LIT for the 118 of the original 141 countries having data for both of these variables. The 49 countries included in Data Set A.1 are indicated with zeros on this Figure. Observe that the countries with values for all seven variables in Data Set A.1 include an extremely disproportionate number of highly literate countries. Conclusions drawn from a study of this data, therefore, must be restricted mainly to countries with high literacy rates. This conclusion applies to the fitting of regression models to this data regardless of whether literacy rate is used as a predictor

---

*When only a few observations are missing from the data base it may be advantageous to adopt statistical procedures to estimate the missing values or adjust for them. Discussions of this approach include Afifi and Elashoff (1967), Elashoff and Afifi (1966, 1969a, 1969b), Cohen and Cohen (1975, Chapter 7), and Cox and Snell (1974, Section 5).

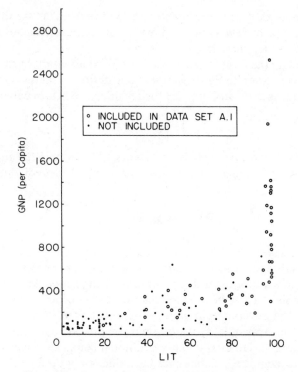

Figure 2.2   Gross national product (GNP) per capita and
literacy rate (LIT) for 118 countries of the world.

variable since the literacy rate of a country could be related to some of the
other measurements in the data set. Unless one can confirm that only variables that have no association with literacy rate are being investigated, the
finding that the countries in Data Set A.1 are highly literate must be
included in any use of fitted models obtained from this data. Again, a simple plot of the data has provided valuable information on an important
characteristic of the data that must be considered prior to fitting regression
models.

2.1.2   Descriptive Statistics

Summary statistics describing characteristics of each variable can provide
valuable insight into data anomalies especially when the amount of data
precludes plots or there are no computer plotting routines available. For
example, suppose one does not have access to a plotting routine so that
Figure 2.2 cannot be constructed other than by hand. While this may not

PRELIMINARY DATA EDITING 27

appear to be much of a problem, recall that there are a total of seven variables under investigation and, consequently, there are many plots of interest. One could select only a few points for plotting and the general trend of Figure 2.2 would be apparent, but would the interesting point in the upper left corner of the graph still be included in the new graph?

One useful device for examining general relationships between variables is the creation of crosstabulations (crosstabs). Crosstabs of two variables can be constructed easily and quickly. While much information on individual data values is inevitably lost if one or both variables are quantitative, trends in the data and errant data points can still be spotted. Table 2.4 displays a crosstab of GNP and LIT for the same 118 countries graphed in Figure 2.2.

Table 2.4. Gross National Product by Literacy Rate, 118 Countries of the World (Countries in Data Set A.1 Indicated in Parentheses).*

		Literacy Rate					
		0-20	20-40	40-60	60-80	80-100	Total
	0-500	31(1)	16(4)	18(7)	18(7)	9(8)	92(27)
	500-1,000	0	0	1(0)	0	13(11)	14(11)
	1,000-1,500	0	0	0	0	8(8)	8(8)
GNP	1,500-2,000	0	0	0	0	2(2)	2(2)
	2,000-2,500	0	0	0	0	0	0(0)
	2,500-3,000	0	1(0)	0	0	1(1)	2(1)
	Total	31(1)	17(4)	19(7)	18(7)	33(30)	118(49)

*All classes include upper bounds.
Source: Loether, McTavish, and Voxland (1974).

It is again evident from this table that the 49 countries included in Data Set A.1 are primarily among the more highly literate of the original countries. Of the 33 countries with 80-100% literacy, 30 (91%) are among the countries in Data Set A.1. Of those with 0-20% literacy, on the other hand, only 1 of 31 (3%) are included in this data set. The L-shaped relationship existing between GNP and LIT is obvious since there are numerous observations in the first row and last column of the table but none in the middle cells. Note also the clear indication of the outlying point in the second cell of the last row.

This technique of substituting a crosstabulation for a two-variable plot will not always indicate trends among the variables as clearly as the graph might. In fact some trends will not be recognizable due to the coarseness of the row and column categories. Gross trends in data are often detectable, however, and in the absence of a plot, the crosstabulation should be constructed.

2.1.3   Graphical Aids

Routinely plotting data was described in the introduction to this chapter as an indispensable aid to properly specifying a regression model and for detecting data anomalies. Several illustrations of the use of graphs have already been presented. At the risk of appearing redundant, we present in this section more examples of the need for plotting in a regression analysis.

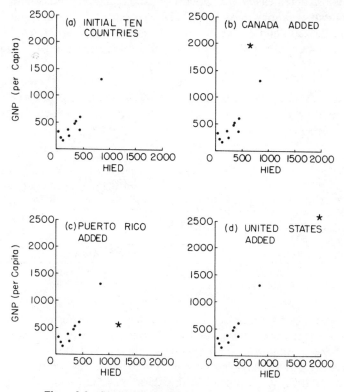

Figure 2.3   Scatter plot of GNP per capita and HIED for
the countries in Table 2.5.

Figure 2.3 depicts four sets of data. Ten of the data points are the first ten observations in Table 2.5, data on GNP and HIED for ten countries in Data Set A.1. Each of the plots (b), (c), and (d) in Figure 2.3 contain one of the additional points listed in Table 2.5. Thus Figure 2.3(b) has the original ten countries and Canada, (c) adds Puerto Rico to the original ten, and the United States is added to the first ten countries to produce (d).

A prediction equation for GNP(Y) using HIED(X) as a predictor variable calculated from the basic ten countries in Table 2.5 is

$$\hat{Y} = 58.39 + 1.22X.$$

The effect of adding Canada can be seen by studying Figure 2.3(b). Canada has an extremely large GNP when compared to the base countries. Visually fitting this data with a straight line suggests that the new slope should be larger than that of the base countries and the Y-intercept should be smaller. The reader might wish to verify these statements by placing a pencil on the graphs through the points in Figures 2.3(a) and (b). These visual impressions are corroborated by fitting the eleven points:

$$\hat{Y} = -24.96 + 1.74X.$$

The implications of so drastic a change in the coefficients due to the addition of a single data point must be weighed carefully. Should the point be deleted? Why (not)?

Table 2.5. Gross National Product and Student Enrollment in Higher Education: 10 Base Countries and 3 Additional Ones

Base Countries	GNP	HIED	Base Countries	GNP	HIED
Bulgaria	365	456	Poland	475	351
El Salvadore	219	89	Romania	360	226
Italy	516	362	USSR	600	539
Jamaica	316	42	Additional Countries		
Mexico	262	258	(a) Canada	1,947	645
New Zealand	1,310	839	(b) Puerto Rico	563	1,192
Nicaragua	160	110	(c) United States	2,577	1,983

Inserting Puerto Rico with the ten base countries also alters the estimated regression coefficients. Visually, the slope in Figure 2.3(c) should be less than that of the base countries and the intercept larger. Indeed, the prediction equation is

$$\hat{Y} = 223.88 + 0.60X.$$

Again one should question whether one data point should be allowed to influence the estimated coefficients so heavily.

Perhaps these two examples lead one to believe that each of the two additional points should be eliminated from the data set prior to estimating the regression coefficients. After all, each point contained an unusually large (compared to the base countries) GNP or HIED value. But the prediction equation is meant to be as general as possible; i.e., we desire that it be applicable to as many countries as possible, even those with large GNP or HIED readings.

Figure 2.3(d) points out one reason why data points should not be arbitrarily deleted because of large data values. Including the United States with the base countries does not appear to appreciably change the slope or the intercept of the prediction equation despite the extremely large measurements of GNP and HIED for the United States. It would be difficult to appreciate this fact without the plot of the data points, but the small change in the prediction equation due to adding the United States is confirmed by calculating the estimates of the coefficients:

$$\hat{Y} = 44.40 + 1.27X.$$

Had we adopted the strategy of arbitrarily deleting a few observations with unusually large (or small) values, we would have deleted each of the three additional countries and severely limited the generality of the resulting prediction equation. The deletion of the United States would be especially unfounded.

The value of graphical displays, then, is to detect trends and extreme measurements. They should not be used to automatically select points for deletion because they look extreme. They can be used—and have been already—to identify potential problems and to alert the data analyst to the effects of including or excluding particular points. One final caution in this regard can be appreciated from Figure 2.3. New Zealand, the data point in the middle of the graph, may visually appear to be an extreme value in Figure 2.3(a). Note that it appears less so in Figures 2.3(b) and (c), and is completely consonant with the other ten countries in Figure 2.3(d).

The most satisfying method of deleting points as outliers is if their presence in the data base can be ascribed to a verifiable error in the data collection or transcribing. For example if in the original data base El Salvadore's GNP and Belgium's LIT were recorded as in Table 2.3, these points could nevertheless be deleted as aberrant. This is because neither value is permissible according to the definitions of the variables. They are clearly errors. The same statement cannot be made for any of the values in Data Set A.1. Let us investigate this data set further.

Figure 2.4 is a graph of GNP and INFD. Three data points have been labeled for consideration: Canada and the United States because of their large GNP values and India due to its extreme INFD value. Should these countries be deleted because they are unrepresentative of the entire data set and may distort the prediction equation? We think not. Observe that all three countries appear to be consistent with a J-shaped relationship between GNP and INFD. That is, although the points do have the largest values for these two variables, they do not abruptly deviate from the apparent relationship between GNP and INFD. They seem to be merely extensions of

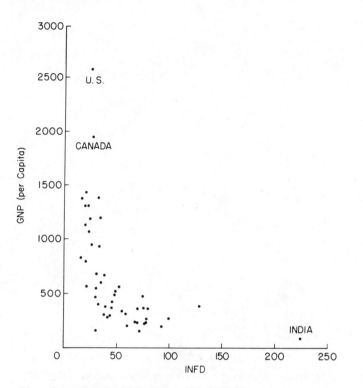

Figure 2.4   Gross national product (GNP) per capita by infant death
rate (INFD) in 49 countries in the world.

this relationship to extreme values of the variables. This does not appear to
be true for two of the data points in Figure 2.5.

PHYS and AGDS have been plotted in Figure 2.5, with Hong Kong and
Singapore labeled as extreme points. If there is any functional relationship
between PHYS and AGDS, it does not seem to include Hong Kong and
Singapore. Other plots of this data do not reveal strong associations
between PHYS and AGDS, with or without these two points; yet the presence of these observations in the data set can severely distort the prediction
equation. With Hong Kong and Singapore included in the data base, the
prediction equation for GNP (using the estimators discussed in Section 5.1)
is

$$GNP = 31.191 - 3.441 \cdot (INFD) + 0.015 \cdot (PHYS) - 0.173 \cdot (DENS)$$

$$+ 0.003 \cdot (AGDS) + 6.312 \cdot (LIT) + 0.690 \cdot (HIED).$$

Without Hong Kong and Singapore the prediction equation becomes

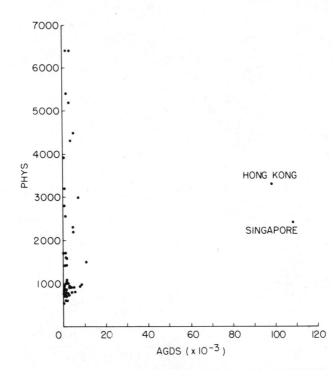

Figure 2.5   Physician ratio (PHYS) by agricultural density (AGDS)
in 49 countries of the world.

$$\text{GNP} = 159.878 - 3.960 \cdot (\text{INFD}) + 0.019 \cdot (\text{PHYS}) + 0.198 \cdot (\text{DENS})$$

$$- 0.050 \cdot (\text{AGDS}) + 5.902 \cdot (\text{LIT}) + 0.681 \cdot (\text{HIED}).$$

Note in particular the change in the estimates of the constant term of the model and in the coefficients of DENS and AGDS.

Hong Kong and Singapore were eliminated from further analysis in Data Set A.1 for reasons similar to those cited above and because they are in reality cities not countries. On other two-variable plots of the variables in the data base these locations were clearly aberrant to the suggested (by the plots) functional relationships. Other countries, notably the United States, were not deleted from the data set despite the presence of outlying values on one or more of the seven variables. This was due to the existence of apparent functional relationships for which the extreme points could be considered extensions.

While we delete these two cities from the data base for any further analysis, we do not intend to ignore them or pretend they do not exist. Just as it will be disclosed that the countries comprising Data Set A.1 are among the more literate of the world, so too the deletion of Hong Kong and Singapore because of their extreme values and potential for distorting the prediction equation will be reported.

Finally, while we have concentrated attention on two-variable graphs in this section, one's ingenuity and insight are the actual limits on the utility of graphs. Creative examples of graphical displays are presented in Gnanadesikan (1977) and Tukey (1977), among more recent publications.

2.2 INITIAL MODEL SPECIFICATION

Previous subsections of this chapter have disclosed how one can be misled by estimating regression coefficients when the data base contains aberrant observations. Incorrect initial definitions of response or predictor variables can also diminish the utility of a fitted model. Predictor variables can erroneously be eliminated from fitted models because they are not properly defined. Confusion and conflicting interpretations can result from an incorrectly specified predictor variable that masks the true contributions of other predictors. Thus care is needed in the *initial* specification of variables in a regression analysis as well as in later transformations and the insertion or deletion of individual predictor variables. Although an in-depth treatise of this topic is beyond the scope of this book, some frequently encountered problems with the specification of variables in regression models are now examined.

2.2.1 Categorical Predictor Variables

A variable that is intrinsically nonnumerical is referred to as a categorical or qualitative variable. Sex, martial status, occupation, and religious preference are examples of strictly nominal variables (labels or descriptions) that are important predictor variables in many regression analyses. Ordinal variables (variables whose values can be ranked or ordered) commonly used in regression models include occupational prestige rankings, social class ratings, attitudinal scales (e.g., strongly disagree, disagree, no opinion, agree, strongly agree), and educational classifications (e.g., no schooling, some elementary education, etc.).

Categorical variables must be assigned numerical codes in a regression analysis. Possible assignments include sex: male = 1, female = 0; marital status: single = 0, married = 1, divorced = 2, separated = 3, widowed = 4; and attitudinal scales: strongly disagree = -2, disagree = -1, ..., strongly agree = 2. These numerical values are completely arbitrary and can be replaced by numerous other selections. Some choices of codes, however, should be avoided.

Consider Data Set A.2 in Appendix A. This is an example of a data base that includes two intrinsically numerical or quantitative variables (academic scores for grade 13 and first year of college) and a categorical one (high school designation). Figure 2.6 shows that the two sets of scores, grade 13 and first year of college, appear to be related linearly, at least as a first approximation. We thus initially allow the grade 13 score to enter the prediction equation for initial year of college average grade without transformation. But what about the high school attended?

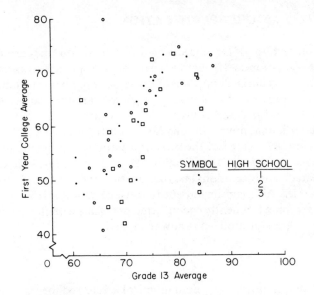

Figure 2.6 Average grades for first-year students at
Queens University (1969) by grade 13 averages for students
from three high schools.

Suppose first that we examine only the data for high schools 1 and 2. One suggestion for incorporating high school designation in a prediction equation along with grade 13 average is to define an "indicator" variable:

$$X_1 = \begin{cases} 0 & \text{for school 1} \\ 1 & \text{for school 2,} \end{cases}$$

i.e., if a student attended high school 1 his observed value for the first predictor variable is $X_1 = 0$, if he attended high school 2 it is $X_1 = 1$. Then letting X_2 denote the grade 13 average, a prediction equation for average academic score for first year of college (Y) is

$$\hat{Y} = -13.17 - 1.10X_1 + 1.05X_2.$$

The coefficient of X_1 indicates that predicted scores for students from high school 2 are 1.10 points lower than those of school 1 for students with the same X_2 scores. For example, a student from high school 1 with an 80 average in grade 13 is predicted to average

$$\hat{Y} = -13.17 - 1.10(0) + 1.05(80) = 70.83$$

in his first year of college, while one from high school 2 is predicted to average

$$\hat{Y} = -13.17 - 1.10(1) + 1.05(80) = 69.73.$$

Another natural definition for high school designation is to let the indicator variable be defined as

$$X_1 = \begin{cases} 1 & \text{if school 1} \\ 2 & \text{if school 2.} \end{cases}$$

The prediction equation then becomes

$$\hat{Y} = -12.07 - 1.10X_1 + 1.05X_2.$$

Observe that only the estimate of the constant term in this prediction equation is changed from the previous one. This is because all the scores for X_1 were changed from the first definition to the second by adding a constant, 1, to them. The slopes of the two predictor variables are unchanged, only the constant term of the model is altered to account for the increase in the scores of X_1. The predicted values from both models are identical and the effect of changing schools (a drop of 1.10 points from high school 1 to high school 2) remains the same.

When a categorical variable has three or more possible values the difficulties with correctly defining the influence of the variable on the response become more acute. In an attempt to incorporate the information provided by all three high schools on college grades, one might decide to specify high school designation as

$$X_1 = \begin{cases} 1 & \text{if school 1} \\ 2 & \text{if school 2} \\ 3 & \text{if school 3.} \end{cases}$$

The prediction equation for average first-year college grades is

$$\hat{Y} = -8.17 - 1.95X_1 + 1.01X_2.$$

The estimated model looks very similar to the previous ones. Even the predicted values are reasonably close to the ones using the previous models: a student with an 80 grade 13 average would be predicted to score 71.48 during his first year of college if he is from school 1 and 69.53 if from school 2. There is, however, an important difference when three schools are incorporated in X_1 rather than two schools.

A tacit assumption underlying the above definition of X_1 for the three schools is that the influence of schools 1 and 3 are extreme and that of school 2 is between them. Using the fitted model given above, the estimated coefficient for X_1, -1.95, indicates that students from school 2 have predicted scores 1.95 points below those of school 1 and those from school 3 have predicted scores 1.95 points below those of school 2 (assuming the same grade 13 scores in all cases). Is this reasonable? Are the school designations merely identifiers or can one truly assert that school 1 produces students with higher scores than school 2 and these latter students have higher scores than school 3? Even if this latter statement is true, are the differences in scores (apart from random fluctuations) constant between schools 1 and 2 and between schools 2 and 3?

By specifying X_1 as having values 1, 2, and 3 we are forcing the schools to have a precise effect on the predicted response. If school 2 is actually the best or worst high school, the specification of X_1 as given above cannot reflect this fact. Visualize the compounding of this problem if one attempts to define a similar school designator for all nine schools in the original data base.

Having raised this question, can we now define an appropriate predictor variable? There are again many possibilities but most useful specifications utilize two, not one, indicator variables. For example, let

$$X_1 = \begin{cases} 1 & \text{if school 1} \\ 0 & \text{otherwise,} \end{cases} \qquad X_2 = \begin{cases} 1 & \text{if school 2} \\ 0 & \text{otherwise,} \end{cases}$$

and X_3 = grade 13 average. Note that if $X_1 = X_2 = 0$, the student must be from high school 3. The advantage of this form of specification is that no a priori ordering is imposed on the schools, nor does this definition require constant differences in the scores between schools. Other possibilities exist, such as

$$X_1 = \begin{cases} 1 & \text{if school 2 or 3} \\ 0 & \text{otherwise} \end{cases} \quad \text{and } X_2 = \begin{cases} 1 & \text{if school 3} \\ 0 & \text{otherwise.} \end{cases}$$

Here school 1 is indicated by $X_1 = X_2 = 0$, school 2 by $X_1 = 1$ and $X_2 = 0$, and school 3 by $X_1 = X_2 = 1$. The prediction equation resulting from this specification is

$$\hat{Y} = -10.22 - 1.04X_1 - 2.89X_2 + 1.01X_3. \qquad (2.2.1)$$

From this predictor one sees that the effect of high school designation on the predicted response is that school 2 decreases the predicted values by 1.04 points over school 1 and school 3 decreases them by an additional 2.89 points. The schools are in fact ordered as the first specification for X_1 requires, but their estimated effects do not change by a constant amount. Using two predictor variables provides the added flexibility required to adequately specify the influence of the schools on the response.

Specifications that require an ordering or constant difference for the values of a variable should be avoided when using categorical variables. Definitions that do not impose conditions on the variable should be sought instead. In general, if a categorical predictor variable can take on any one of k different values, one should define $k - 1$ indicator variables like those used above rather than utilizing a single predictor variable with an arbitrary assignment of k values for it.

2.2.2 Interactions

Interaction terms in a regression model are products of two or more predictor variables. They are useful when it is believed that the effect of a predictor variable on the response depends on the values of other predictor variables. For example, in the two-variable model

$$Y = \alpha + \beta_1 X_1 + \beta_2 X_2 + \varepsilon,$$

the effect of X_1 on the response is measured by the magnitude of β_1 regardless of the value of the second predictor variable, X_2. In other words, given any fixed value of X_2, the response varies linearly with X_1 (apart from the random error term) with a constant slope, β_1; e.g., if $X_2 = 1$,

$$Y = (\alpha + \beta_2) + \beta_1 X_1 + \varepsilon$$

or if $X_2 = 10$,

$$Y = (\alpha + 10\beta_2) + \beta_1 X_1 + \varepsilon.$$

Consider now the effect of adding an interaction term, $\beta_3 X_1 X_2$, to the original model. The new model is

$$Y = \alpha + \beta_1 X_1 + \beta_2 X_2 + \beta_3 X_1 X_2 + \varepsilon.$$

If $X_2 = 1$,

$$Y = (\alpha + \beta_2) + (\beta_1 + \beta_3)X_1 + \varepsilon$$

and if $X_2 = 10$,

$$Y = (\alpha + 10\beta_2) + (\beta_1 + 10\beta_3)X_1 + \varepsilon$$

so that both the constant term and the slope are affected by the specific choice of X_2. Thus the effect of X_1 on the response depends on the value of X_2. Similarly, the effect of X_2 on the response depends on the value of X_1.

As an illustration of the added flexibility that accrues when interaction terms are added to regression models, let us return to the prediction of average college grades for high schools 1 and 2. If one does not wish to assume that the grade 13 averages have the same influence on the first year of college averages for both high schools, an interaction term can be added to the original model resulting in the following fit:

$$\hat{Y} = -25.97 + 19.60X_1 + 1.23X_2 - 0.29X_1X_2.$$

Using the prediction equations derived in the previous subsection without an interaction term, a student from high school 1 would be predicted to average 1.10 points higher than a student from high school 2 who has the same grade 13 average, regardless of the value of the grade 13 average. By adding the interaction term, the difference between the predicted scores for students from the two high schools with the same grade 13 average does depend on the grade 13 score: e.g., the difference is 3.60 points (72.43 vs. 68.83) for students with grade 13 averages of 80 but only 0.70 points (60.13 vs. 59.43) for students with grade 13 averages of 70.

Interaction terms clearly allow more opportunity for individual predictor variables to exhibit joint effects with other predictor variables. Several interaction terms involving two or more predictor variables can be included in regression models but they should not be inserted routinely for several reasons. First the number of possible interaction terms can be large for regression models with several predictor variables. With only 5 predictor variables there are 10 possible two-variable interaction terms, 10 three-variable interaction terms, 5 four-variable interaction terms, and 1 five-variable interaction term. Use of all predictor variables and their interactions could result in a complicated model with 36 terms in it. Many or all of the interaction terms would not substantially improve the fit to the data, only add to the complexity of the prediction equation.

Another reason for not arbitrarily adding interaction terms to regression models is that interaction terms sometimes repeat information provided by the individual predictor variables. If the redundancy induced by the interaction terms is too strong, coefficient estimates for the individual predictor variables can be distorted. The drastic changes in the coefficients for the above prediction equations with and without the interaction term could be reflecting such a strong redundancy. This type of redundancy is referred to as a "multicollinearity" and will be discussed in greater detail in Chapter 4.

Despite these admonitions, interaction terms are valuable additions to the specification of regression models. Whenever it is reasonable to believe that predictor variables have a joint influence on the response variable, specific interaction terms should be considered for inclusion in the model. This belief can be reinforced through theoretical arguments, plots of Y versus $X_1 X_2$ if both predictor variables are numerical, or plots of Y versus X_2 for each value of X_1 if X_1 is an indicator variable. If X_1 and X_2 are numerical, plots of Y versus $X_1 X_2$ should show a linear trend if the interaction term is an important predictor variable. Plots of Y versus X_2 for each value of X_1 should reveal linear associations with different slopes and intercepts if the interaction of a numerical (X_2) and an indicator (X_1) variable is an important additional predictor variable. As always, the plots are not foolproof but they can provide valuable insight into the use of interaction terms.

2.2.3 Smoothing

Reluctance to devote the time necessary for properly specifying regression models stems in part from a concern that this is a confusing and complicated process. Confusion arises not only because of the wide variety of variable specifications that are available but also because plots of the data do not always clearly indicate which specifications might be useful. Data smoothing techniques aid in specifying the form of response or predictor variables by reducing the variability of plotted observations and enabling trends to be more clearly recognized.

One thing to remember when examining plots of data points is to focus on gross, overall trends in the observations rather than concentrating on small perturbations of, say, three or four points. Partly with this in mind we concluded that the grade 13 averages in Figure 2.6 should enter a prediction equation for first year of college averages as a linear term, without any transformation or reexpression. Let us now reconsider this conclusion.

The 16 pairs of scores for students from high school 3 in Data Set A.2 have been plotted in Figure 2.7(a). It appears from this graph that a linear term should be adequate to relate the two sets of grades, although the plotted points exhibit a great deal of variability. Some of this variability can be eliminated by smoothing the data as follows. The pairs of observations are first ordered from smallest to largest according to the magnitudes of one of the variables, say grade 13 averages. The first two columns of Table 2.6 exhibit this rearrangement. Next, the first and last first-year averages (65.0 and 63.2, respectively) are rewritten in the third column of the table. The second first-year average (45.0) is replaced by the median of the first,

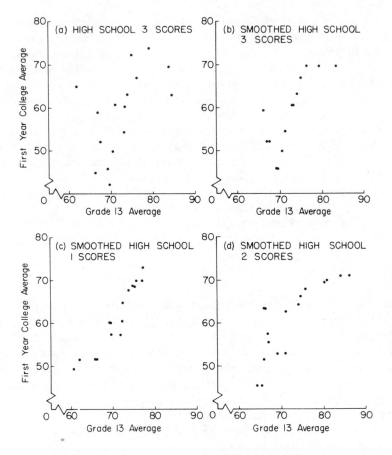

Figure 2.7   Smoothed first-year college scores by grade 13 averages.

second, and third scores (i.e., the middle score after the three have been
ordered in magnitude). The median of 65.0, 45.0, and 59.0 is 59.0. Then the
third first-year average (59.0) is replaced by the median of the second, third,
and fourth averages (52.2, the median of 45.0, 59.0, and 52.2). Each first-
year average, then, is replaced in column three of Table 2.6 by the median
of the three scores including the ones immediately before and after it.
Smoothing can be repeated until no changes occur in the smoothed values;
for the scores from high school 3, only two smoothings are needed. The
smoothed first-year averages are then plotted against the grade 13 average
as in Figure 2.7(b). Since we are unable to smooth the first and last points
with this procedure, we do not plot those two points [Tukey (1977) suggests
several alternative approaches for dealing with end points].

Table 2.6. Smoothing the First Year of College Average Grades for Students from High School 3

Initial Scores		Smoothed First-Year Averages	
Grade 13 Average	First-Year Average	First Smoothing	Second Smoothing
61.8	65.0	65.0	65.0
66.6	45.0	59.0	59.0
67.0	59.0	52.2	52.2
67.6	52.2	52.2	52.2
69.0	46.0	46.0	46.0
69.2	42.0	46.0	46.0
70.8	50.0	50.0	50.0
71.5	61.2	54.3	54.3
73.2	54.3	60.4	60.4
73.4	60.4	60.4	60.4
74.0	63.2	63.2	63.2
75.2	72.4	67.0	67.0
76.6	67.0	72.4	69.5
79.3	73.6	69.5	69.5
83.7	69.5	69.5	69.5
84.5	63.2	63.2	63.2

Source: Ruben and Stroud (1977). (See Appendix, Data Set A.2.)

The plot of the smoothed first-year average is somewhat shocking. Contrary to expectation, the smoothed plot is not a straight line; in fact, the points appear to lie very close to a smooth curve that initially decreases, then rises sharply, and finally flattens out. Unfortunately this curve is almost too exact. The apparent preciseness of the smoothed points has caused us to concentrate on minor fluctuations rather than gross trends in the data. Indeed, the initial decrease and final flattening are based on, respectively, three and two smoothed data points. While these shifts in the curve should arouse our curiosity, we must be careful that they do not lead us to specify a model primarily to fit these few points. If we examine the plot for overall rather than localized tendencies, it appears that a straight line does fit most of the data (this will be easier to see if the first and last plotted points are covered and a pencil is laid over the remaining points).

Figures 2.7(c) and (d) are plots of the smoothed first-year average grades for high schools 1 and 2 (again removing the nonsmoothed first and last scores for each school). The graph for high school 1 looks linear but that of high school 2 indicates that a term accounting for curvature might be needed. Here again, we are possibly being deceived by the last few points, but the inclusion of both a linear and a quadratic term for grade 13 averages could account for much of the tendency for curvature in both Figures 2.7(b) and (d).

In summary, smoothing can clear up much of the confusion about the need for transformations and which ones might be appropriate. When several predictor variables are included in regression models, two-variable plots can be highly erratic. Smoothing can reduce this variability but one must be careful not to be overly concerned with localized trends in the smoothed values. Finally, median smoothing has been presented in this subsection due to its simplicity. Other types, such as moving averages and exponential smoothing [e.g., Boardman and Bryson (1978)], are also available and just as effective.

2.3 REEXPRESSING VARIABLES

When two-variable plots or plots of smoothed data values suggest that the relationship between the response variable and one or more predictor variables is not linear, as in Figures 2.4 and 2.7(d), some form of reexpression (transformation) of the variables is required. If only one or two of the plots suggest the need for reexpression, it is generally preferable to transform the one or two predictor variables that are involved in the nonlinearity rather than the response variable. A transformation of the response variable in such circumstances could induce nonlinear relationships with the other predictor variables which previously appeared to be linearly related to the response. On the other hand, if several of the plots indicate the need for reexpression—particularly if the same nonlinear functional relationship appears to be present in most of the plots—it is generally preferable to transform the response variable. Of course, knowledge of any theoretical relationships between response and predictor variables should always be incorporated when selecting transformations since theoretical relationships can dictate the precise transformation needed.

In this section we discuss transformations of response and predictor variables. As a preliminary we first briefly review some of the more common and useful functional relationships between two variables. Because of their importance in regression modelling, readers who are unfamiliar with polynomials and exponential and logarithmic functions might wish to consult other references for more extensive coverage of these topics.

2.3.1   Special Functions

Polynomials, equations of the form

$$Y = \alpha + \beta_1 X + \beta_2 X^2 + \ldots + \beta_k X^k,$$

are among the most useful specifications of relationships between response and predictor variables in a regression analysis. Polynomials can approximate a wide variety of curved relationships and polynomial terms for two or more predictor variables are easily incorporated in a single regression model. Negative powers of X (such as $1/X$, $1/X^2$, ..., $1/X^k$) are also easily included.

In order to appreciate the variety of curves that can be approximated with polynomials, Figure 2.8(a)-(d) graphs a response variable Y as, respectively, a linear, quadratic, cubic, and inverse function of a predictor variable X.

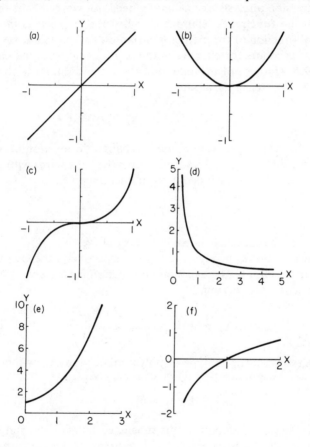

Figure 2.8 Useful relationships between two variables: (a) $Y = X$, (b) $Y = X^2$, (c) $Y = X^3$, (d) $Y = 1/X$, (e) $Y = \exp(X)$, (f) $Y = \ln(X)$.

Changes of location (center of the graph) and scale (the rate at which the curve changes) are accommodated by minor changes in the basic form. For example, the quadratic function

$$Y = 4 + 2(X - 6)^2$$

shifts the curve in Figure 2.8(b) from being centered at the point $X = 0$, $Y = 0$ to the point $X = 6$, $Y = 4$ and doubles the rate at which the curve increases as X moves in either direction from the center of the curve. A negative coefficient on any of the polynomial graphs in Figure 2.8(a)-(d) will result in the "mirror image" of the curve reflected around the X-axis.

In a regression analysis one generally need not worry about the location or scale of the functional relationship when the response is modeled as a polynomial function of the predictor variables. Estimation of the regression coefficients accounts for changes in the relationship from the basic graphs in Figure 2.8. If a plot of the data indicates a trend similar to the cubic one of Figure 2.8(c), fitting the model

$$Y = \alpha + \beta_1 X + \beta_2 X^2 + \beta_3 X^3 + \varepsilon$$

will allow for automatic accounting of changes in location or scale. Note that both the linear and the quadratic terms (i.e., the terms with powers less than the cubic) must be included in the model since a fit of

$$Y = \beta_3 X^3 + \varepsilon$$

only allows for a scale change from the basic curve.

The last two curves in Figure 2.8 are, respectively, the exponential and logarithmic functions. The exponential function [written e^X or $\exp(X)$] can be defined as an infinite sum

$$e^X = 1 + X + \frac{X^2}{2} + \frac{X^3}{3 \cdot 2} + \frac{X^4}{4 \cdot 3 \cdot 2} + \dots$$

and is, therefore, a special type of polynomial. Many growth processes are modelled as an exponential function of time; e.g.,

$$Y = \alpha + \beta_1 e^t + \varepsilon$$

where Y is the size or amount of the substance being grown and t is the time the substance has been growing. The logarithmic function is the inverse of the exponential function, i.e., if $Y = e^t$ then $t = \ln(Y)$. The exponential and logarithmic functions are extensively tabulated and virtually all computer libraries contain algorithms to calculate both functions, as do many hand calculators. [Note: Transformations such as $-\ln(X)$ and $\exp(-X)$ enable the logarithmic and exponential functions to decrease similar to the curve in Figure 2.8(d) rather than increase as in Figures 2.8(e) and 2.8(f).]

2.3.2 Predictor Variable Transformations

One reason cited earlier for preferring to transform predictor variables when only a few two-variable plots show nonlinear relationships with the response is that respecification of the response variable could produce linear relationships with some predictor variables while simultaneously destroying its linearity with others. A second reason is that the response variable is often selected for analysis because it is the variable of interest whereas the predictor variables might be chosen only because of their presumed association with the response. The main objective in these situations is to formulate a prediction equation for the response, not a transformation of the response.

As an illustration of how to respecify a predictor variable, let us return to the two-variable plot of GNP versus INFD, Figure 2.4. The plot resembles either the curve in Figure 2.8(d), $Y = 1/X$, or an inverted Figure 2.8(f), $Y = -\ln(X)$.

Figure 2.9 is a plot of GNP versus ln(INFD). There is still a great amount of curvature evident in the plot. Since the graph of $Y = 1/X$ curves more sharply in Figure 2.8 than does that of $Y = \ln(X)$, perhaps the reciprocal of

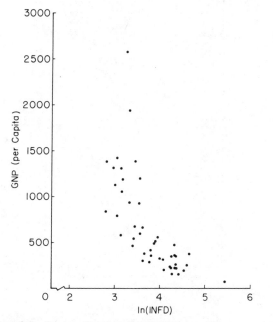

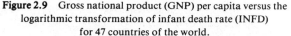

Figure 2.9 Gross national product (GNP) per capita versus the logarithmic transformation of infant death rate (INFD) for 47 countries of the world.

INFD, 1/INFD, or 1/(INFD)^2 might eliminate more of the curvature of GNP with INFD than did the logarithmic transformation. Figures 2.10 and 2.11 confirm this observation. (The fact that some transformations of INFD cause the scatter of points to decline while others cause the points to rise should not cause the reader to be concerned. The purpose of the transformation is to obtain a linear association between the variables; whether the transformation causes the points to rise or fall can be reversed merely by using the negative of the transformation, e.g., $Y = -\ln(X)$ or $Y = -1/X$.) The scatter of points in Figure 2.10 is clearly more linear than the original scatter and the plot of the logarithmic transformation. There does appear to remain a slight quadratic curvature evident in the figure. Perhaps we are again being deceived by looking too closely at a few points; nevertheless, a further transformation can be made to see what effect it has.

Figure 2.11 is a plot of GNP versus 1/INFD^2. Comparing Figure 2.10 and 2.11 it appears this last transformation does result in a more linear plot. There remain three or four straggler points in Figure 2.11 but the bulk of the data points do indicate a linear association.

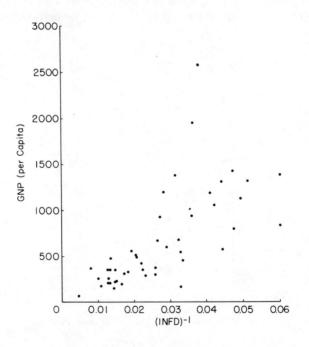

Figure 2.10   Gross national product (GNP) per capita versus the reciprocal of infant death rate (INFD) for 47 countries of the world.

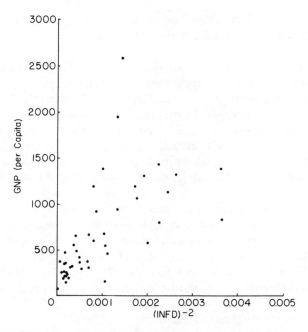

Figure 2.11 Gross national product (GNP) per capita versus the square of the reciprocal of infant death rate (INFD) for 47 countries of the world.

Whether ln(INFD), 1/INFD, or 1/INFD^2 is used as a predictor of GNP depends on two things: (i) is there a theoretical model which, *apart from this data*, dictates that one of these respecified infant death rates is related to gross national product; or (ii) does either simplicity of the fitted model or accuracy of prediction dominate when weighing the choices? These questions should be carefully considered before a final choice for the transformation is made. Further discussion of these points will follow in Section 2.3.4; for now, we are only attempting to point out the benefits of reexpressing predictor variables in terms of producing as linear a scatter plot as possible.

2.3.3 Linearization of the Response

The previous section expounded the use of transformations on predictor variables rather than the response variable when associations with the response are nonlinear. There are situations, however, when it is more desirable to transform the response. For example, if ln(Y) is linearly related to a set of prediction variables, X_1, X_2, ..., X_p, two-variable plots of Y versus the predictor variables will produce several plots with the appearance of the curve in Figure 2.8(e). Observing several plots of the same general relationship, therefore, should suggest the need to transform the response variable.

Figure 2.8 can guide the data analyst in selecting transformations of the response variable as it did reexpressions of the predictor variables. One only needs to make the inverse of the transformation indicated. If, for instance, cubic relationships are observed between the response and predictor variable, reexpressing the response as $Y^{1/3}$ should linearize the plots. Analysis of relationships such as is indicated in Figure 2.4 should suggest transformations of $1/Y$ or $1/Y^{1/2}$. The process of selecting the most appropriate respecification of the response is, as demonstrated in the last subsection, often one of trial and error.

As with transformations of the predictor variables, there are numerous respecifications of the response variable that can be advantageous. Hoerl (1954) provides an extensive series of graphs similar to Figure 2.8 that can be of great aid in selecting an appropriate transformation. Tukey (1977, Chapters 5-6) and Mosteller and Tukey (1977, Chapters 4-6) also discuss and give examples of many of the more important variable reexpressions.

2.3.4 Prediction Versus Model Building

Throughout much of Chapter 1 care was taken to differentiate the construction of regression models for prediction purposes from the formulation of theoretical models. The latter situation requires that not only must the response and predictor variables be chosen in advance of the data analysis but their functional relationships must be specified as well. This can be accomplished through theoretical arguments or as a result of previous empirical analyses. The reasons for insisting on this course of action for model building are somewhat clearer now that transformations have been discussed.

Consider again Figure 2.7(b). Had we allowed the smoothed plot of these data points to dictate a model to be fit, a quite complicated one would have resulted. We could indeed have selected one that would fit all the data points quite well. Then it would have been tempting to ascribe reasons for the initial decline and subsequent leveling off of the first year of college scores as a function of the increasing grade 13 scores. We could thereby have proposed a theory concerning the assessment of grade 13 scores as they relate to first-year college scores. On the basis of Figure 2.7(b) and the model fit to the data, we might claim that it is reasonable to admit students from high school 3 who have grade 13 scores between 60 and 65 and above 72!

Ignoring the difficulty of defending such an argument, conclusions are often drawn in this fashion. Based on the analysis of a single data set with no predetermined model stipulated, a regression model is fit to the data. The specification of the model might not employ plotting techniques but the inclusion or exclusion of predictor variables and whether quadratic or inter-action terms are inserted into the prediction equation might be determined by several analyses of the same set of data. Then the conclusions "implied" by the fitted model are drawn. The dangers inherent in this set of procedures are well-illustrated by further analysis of the grade prediction example.

First, the trends in Figure 2.7(b) are not evident in Figures 2.7(c) and (d), data on two other high schools in the same city whose students matriculated at Queen's University during the same year as those of high school 3. Thus, this trend must be peculiar to high school 3. Next, if the first-year college scores are smoothed instead of the grade 13 scores (left to the reader as an exercise), the trend disappears. There is no initial decline in the scores or a flattening out at the end.

By comparing the data for high school 3 with high schools 1 and 2, we are actually examining a proposed theoretical model with other data sets and finding that the proposed model is invalid. By comparing the smoothed values of Figure 2.7(b) with those obtained by smoothing the first year of col-lege averages, we are in the fortunate position of rejecting the proposed model because it does not allow for internal consistency within the same set of data; i.e., we would like a theoretical model to be consistent with a set of data regardless of how it is smoothed since smoothing aims to reduce the variabil-ity in the data and not destroy true relationships.

What if our goal, however, is not model building but only prediction? When one desires only that a good prediction equation be found it is not always crucial that correct models relating response and predictor variables be formulated. If one desires to predict first year of college average grades from the grade 13 averages for students with grade 13 averages between 70 and 80, it appears that fitting a straight line to the data from any of the high schools (or all three combined in a single data base with additional variables designating the high schools) will provide adequate prediction. This does not imply that the two scores are functionally related by a straight line, only that over this restricted range of grade 13 averages a straight line is consistent with all the data sets.

If one desires to predict outside the 70-80 range on grade 13 averages, how-ever, more is demanded of the fitted model. The plot for high school 2 hints at a curved relationship, perhaps quadratic, as one views the entire range of the data. A curved relationship is not inconsistent with the flattening in Figure 2.7(b) either. Hence in order to predict at the extreme ends—certainly if one desires to extrapolate beyond the range of this data—a theoretical model must be estimated. But, as argued above, the theoretical model should not be specified and verified just by an examination of this one data base.

EXERCISES

1. Construct a crosstab of GNP and PHYS, and GNP and HIED including all countries in Data Set A.1 except Hong Kong and Singapore. What trends appear to be exhibited in these two crosstabulations? Confirm the trends by plotting scatter diagrams.

2. Add Hong Kong and Singapore to the crosstabs and plots constructed in Exercise 1. Do these plots appear to be outliers? Do they appear to be outliers in the relationship between GNP and DENS?

3. Define indicator variables appropriate for specifying college classification: freshman, sophomore, junior, and senior. Is this an appropriate specification if one wishes to relate a student's summer earnings (Y) to the number of years of college completed (X)?

4. Figure 2.6 is a plot of first-year college average versus grade 13 average for each high school. Does this plot suggest that interaction terms involving the high school and grade 13 averages would be beneficial in predicting first-year college average?

5. A quick method for approximating the slope of a scatter of points is to pick a representative point at each end of the swarm of points and divide the difference of the Y values by the corresponding difference in the X values (see the Introduction to Chapter 3). Approximate the slopes for the smoothed points in Figure 2.7(b), (c), and (d). Do the calculated slopes tend to confirm or dispute the answer to Exercise 4?

6. Construct a table similar to Table 2.6 for each high school by smoothing the grade 13 average instead of the first year of college averages. Plot the smoothed grade 13 average versus the original first year of college averages. Do the same trends appear in these plots as appeared in Figures 2.7(b), (c), and (d)?

7. Let c denote any nonzero constant. Expand the algebraic expression $\beta_3(X - c)^3$, thereby showing that the inclusion of linear and quadratic terms along with a cubic term in the specification of a regression model can account for shifts in the location of the cubic relationship between a response and predictor variable.

8. Using a hand calculator or tables of the exponential function calculate $\exp(GNP)$ for the 47 countries plotted in Figure 2.9. Make graphs of

exp(GNP), 1/GNP, and 1/GNP$^{1/2}$ versus INFD. Compare these graphs with Figures 2.9 to 2.11. If it is more desirable to transform the response rather than the predictor variable, which transformation would you suggest?

9.   From the plots already constructed of GNP versus each of the predictor variables in Data Set A.1 (see Figures 2.2, 2.4, 2.5, and Exercises 1 and 2) does it appear that GNP or a subset of the predictor variables should be transformed before attempting to fit a regression model?

CHAPTER 3

SINGLE-VARIABLE LEAST SQUARES

Theoretically the fitting of a straight line poses no great difficulty. By selecting any two points on the line $Y = \alpha + \beta X$, both the slope (β) and the intercept (α) coefficients are readily calculated. For example, from any two points (X_1, Y_1) and (X_2, Y_2) on the line $Y = 5 + 2X$ the slope can be found by the equation

$$\beta = \frac{Y_2 - Y_1}{X_2 - X_1} \ ,$$

the familiar "rise" divided by "run" expression. Once β has been obtained, the intercept can be found using either point by solving for α in the equation of the straight line:

$$\alpha = Y - \beta X.$$

Using the points $(1,7)$ and $(4,13)$, both of which lie on the line $Y = 5 + 2X$,

$$\beta = \frac{13 - 7}{4 - 1} = 2$$

and $\alpha = 7 - (2)(1) = 5$ or $\alpha = 13 - (2)(4) = 5$. Any other pair of points can also be used to solve for exactly the same values of the model coefficients.

While this procedure provides exact, correct coefficients using points that lie on a straight line, experimental points rarely (if ever) do so. This is indicated by the expression generally used to denote a regression model with a single predictor variable:

$$Y = \alpha + \beta X + \varepsilon.$$

The additional term in this model, ε, signifies that for a fixed value of the predictor variable the observed value of the response will not lie exactly on the line $Y = \alpha + \beta X$, but will be in "error" by an amount ε. Several observations for the same value of X, moreover, all have different errors associated with them so that Y tends to fluctuate around the value $\alpha + \beta X$.

Because the responses are subject to errors, fitting a theoretical model is no longer a simple task. Examine the data listed in Table 3.1. These figures

Table 3.1. Average Charges for Full-Time Undergraduate Resident Students in Institutions of Higher Education, Selected Years from 1964-65 ($t = 1$) to 1973-74 ($t = 10$)

Year (t)	Tuition and Required Fees*
1	$403
3	433
5	429
8	473
10	494

*in 1974-75 dollars.
Source: Adapted from U. S. Department of Health, Education, and Welfare, National Center for Education Statistics (1976), Table 44, p. 104.

represent the average expenses of full-time resident undergraduate students in institutions of higher education for selected years from 1964-65 ($t = 1$) to 1973-74 ($t = 10$). One of the goals in analyzing this data is to be able to estimate (interpolate) the expenses for those years between 1964-65 and 1973-74 for which data are not available and to project (extrapolate) beyond 1973-74. Figure 3.1 is a plot of this data. There is no apparent curvature and no transformation seems necessary.

One theoretical model postulated between these two variables based on Figure 3.1 would be a straight line; i.e., if Y represents tuition and required fees and t indicates time, a theoretical model could be formulated as $Y = \alpha + \beta t + \varepsilon$. Due to the errors associated with the responses it is clear from

Figure 3.1 that the data points do not all lie on a straight line. In fact, using the slope and intercept formulas presented above leads to confusion because the results are not unique. Table 3.2 lists estimates of the slope and intercept parameters that are calculated from all ten possible pairs of points from the original five points in Table 3.1.

It should be apparent from Table 3.2 why the slopes and intercepts that are calculated from a data set are referred to as "estimates" and not

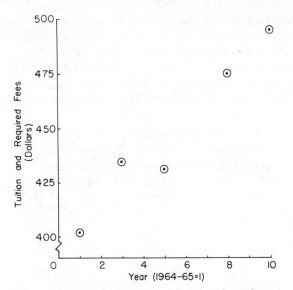

Figure 3.1   Average charges per full-time undergraduate resident, public institutions of higher education.

Table 3.2. Estimated Regression Coefficients Using Pairs of Points from Table 3.1

Time Points	Estimated Coefficients	
	Intercept	Slope
1,3	$388.00	$15.00
1,5	396.50	6.50
1,8	393.00	10.00
1,10	392.89	10.11
3,5	439.00	–2.00
3,8	409.00	8.00
3,10	406.87	8.71
5,8	355.65	14.67
5,10	364.00	13.00
8,10	389.00	10.50

"parameters." Unlike the example at the beginning of this chapter, when the responses are subject to unknown errors, unique solutions for the parameters cannot be obtained from pairs of points. The values in Table 3.2 must, therefore, be considered estimates of the true parameters. Yet these estimates vary greatly and their effect on the prediction equation can be quite different, as illustrated in Figure 3.2.

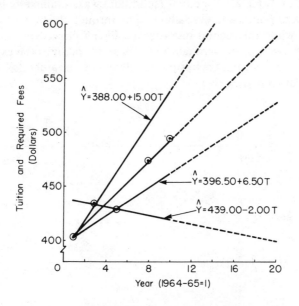

Figure 3.2 Estimated ($t = 1$ to $t = 10$) and projected ($t = 10$ to $t = 20$) average charges per full-time undergraduate resident, public institutions of higher education.

The lines plotted in Figure 3.2 include three of the extreme sets of parameter estimates from Table 3.2, those based on the pairs of points (1,3), (1,5), and (3,5). None of these fitted lines appear to approximate the five data points very well despite the fact that each passes through two of the five points. The unlabeled line in Figure 3.2, on the other hand, seems to approximate the points reasonably well although it does not pass exactly through any of the points. This line is similar, but not equal to, the lines that one would obtain using the estimates for either of the pair of points (1,8) or (1,10). It is the "least squares" fitted line,

$$\hat{Y} = 393.80 + 9.74t$$

which we will discuss shortly.

From the discussion of this simple example, one might be tempted to conclude that a decent fit to the points in Table 3.2 can be obtained by "eyeballing" a line through the points in Figure 3.2. Perhaps this is true for so small and well-behaved a data set. But how does one "eyeball" a more complicated data set such as the GNP data in Figure 2.4 or the high school grade data in Figure 2.6. Figure 3.3 shows two "eyeball" fits to the relatively simple data set of Table 2.1. The first (solid line) was obtained by ignoring the first three data points and eyeballing a line through the bulk of the remaining points, while the second one (dashed line) followed the same process only the last two points were ignored. These are perhaps the extremes that would be attempted, but each is potentially some researcher's "best" fit as is any line between these two.

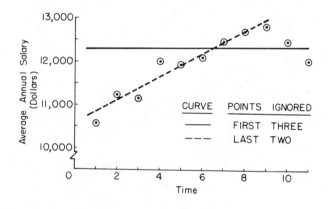

Figure 3.3   Visual fits to salary data.

Crude model fitting techniques such as this suffer not only from subjectivity but also from the lack of desirable statistical properties. Least squares parameter estimators have long been advocated because of their optimal statistical properties. Following a discussion of the appropriateness of regression models containing a single predictor variable, the least squares estimators will be introduced.

3.1   APPROPRIATENESS

Undoubtedly the simplest regression models to analyze are those that contain a single predictor variable. With this thought in mind, many authors refer to such models under the title of "Simple Linear Regression." Yet when one considers the uses and interpretations of single-predictor-variable models, the inherent problems associated with the models are by no means "simple" to overcome. Perhaps the nomenclature derives most from the relative ease of the computations.

Before discussing the fitting of regression models containing one predictor variable, we wish to touch on some of the considerations that should be made prior to deciding that such a model is appropriate for the problem being undertaken. These considerations include both theoretical and empirical facets; i.e., is there a theoretical basis for utilizing only one predictor variable and does the fitted model adequately approximate the response variable? Negative answers to either of these questions could cause one to reevaluate the assumed model.

3.1.1 Theoretical Validity

In the mathematical and physical sciences theoretical models are commonplace. Few would question the validity of models such as

$$d = vt,$$

relating the distance (d) an object travels to the time (t) it travels and the constant velocity (v) of the object; nor would many object to the model

$$F = ma$$

relating the force (F) exerted in moving a mass (m) with a constant acceleration (a). Even these well-defined theoretical models are never exact in practice for it is virtually impossible to travel at a constant velocity or exert a constant acceleration just as it is not possible to measure distance, time, force, and mass without error.

An analog to these theoretical models which incorporates errors is constructed by adding an additional term to the models:

$$d = vt + \varepsilon \quad \text{and} \quad F = ma + \varepsilon. \tag{3.1.1}$$

Presumably, the errors are small and random relative to the responses being measured, d and F. Without complicating the situation any further, one basic principle that is crucial to the understanding and correct interpretation of fitted regression models can now be appreciated: *all theoretical regression models are approximations.*

Despite the errors associated with the models in eqn. (3.1.1), we are generally quite pleased when we have theoretical arguments from which to construct regression models. This is because the theory surrounding the model has confirmed that all the relevant constituents (predictor variables) have been included. Or have they?

Empirical studies usually do not operate under the type of idealized conditions needed for models such as (3.1.1) to be valid unless the experiments are performed in well-equipped laboratories. Ordinarily factors other than the theoretical ones influence the response variable. Wind or surface friction, air resistance, temperature and humidity, inaccuracies in the measuring instruments—all contribute errors that may be nonnegligible relative to the responses. In some circumstances a more appropriate model for distance traveled might be

$$d = vt + uf + wr + \varepsilon, \tag{3.1.2}$$

where f and r represent friction and air resistance, respectively, and u and w are constants. The danger in basing regression models solely on theoretical considerations is that model (3.1.1) will be assumed without regard to potentially important intervening factors such as those indicated in eqn. (3.1.2). Lest the reader feel that we are downgrading the use of theoretical models, we wish to stress that our intent is to point out (i) that theoretical models are indeed of great value since they identify the necessary constituent terms of a regression model, but (ii) the suggested model may be adequate only under restricted conditions. Therefore, theoretical models should be examined carefully for their empirical validity.

In many research studies not only are there no theoretical models to guide the investigations but the factors or variables of interest are not unique. Both a multiplicity of variables influencing a response and a multiplicity of representations of the response and predictor variables increase the complexity of developing theoretical models in such studies. Advocating a single predictor variable for a theoretical regression model in these circumstances should be done with extreme care.

To illustrate these latter points, consider the data on homicide rates and firearm availability listed in Table 3.3. One intent of the study from which the data are taken was to investigate the role of firearms in accounting for the homicide rate in Detroit. No exact accounting can be made of the availability of all firearms since most weapons are not under state or federal regulation. Handguns, however, must be registered and one must obtain a license to purchase a handgun, both Michigan state statutes.

Neither of these predictor variables can be said to adequately measure the availability of firearms, as discussed by Fisher (1976). Only handguns are required to be licensed and registered, and many that are illegally obtained are not. Moreover, those firearms owned prior to 1961, while available for use as weapons, are not included in the licenses or registrations for 1961-1973. Also, the two measures of firearm availability cannot be considered equivalent. Individuals may obtain licenses to purchase handguns and not

Table 3.3. Detroit Homicide Rate and Two Indicators of Firearm
Availability: 1961-1973

| Year | Homicide* Rate | Firearm Availability* | |
		Licenses Issued	Handguns Registered
1961	8.60	178.15	215.98
1962	8.90	156.41	180.48
1963	8.52	198.02	209.57
1964	8.89	222.10	231.67
1965	13.07	301.92	297.65
1966	14.57	391.22	367.62
1967	21.36	665.56	616.54
1968	28.03	1,131.21	1,029.75
1969	31.49	837.60	786.23
1970	37.39	794.90	713.77
1971	46.26	817.74	750.43
1972	47.24	583.17	1,027.38
1973	52.33	709.59	666.50

*Number of homicides, licenses, and registrations per 100,000 population.
Source: Fisher (1976). (See Appendix, Data Set A.3.)

eventually do so, or they may fail to register the handgun. It is also possible that some individuals may fail to obtain licenses for handguns but register them at a later date; e.g., if a resident moves into Detroit from another state. Finally, the time lag between the acquiring of a license and the registration of the handgun is unknown. This could affect the relative counts for each year.

Suppose now that two researchers theorize that the homicide rate (Y) is directly related to the availability of firearms (X) in a linear fashion:

$$Y = \alpha + \beta X + \varepsilon.$$

Not only does the use of this model postulate that there are no other intervening predictor variables, it also presumes a unique definition of the predictor variable, X. If one researcher decides to use the rate of license issuance (X_1) as the predictor variable and the other believes that the rate of registrations (X_2) is the correct formulation of the model, the resulting prediction equations for the homicide rate are, respectively,

$$\hat{Y} = 4.91 + 0.0376X_1 \quad \text{and} \quad \hat{Y} = 1.66 + 0.0430X_2.$$

While similar, these prediction equations are not identical and it is apparent from Figure 3.4 that neither fit is extremely good.

Figure 3.5 shows two good fits to the homicide data, each using a prediction equation with one predictor variable. *Subsequent* to obtaining fits like

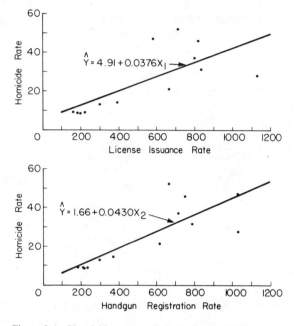

Figure 3.4   Homicide-rate prediction using two indicators
of the availability of firearms.

these, researchers sometimes postulate models and contend that the data
support the model. This is a true statement, but the data actually support
several possible models. Again, if two researchers were to adopt this philos-
ophy of specifying theoretical models, two completely different theoretical
models could be "proven." The two predictor variables (size of the police
force, rate of clearance of homicides) measure different characteristics and
the prediction equations cannot be considered equivalent despite the fact
that they both approximate the homicide rate well. One could argue that the
full-time police and clearance rates should be influencing the homicide rate
but it is not possible to defend a causal relationship between homicide rate
and either predictor variable on the basis of this data alone. Figure 3.5
merely reveals that both predictor variables change similarly to the homicide rate over these 13 years.

   The admonition that this example is intended to produce is that both
theoretical models and empirical verification are necessary for the successful specification of a theoretical regression model whose purpose is to relate
a response variable with predictor variables. Because of the close concommitant variation between the homicide rate and both of the variables in

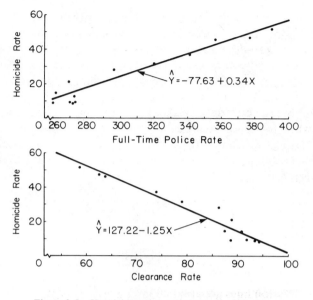

Figure 3.5   Homicide-rate prediction using full-time
police rate and clearance rate.

Figure 3.5, either of them could be used to predict the homicide rate—if that is one's only goal. As discussed in Section 1.4, however, much more is demanded if one wishes to infer cause-effect from a regression model. In this case one would need theoretical justification for the inclusion of the predictor variables, empirical verification that the fitted model is an adequate approximation to the responses, and substantiation of the model under a wide range of conditions.

3.1.2   Approximations in Restricted Regions

A second reason for considering fitted or theoretical regression models to be approximations is that one often does not collect data on the response variable over the entire range of interest of the predictor variables. An obvious example of this is when one wishes to predict into the future as was discussed in Section 1.4.1. Here time would be used as a predictor variable but one can only collect data up to the present. Conceivably one could construct a theoretically correct regression model for the time-span of available data, but the model could change in the future.

Similarly, due to a limited range of available data one could approximate a response variable well yet not have a correct theoretical model. Consider, for example, the fitting of suicide rate data. Many theoretical models have been proposed to "explain" suicides, as well as merely to describe trends in suicidal behavior. Let us oversimplify the problem and suppose that one is interested in describing suicide rates as they vary with age. A plausible hypothesis is that suicide rates increase in a steady, perhaps linear, fashion with age up to, say, age 65 when they may rise sharply due to forced retirement.

Now suppose the only data available to study this question are the suicide rates for white males and females from the 25-29 to 50-54 age brackets in Table 3.4. Using only this data, the two fitted models for the suicide rate for males ($\hat{Y}_1$) and females ($\hat{Y}_2$) as a function of the midpoint (X) of each age group are, respectively,

$$\hat{Y}_1 = -13.90 + 0.93X$$

and

$$\hat{Y}_2 = -1.02 + 0.23X.$$

The fitted regression lines are graphed in Figure 3.6 along with the observed responses and both fits appear to be excellent.

Knowing the dangers of extrapolation, we should be careful not to attempt to predict very far outside the 25-29 and 50-54 age groups. It does

Table 3.4. Average Annual Suicide Rates, 1959-1961 White Males and Females, by Age

Age	Suicide Rate*	
(Years)	Male	Female
15-19	5.6	1.6
20-24	11.4	3.1
25-29	13.2	5.1
30-34	16.0	6.5
35-39	19.6	7.6
40-44	24.5	8.4
45-49	29.9	9.8
50-54	36.4	10.9
55-59	40.1	10.5
60-64	41.0	10.6
65-69	40.4	9.5
70-74	46.5	9.1
75-79	54.1	8.8
80-84	65.1	7.8

*Number per 100,000 population.
Source: Gibbs (1969). ©1969 by The University of Chicago. Reprinted with permission.

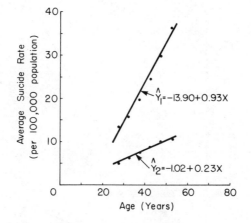

Figure 3.6 Suicide-rate prediction equations, U. S. white males ($\hat{Y}_1$) and females ($\hat{Y}_2$), 25-29 to 50-54.

seem reasonable, however, in view of the good fits, that one could predict adequately for a few age categories beyond those used to calculate the fits. It also seems reasonable to conclude that the original hypothesis is supported by this analysis. Combining these two conclusions, one could summarize by inferring that the suicide rates for both white males and white females is linearly related with age, at least for age groups, say 15-19 to 60-64. One might even acknowledge that this statement cannot be considered exactly true, particularly for the age brackets outside 25-29 to 50-54, but the approximation should be adequate for most purposes.

The degree to which the approximation is truly adequate can be gauged from Figure 3.7. It is apparent from this figure that the two predictors do not yield accurate prediction outside the span of ages for which they were calculated. Within the age groups 25-29 to 50-54 the curves are well-approximated by straight lines but outside these ages both suicide rates are decidedly nonlinear. Once again the danger of extrapolation is clear: the fitted model is only an approximation to the suicide rates and a poor approximation outside the age brackets used to obtain the fit.

3.1.3   No-Intercept Models

The final topic we wish to discuss prior to actually fitting single-variable models concerns the question of whether an intercept parameter should be included in the model specification. For a single predictor variable the two choices of a theoretical model are

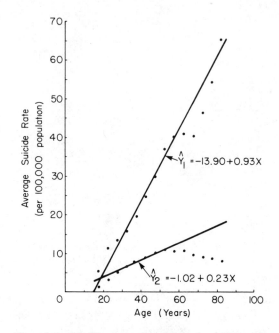

Figure 3.7 Suicide rate prediction equations, U. S. white
males ($\hat{Y}_1$) and females ($\hat{Y}_2$) computed from 25-29 to
50-54 age groups, compared with suicide rates
from 15-19 to 80-84.

$$Y = \alpha + \beta X + \varepsilon \quad \text{and} \quad Y = \beta X + \varepsilon,$$

the latter model being referred to as a "no-intercept" model. The first point
to note about no-intercept models is that, apart from random error, they
require the response to be zero when the predictor variable is zero. Fitted
models, $\hat{Y} = \hat{\beta}X$ where $\hat{\beta}$ is an estimate of β, likewise have this property. If
theoretically the response variable is not required to be zero (again, apart
from random error) when the predictor variable is zero, models without intercepts should not be employed.

Even if the response variable is theoretically zero when the predictor variable is, this does not necessarily mean that the no-intercept model is appropriate. If the specified model is actually only an approximation to the true
(unknown) model, especially if the model is fit over a limited range of the
predictor variable, then forcing the approximation through the origin can
result in a very poor fit.

Let us examine the fit to the male suicide rates for a no-intercept model.
Theoretically one could argue that no suicide can occur at age zero. Consequently, a straight-line regression of suicide rate on age should pass through

the origin; i.e., a should be zero. Assuming a no-intercept model, a prediction equation for the male suicide rates for the age groups 25-29 to 50-54 is

$$\hat{Y} = 0.60X.$$

The estimated slope coefficient is now about two-thirds of its value for the prediction equation containing an intercept term. Figure 3.8 reveals that the no-intercept prediction equation fits very few of the observed suicide rates; it even predicts the suicide rates for the 25-29 to 50-54 age groups poorly.

The added restriction of no intercept term in a prediction equation can, as it did here, ruin the approximation to the responses because it requires that the line pass through the origin. The true relationship may be curved as in the nonlinear fit drawn in Figure 3.8. Despite the curvature, over a large

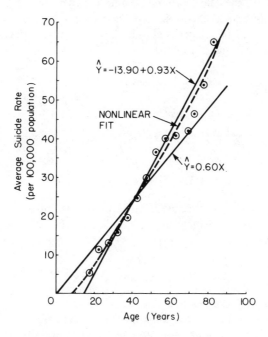

Figure 3.8 Comparison of fits to the male suicide rates.

region of age groups a straight line (with an intercept) approximates the male suicide rates well. If the intercept model is calculated over the entire range of age groups in Table 3.4, the linear approximation may be sufficiently accurate for a fairly wide age spread. But if the line is forced through the origin, the approximation is ruined. If one insists on a no-intercept

model for the male suicide rates, data transformations or nonlinear models
must be tried in order to obtain an adequate fit.

3.2 LEAST SQUARES PARAMETER ESTIMATION

Among the many possible estimators of coefficients in a linear regression
model, least squares parameter estimation is overwhelmingly the most popular. Several characteristics of least squares estimation contribute to its
popularity. Least squares estimators are comparatively easy to compute.
They have known probability distributions. They are unbiased estimators of
the regression parameters and, while they are not the only unbiased estimators, they have the smallest variances of all unbiased linear functions of the
responses under some relatively mild assumptions on the true model. These
characteristics of the least squares estimators will be explored further in
subsequent chapters. At this point suffice it to say that least squares estimation historically has been accepted almost universally as the best estimation
technique for linear regression models.

3.2.1 The Principle of Least Squares

If one wishes to predict responses for the model

$$Y = \alpha + \beta X + \varepsilon, \tag{3.2.1}$$

it is natural to select a prediction equation of the same form:

$$\hat{Y} = \hat{\alpha} + \hat{\beta} X, \tag{3.2.2}$$

where $\hat{\alpha}$ and $\hat{\beta}$ are estimates of the true intercept and slope coefficients α
and β, respectively. A measure of how well $\hat{Y}$ predicts the response variable
Y is the magnitude of the residual, r, defined as the difference between Y
and $\hat{Y}$:

$$r = Y - \hat{Y} = Y - \hat{\alpha} - \hat{\beta} X. \tag{3.2.3}$$

One would like all the residuals to be as small as possible, ideally zero.
Unless $\varepsilon = 0$ in eqn. (3.2.1) for all responses, all the observed responses will
not lie exactly on a straight line and no prediction equation of the form
(3.2.2) will fit the data points exactly; hence, no linear prediction equation
can make all the residuals zero.

Returning to the tuition costs in Table 3.1, Figure 3.9 shows the residuals
for two possible fits to the data. In the first fit, the last three points have
been fit extremely well (intentionally). The first two points, however, have
large residuals. This is characteristic of attempts to force some of the residuals to be very small—others simultaneously can be forced to be large. The

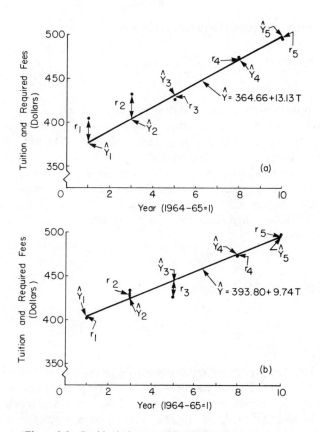

Figure 3.9 Residuals from two fits to the data in Table 3.1:
(a) Precise fit to last three points, (b) Least squares fit.

second fit, the least squares prediction equation, compromises by making the average size (in a sense to be defined shortly) of the residuals small. None of the residuals are zero, but neither are any of the residuals as large as the first two in the previous fit.

Least squares estimators of the regression coefficients minimize the average squared residual*

$$n^{-1} \sum_{i=1}^{n} r_i^2 = n^{-1} \sum_{i=1}^{n} (Y_i - \hat{\alpha} - \hat{\beta} X_i)^2 \,,$$

---

*Unless specifically stated otherwise all summations in this text are from $i = 1$ to $i = n$; i.e., the following three expressions are equivalent:

$$X_1 + X_2 + \ldots + X_n = \sum_{i=1}^{n} X_i = \Sigma \, X_i \,.$$

or, equivalently, the sum of squared residuals

$$\sum_{i=1}^{n} r_i^2 = \sum_{i=1}^{n} (Y_i - \hat{\alpha} - \hat{\beta}X_i)^2 . \tag{3.2.4}$$

The least squares estimators, $\hat{\alpha}$ and $\hat{\beta}$, are thus chosen so that the sum of the squared residuals (or the average squared residual) is as small as possible. At first this might seem like an unusual criterion. Why not simply make the average residual as small as possible?

An answer to this latter query is that one can always force the average residual to be zero, yet there are many choices of α and β that will do so and none of these choices may make *individual* residuals small. For instance if we let $\hat{\alpha} = \overline{Y}$, the average response, and $\hat{\beta} = 0$, then each of the n responses will be predicted by

$$\hat{Y}_i = \overline{Y}.$$

The average residual for this estimator is

$$n^{-1}\sum_{i=1}^{n} r_i = n^{-1}\sum_{i=1}^{n} (Y_i - \overline{Y}) = \overline{Y} - \overline{Y} = 0 . \tag{3.2.5}$$

In Figure 3.9, this prediction equation would be represented by a horizontal line at $\hat{Y} = 446.40$. The smallest residual (ignoring signs) for the five data points using this prediction equation is 13.40 and the largest is 47.60, whereas the smallest residual for the least squares prediction equation is 0.54 and the largest one is 13.50. Hence the largest residual from the least squares predictor (obtained by minimizing the average squared residual) is approximately the same size as the smallest residual using the constant predictor (obtained by minimizing the average residual).

The reason minimizing the average residual does not guarantee good prediction is that very large positive residuals can cancel numerically with very large negative residuals in the summation in eqn. (3.2.5). The five residuals for the tuition cost data using $\hat{Y} = 446.40$ are:

$$-43.40, -13.40, -17.40, 26.60, 47.60.$$

None of these residuals is small but the negative residuals add up to -74.20 and the positive ones sum to $+74.20$, the average residual being zero.

In minimizing the sum of the squared residuals with least squares estimators, the magnitudes of the individual residuals are made small since any large residual, positive or negative, results in a large squared residual and a large sum of squared residuals, eqn. (3.2.4). The five residuals for the tuition cost data using the least squares prediction equation are:

$$-0.54, 9.98, -13.50, 1.28, 2.80.$$

Surprisingly, the sum of the positive least squares residuals almost equals the sum of the negative ones, 14.04. If enough decimal places were carried in all calculations, the sum of the positive residuals would exactly equal the sum of the negative ones. The least squares estimator, it turns out, minimizes both the sum of the residuals and the sum of the squared residuals.

The values of $\hat{\alpha}$ and $\hat{\beta}$ that minimize the sum of squared residuals, eqn. (3.2.4), are*

$$\hat{\beta} = \frac{\sum_{i=1}^{n} X_i Y_i - n\overline{X}\overline{Y}}{\sum_{i=1}^{n} X_i^2 - n\overline{X}^2} \qquad \text{and} \qquad \hat{\alpha} = \overline{Y} - \hat{\beta}\overline{X}. \qquad (3.2.6)$$

Using these formulae for the estimators of α and β, the principle of least squares guarantees that the sum of squared residuals will be as small as possible for all prediction equations of the form (3.2.2); i.e., $\hat{Y} = \hat{\alpha} + \hat{\beta}X$. This is the principle of least squares as applied to model (3.2.1).

All the prediction equations employed in previous sections, except the no-intercept predictors, were obtained by estimating the model parameters using eqn. (3.2.6). For example, the prediction of average annual educational costs for the data in Table 3.1 has the following summary statistics:

$$\overline{X} = 5.40 \qquad \Sigma X_i^2 = 199.00$$

$$\overline{Y} = 446.40 \qquad \Sigma X_i Y_i = 12,571.00 .$$

Hence the least squares estimates of α and β are

$$\hat{\beta} = \frac{12,571.00 - 5(5.40)(446.40)}{199.00 - 5(5.40)^2} = \frac{518.20}{53.20} = 9.74$$

---

*As with many technical derivations in this book, the derivations of these estimators is relegated to the chapter appendix so the continuity of the presentation will not be impaired. The derivation employs calculus; some readers may wish to skip the appendix and can do so without loss of continuity or essential detail.

and

$$\hat{\alpha} = 446.40 - (9.74)(5.40) = 393.80,$$

so that the least squares prediction equation is $\hat{Y} = 393.80 + 9.74X$.
   If one desires to use a no-intercept model,

$$Y = \beta X + \varepsilon, \tag{3.2.7}$$

a prediction equation of the form

$$\hat{Y} = \hat{\beta} X \tag{3.2.8}$$

is required. The least squares estimator of β in eqn. (3.2.8) is not the same as
is given in eqn. (3.2.6) for intercept models. This is because the sum of
squared residuals for the no-intercept model, namely

$$\Sigma r_i^2 = \Sigma (Y_i - \hat{\beta} X_i)^2, \tag{3.2.9}$$

only needs to be minimized by the choice of $\hat{\beta}$ and not by the simultaneous
choice of $\hat{\alpha}$ and $\hat{\beta}$ as for intercept models. The least squares estimator
obtained by minimizing eqn. (3.2.9) is

$$\hat{\beta} = \frac{\displaystyle\sum_{i=1}^{n} X_i Y_i}{\displaystyle\sum_{i=1}^{n} X_i^2}. \tag{3.2.10}$$

   The no-intercept fit to the male suicide rates for age groups 25-29 to 50-54
in Table 3.4 (see Figure 3.8) is computed as follows (using class midpoints
of 27.5, 32.5, ..., 52.5 for the six age groups):

$$\Sigma X_i Y_i = 5{,}990.50 \qquad\qquad \Sigma X_i^2 = 10{,}037.50$$

$$\hat{\beta} = \frac{5{,}990.50}{10{,}037.50} = 0.60 \,.$$

The six least squares residuals are

$$-14.3, \; -3.50, \; -2.90, \; -1.00, \; 1.40, \; 4.90$$

and they sum to -15.4. The residuals from least squares models with no intercepts do minimize the sum of the squared residuals, eqn. (3.2.9), but they do not in general minimize the sum of the residuals (whose minimum value is zero). Hence the principle of least squares only guarantees that the sum (average) of squared residuals is minimized. For some models, such as linear regression models with intercepts, the sum (average) of the residuals is also minimized.

3.2.2 Model Assumptions

Since virtually all regression models are approximations, it is important to inquire under what assumptions the least squares prediction equation is derived. Must an assumed model be the correct theoretical model before the estimators can be effectively used in prediction equations? Can the assumed model be misspecified and the estimated coefficients still yield adequate prediction equations? Can the estimated coefficients be interpreted similarly to the true parameters?

These and other questions will be answered in some detail in Chapter 6. We wish to stress one important point at this time: *no* assumptions about the correspondence between the specified models (3.2.1) or (3.2.7) and the true functional relationship between the response and predictor variable have been made. In order to derive the estimators (3.2.6) and (3.2.10) we did not have to assume that the specified models were the true functional or causal ones; we did not have to assume anything about the errors, ε; and we did not even have to assume that the single predictor variable, X, is the only one that influences the response, Y, or that it is the best single predictor variable. We have made no assumptions at all!

This point is at the heart of many missapplications of regression analysis. The prediction equations can be computed regardless of whether the true model has been completely and properly specified. Yet frequently—all too frequently—prediction equations are computed and inferences are drawn on estimated regression coefficients without regard to whether the prediction equation fits the data, much less any discussion of other possible predictors or alternative specifications of the predictor variables. Occasionally causality is inferred from a single-variable prediction equation with no stronger basis than the fact that the prediction equation can be calculated.

A linear prediction equation can be found for the female suicide rates in Table 3.4:

$$\hat{Y} = 2.92 + 0.10X.$$

Some would interpret this prediction equation as indicating that the female suicide rate increases linearly with age at about 0.10 per 100,000 population per year. From Figure 3.7 this is obviously not true for all age groups; nevertheless, simple plots are often not made and nonlinear trends such as with the female suicide rates are not detected. Assessment of the adequacy of the fit is also frequently ignored, perhaps due to a feeling that the principle of least squares insures that the best possible fit will be obtained. Least squares estimation merely guarantees that the sum of squared residuals will be minimized by the estimates of the parameters *for the model used* to obtain a predictor. It neither guarantees that the model used is the correct one nor that the sum of squared residuals for the resulting prediction equation will be small (only that they will be as small as possible *for the model used*).

3.2.3 Standardization

Transformation of the response and predictor variables are often used not only to linearize a theoretical or empirical relationship, but also to enable estimated regression coefficients to be more directly comparable. If one desires to predict gross national product (GNP) using infant death rate (INFD) as a single predictor variable, the magnitude of the estimated regression coefficient depends in part on the scale of measurement for both GNP and INFD. In Data Set A.1, INFD is measured as the number of infant deaths per 1,000 live births. What would happen to $\hat{\beta}$ if INFD is measured per 100 or per 10,000 live births? From the expression for $\hat{\beta}$ in eqn. (3.2.6), if one multiplies all the values of the predictor variable by a constant, one can verify that its estimated regression coefficient is divided by the same constant. One can also verify that $\hat{\alpha}$ is unchanged under this transformation.

Similarly, if all GNP values are multiplied by a constant, the estimated coefficients for both α and β are multiplied by the same constant. Adding a constant c to all X values does not change $\hat{\beta}$ but does change the estimator of α to $\hat{\alpha} - \hat{\beta}c$. Adding a constant to all Y values likewise does not alter the estimator of β but changes that of α to $\hat{\alpha} + c$.

In all these transformations of Y and X the underlying quantities have not been altered, only the numerical scales on which they are measured have been changed. If regression equations are calculated by different researchers who use different scales of measurement on the same variables, the resulting coefficient estimates are not directly comparable. In other words, unless one makes appropriate transformations so that each of the researchers measure the response variable on the same scale and, likewise, all the predictor variables are measured on the same scale, the magnitudes of the intercept and slope coefficient estimates cannot be meaningfully compared.

A similar problem arises if each researcher uses the same scales of measurement but characteristics of the individual samples indicate that the distributions of predictor or response variables are dissimilar. For example, if one were to assess the effectiveness of using I.Q. scores to predict a particular achievement score for two groups of students, the estimated coefficients for the two prediction equations could appear dissimilar if the *average* I.Q. scores differed appreciably for the two groups (similar to adding a constant to one set of I.Q. scores to obtain the other) or if the *variability* of the two sets of I.Q. scores differed greatly (similar to multiplying one set of scores by a constant to obtain the other). Here it is not the definitions of the variables that make the estimated coefficients difficult to compare but the distribution of scores of the predictor variables.

Conceptually an even more difficult problem arises when one wishes to determine the relative importance of several predictor variables in a multiple regression model. How can coefficients for all the predictor variables in Data Set A.1 be directly compared? All the predictor variables have different units of measurement. The predictor variable with the largest estimated coefficient might not be the most influential predictor variable. Its magnitude could be due mainly to the scale in which it was measured.

To overcome these problems data analysts often choose to standardize predictor variables when it becomes necessary to compare the magnitudes of estimated coefficients. Standardization is merely a transformation on a response or predictor variable that (i) eliminates all units of measurement and (ii) forces the standardized variables to have the same average and the same amount of variability (defined to be equal standard deviations).

Figure 3.10 contains plot of the distribution of observations on DENS and AGDS, two measures of the population density for the countries in Data Set A.1. As can be seen from the plot and is confirmed by the sample means and sample standard deviations recorded thereon, the distributions of the observations on these two variables are quite dissimilar. Predicting GNP(Y) using single-variable prediction equations for DENS(X_1) and AGDS(X_2) results in the following two least squares prediction equations:*

$$\hat{Y} = 728.64 - 0.51X_1 \quad \text{and} \quad \hat{Y} = 776.27 - 0.05X_2.$$

Since the distributions in Figure 3.10 are so dissimilar, it is not possible to infer much about the relative magnitudes of the coefficients in the two prediction equations.

In order to make these estimates more comparable, some form of standardization of the predictor variables is necessary. There are many possibilities. One choice is to mimic standard normal deviates; i.e., let

---

*In this analysis and all which follow, unless explicitly noted to the contrary, Hong Kong and Singapore are removed from Data Set A.1.

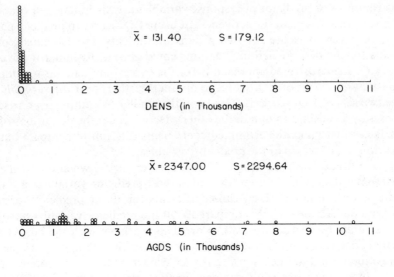

Figure 3.10 Distribution of observations for two measures of population density (truncated to multiples of 100 for ease of plotting).

$$Z_i = \frac{X_i - \overline{X}}{S_X},$$

(3.2.11)

where $\overline{X}$ is the sample mean of the population variable scores and S_X is the sample standard deviation of the scores:

$$\overline{X} = n^{-1} \sum X_i \qquad \text{and} \qquad S_X = \sqrt{\frac{\sum (X_i - \overline{X})^2}{n-1}}.$$

If this is done to all predictor variables of interest, then for each of them

$$\overline{Z} = 0 \qquad \text{and} \qquad S_Z = 1.$$

Estimates of the coefficients α^* and β^* for the transformed model

$$Y_i = \alpha^* + \beta^* Z_i + \varepsilon_i$$

can be found by applying eqn. (3.2.6) to the standardized predictor variable, yielding (see Appendix 3.B for details)

$$\hat{\alpha}^* = \overline{Y} \quad \text{and} \quad \hat{\beta}^* = \frac{\sum_{i=1}^{n} Z_i Y_i}{n-1} . \tag{3.2.12}$$

Coefficient estimates for the original (nonstandardized predictor-variable) model can be obtained from the standardized estimates through the relationships

$$\hat{\beta} = \frac{\hat{\beta}^*}{S_X} \quad \text{and} \quad \hat{\alpha} = \hat{\alpha}^* - \hat{\beta}\overline{X}.$$

If in addition to the predictor variable the response variable is also standardized by

$$Y_i^0 = \frac{Y_i - \overline{Y}}{S_Y} ,$$

least squares estimators of the parameters of the transformed model

$$Y_i^0 = \alpha^0 + \beta^0 Z_i + \varepsilon_i^0 ,$$

where $\varepsilon_i^0 = \varepsilon_i / S_Y$, are

$$\hat{\alpha}^0 = \overline{Y}^0 = 0 \quad \text{and} \quad \hat{\beta}^0 = \frac{\sum_{i=1}^{n} Z_i Y_i^0}{n-1} . \tag{3.2.13}$$

The estimated coefficients for this type of a regression model are referred to in the statistic and economic literature as "beta weights" and are related to the estimates from the original model by

$$\hat{\beta} = \frac{\hat{\beta}^0 S_Y}{S_X} \quad \text{and} \quad \hat{\alpha} = \overline{Y} - \hat{\beta}\overline{X}.$$

For reasons that will become clear in the next few chapters, a standardization of the predictor variables that is slightly different from eqn. (3.2.11) has become quite popular:

$$W_i = \frac{X_i - \overline{X}}{d_X} , \tag{3.2.14}$$

where

$$d_X = S_X \sqrt{n-1} = \sqrt{\Sigma (X_i - \overline{X})^2}.$$

The average $\overline{W}_i$ is zero ($W = 0$) and $d_W = 1$. The least squares estimators of α^* and β^* in the transformed model

$$Y_i = \alpha^* + \beta^* W_i + \varepsilon_i$$

are

$$\hat{\alpha}^* = \overline{Y} \quad \text{and} \quad \hat{\beta}^* = \sum_{i=1}^{n} W_i Y_i \qquad (3.2.15)$$

and are related to the original parameter estimates by

$$\hat{\beta} = \frac{\hat{\beta}^*}{d_X} \quad \text{and} \quad \hat{\alpha} = \hat{\alpha}^* - \hat{\beta}\overline{X}.$$

If both response and predictor variables are standardized similar to eqn. (3.2.14), the transformed model is

$$Y_i^+ = \alpha^+ + \beta^+ W_i + \varepsilon_i^+$$

where $Y_i^+ = (Y_i - \overline{Y})/d_Y$ and $\varepsilon_i^+ = \varepsilon_i/d_Y$. It is readily verified that

$$\hat{\beta}^+ = \frac{\hat{\beta}d_X}{d_Y} = \frac{\hat{\beta}S_X}{S_Y} = \hat{\beta}^0$$

and

$$\hat{\alpha}^+ = \frac{(\hat{\alpha}^* - \overline{Y})}{d_Y} = 0 = \hat{\alpha}^0.$$

The beta weights, therefore, are the same for both standardizations, despite the fact that the numerical values of the response and predictor variables differ for the two.

Prediction equations for GNP using standardized DENS and AGDS predictor variables are, respectively,

$$\hat{Y} = 661.17 - 91.97Z_1 \quad \text{and} \quad \hat{Y} = 661.17 - 112.53Z_2$$

using the normal deviate standardization and

$$\hat{Y} = 661.17 - 623.74\,W_1 \quad \text{and} \quad \hat{Y} = 661.17 - 763.21\,W_2$$

using the unit length standardization. In either set of prediction equations the coefficients for DENS and AGDS are similar, much closer than the order of magnitude difference for the coefficients of the original variables (–0.51 for X_1 and –0.05 for X_2). For both sets of standardized prediction equations the ratio of the standardized coefficient of AGDS to that of DENS is 1.22.

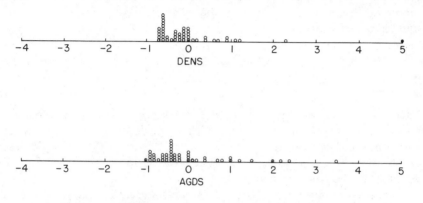

Figure 3.11   Distribution of standardized observations for two measures of population density (truncated to nearest tenth).

Visually the effect of standardization of predictor variables can be seen from Figure 3.11. Standardized observations for DENS and AGDS are plotted using transformation (3.2.11) (a plot for transformed values using eqn. (3.2.14) would appear virtually identical since each point would be multiplied by the same constant, $\sqrt{n-1} = \sqrt{46} = 6.78$). By comparison with Figure 3.10, the standardized densities are much more comparable than the two sets of original densities. It is, therefore, more appropriate to compare the standardized coefficient estimates than the ones from the unstandardized variables.

3.3   ADDITIONAL MEASURES OF FIT

Since a least squares prediction equation can always be calculated for single-variable models regardless of whether the assumed model is the correct representation of the true relationship between the response and the predictor variable, measures of the adequacy of the fit should be examined prior to an attempt to interpret the prediction equation.

A systematic approach to the assessment of the fit of a prediction equation is begun in this section. Basically, the techniques proposed in the following subsections attempt to summarize a few key pieces of information on the fit and enable the data analyst to obtain a quick intuitive feel for how well the prediction equation approximates the observed responses.

3.3.1 Partitioning Variability

Statistical analysis of data occurs because responses are variable. One of the simplest measures of the variability of a set of observations is the sum of the squared observations:

$$\sum_{i=1}^{n} Y_i^2 \,. \tag{3.3.1}$$

This is an appropriate measure of the variability of a set of observations because it is zero if and only if each and every observation is zero; otherwise, this quantity is some positive number. The sum of the squares of a set of observations, eqn. (3.3.1), is often referred to as the Total Sum of Squares (TSS) of the observations.

If the total sum of squares is not zero, it is natural to inquire why there is variability in the data. One possibility is that all the observations are equal to some nonzero constant. In this case there is no variability per se in the data, each data value is merely nonzero. A better measure of the variability in the data would be one that is zero when all the observations are constant, regardless of whether that constant is zero. Such a measure can be constructed from eqn. (3.3.1) as follows (see Appendix 3.C):

$$\sum_{i=1}^{n} Y_i^2 = \sum_{i=1}^{n} (Y_i - \overline{Y})^2 + n\overline{Y}^2 \,. \tag{3.3.2}$$

The first term on the right side of eqn. (3.3.2) is called the Adjusted Total Sum of Squares, or TSS(adj), and is zero only when each observation is equal to a constant. (Note: If $Y_i = c$ for $i = 1, 2, \ldots, n$, then $\overline{Y} = c$ and $Y_i - \overline{Y} = 0$.) TSS(adj) is, therefore, a true measure of the variability of a sample and is the numerator of the sample variance, S_Y^2, of the responses. The second term on the right side of eqn. (3.3.2) is the Sum of Squares due to the Mean of the sample, SSM, and is zero only if the sample mean, $\overline{Y}$, is identically zero. Symbolically eqn. (3.3.2) is representable as

$$\text{TSS} = \text{TSS(adj)} + \text{SSM}$$

and the total sum of squares of the responses is zero if and only if (i) each observation is equal to the sample mean, and (ii) the sample mean is zero. If TSS is not zero, eqn. (3.3.2) partitions the total variability in the sample into two components: variability attributable to deviations of the observations from the sample mean and variability due to a nonzero sample mean. TSS can be computed for the responses in Table 3.1 as follows:

$$SSM = 5 \cdot (446.4)^2$$

$$= 996,364.8$$

$$TSS(adj) = (-43.4)^2 + (-13.4)^2 + (-17.4)^2 + (26.6)^2 + (47.6)^2$$

$$= 5,339.2$$

$$TSS = SSM + TSS(adj) = 1,001,704.0.$$

The partitioning of TSS can be carried further when the response is modeled as a linear function of a single predictor variable (and also for multiple linear regression models as we shall see in Chapter 5):

$$\sum_{i=1}^{n} Y_i^2 = n\overline{Y}^2 + \sum_{i=1}^{n} (\hat{Y}_i - \overline{Y})^2 + \sum_{i=1}^{n} r_i^2 \tag{3.3.3}$$

This partitioning is derived in Appendix 3.C.

The first term on the right side of eqn. (3.3.3) is SSM. The second term measures the variability of the predicted values from the sample mean and is referred to as the Sum of Squares due to the Regression of Y on X, abbreviated SSR. From eqn. (3.2.6), $\hat{Y}_i = \hat{\alpha} + \hat{\beta}X_i$ differs from $\overline{Y}$ only if $\hat{\beta} \neq 0$; i.e., only if the regression of Y on X has a nonzero estimated regression coefficient. If X is of little value in predicting the response, $\hat{\beta}$ and SSR should be small. The final term in eqn. (3.3.3) is the Residual Sum of Squares or Sum of Squares due to Error, abbreviated SSE. It measures the variability in the responses that cannot be attributed to a nonzero mean or to the responses all falling on the fitted regression line. Ideally, SSE would be very small relative to SSR; i.e., the adjusted variability in the responses, TSS(adj), is mainly due to a nonzero estimated regression coefficient and not due to the failure of the responses to lie on the fitted line. If the fit is extremely good, the residuals are all small and so is SSE. Symbolically, eqn. (3.3.3) is

$$TSS = SSM + TSS(adj)$$
$$= SSM + SSR + SSE. \tag{3.3.4}$$

As with the estimation of regression coefficients, no assumptions about the true model are required in order to partition the total sum of squares as in eqn. (3.3.4). The derivation is, as shown above, merely an algebraic manipulation. The various terms in eqn. (3.3.4) allow the total variation in the responses to be assigned to three sources: a nonzero mean, differences in predicted values from a constant mean (i.e., a nonzero estimated regression coefficient, $\hat{\beta}$), and unexplained variation about the estimated regression line (i.e., residual variation).

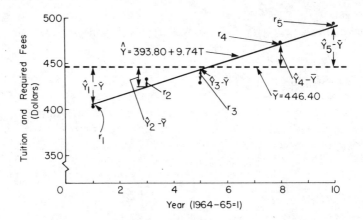

Figure 3.12   Partitioning responses into components for the mean, deviations of predictions from the mean, and residuals.

Figure 3.12 partitions each of the responses in Table 3.1 into the three components just described. Apart from the large nonzero sample mean, $\overline{Y} = 446.4$, most of the variation about the mean, TSS(adj), is attributable to a nonzero estimated regression coefficient; i.e., most of the variation in $Y_i - \overline{Y}$ values is ascribable to variation in $\hat{Y}_i - \overline{Y}$ values rather than in residual variation, $r_i = Y_i - \hat{Y}_i$. This visual impression can be summarized using eqn. (3.3.4):

$$\text{SSM} = 5 \cdot (446.4)^2$$

$$= 996{,}364.80.$$

$$\text{SSR} = (-42.86)^2 + (-23.38)^2 + (-3.90)^2 + (25.33)^2 + (44.80)^2$$

$$= 5{,}047.46$$

$$SSE = (-0.54)^2 + (9.98)^2 + (-13.50)^2 + (1.28)^2 + (2.80)^2$$

$$= 291.62,$$

and

$$TSS = SSM + SSR + SSE = 1,001,703.88$$

which differs from the exact value 1,001,704 only because of roundoff error. From these figures it is clear that the variation attributable to the prediction equation accounts for most of the variability of the responses, TSS(adj).

Computationally the calculation of the partitioning of TSS can be arduous using the formulae in eqn. (3.3.3) if the data set is not extremely small. First one must determine all n predicted values, $\hat{Y}_i = \hat{\alpha} + \hat{\beta}X_i$. Next the deviations, $\hat{Y}_i - \overline{Y}$, and residuals $r_i = Y_i - \hat{Y}_i$, need to be determined. Finally, each of the deviations and residuals must be squared and summed, yielding, respectively, SSR and SSE. If a computer is available this is no real problem; if one desires to use a calculator, it is.

Fortunately, there is a simpler computing formula for SSR. Recalling that $\hat{Y}_i = \hat{\alpha} + \hat{\beta}X_i = (\overline{Y} - \hat{\beta}\overline{X}) + \hat{\beta}X_i$,

$$\hat{Y}_i - \overline{Y} = \hat{\beta}(X_i - \overline{X}) .$$

Consequently,

$$SSR = \hat{\beta}^2 \sum_{i=1}^{n} (X_i - \overline{X})^2 = \hat{\beta}^2 d_X^2 .$$

The partitioning of TSS can now be accomplished using the following formulae:

$$TSS = \sum Y_i^2, \quad SSM = n\overline{Y}^2, \quad SSR = \hat{\beta}^2 (\sum X_i^2 - n\overline{X}^2) \qquad (3.3.5)$$

$$SSE = TSS - SSM - SSR .$$

For the data in Table 3.1,

$$TSS = 1,001,704.00 \qquad SSM = 996,364.80$$

$$SSR = (9.7406)^2 [199 - 5(5.4)^2] = 5,047.58$$

$$SSE = 291.62,$$

as was previously computed using the formulae in eqn. (3.3.3). Note that in this computation we have carried more decimal places for $\hat{\beta}$ to reduce roundoff error. In all future partitionings of TSS, we will use the computational formulae in eqn. (3.3.5) rather than the more laborious ones in eqn. (3.3.3). Both are theoretically correct and will yield exactly the same partitioning if enough decimal places are retained in all intermediate calculations.

3.3.2 Coefficient of Determination

When determining the strength of a relationship between two variables, one often desires to compute a measure of association between them. Since we can write

$$Y_i = \hat{Y}_i + r_i \qquad i = 1, 2, ..., n,$$

zero residuals imply that $(Y_i, \hat{Y}_i)$ points all lie on a straight line: $Y = \hat{Y}$. An appropriate measure of the strength of the association between observed and predicted responses is Pearson's product moment correlation coefficient, r. It is well known that Pearson's r obtains its largest values, ± 1, when all points in a data set lie on a straight line. Its minimum value, 0, occurs when the data points are essentially randomly distributed throughout a two-variable plot or appear in certain types of nonlinear patterns. For observed and predicted responses, Pearson's r can be written as

$$r = \frac{\Sigma (Y_i - \overline{Y})(\hat{Y}_i - M)}{[\Sigma (Y_i - \overline{Y})^2 \cdot \Sigma (\hat{Y}_i - M)^2]^{1/2}} \tag{3.3.6}$$

where M is the average value of all the predicted responses. We are not concerned with whether the association is positive or negative but only how close Y_i and $\hat{Y}_i$ values are to one another; therefore, we can use r^2 instead of r. If the residuals are negligible, r^2 will be close to 1.0; if they remain very large after the regression model has been fitted, Y_i and $\hat{Y}_i$ pairs will not form a straight line and r^2 will be close to zero.

The square of Pearson's r is called the "Coefficient of Determination" of a fitted regression model and is denoted by R^2. Instead of calculating eqn. (3.3.6) and squaring it, a computational simplification is possible (Appendix 3.D):

$$R^2 = \frac{SSR}{TSS(adj)} = \frac{SSR}{SSR + SSE}. \tag{3.3.7}$$

In this form R^2 can be seen to be the proportion of the adjusted variation in the responses that is attributed to the estimated regression line. If the residuals are negligible $R^2 \approx 1.0$, while if they are large R^2 can be very small.

For the example of the previous subsection,

$$R^2 = \frac{5,047.58}{5,339.20} = .945 \ .$$

As a second example, the coefficient of determination for the female suicide rates is

$$R^2 = \frac{12.77}{34.62} = .369.$$

Thus in the first set of data approximately 94.5% of the variability of the responses can be attributed to the regression of cost on time, while in the latter one fully 63% of the variability is still left unexplained (most of which is due to the need for a quadratic term in the prediction equation).

Simple interpretations of R^2 such as these coupled with the theoretical appeal of its being a function of Pearson's r make the coefficient of determination one of the most important measures of the adequacy of prediction equations. Before reasonable interpretations of prediction equations can be made—indeed before confidence can be placed in predictions from a fitted model—R^2 must be considered sufficiently large by the data analyst. A large value of R^2 does not necessarily guarantee accurate prediction (a point we shall return to later), but it should be required before undue claims are made about the fitted model.

3.3.3 Residual Variability

The unexplained variability of the responses, SSE, provides more information about the adequacy of the fit of a prediction equation than only its function in the coefficient of determination. If we consider the predictor variable X to be nonrandom, as we have tacitly done up to this point, then all the variability in the responses is due to the random error term of the model, ε. Theoretically, the variability of a random variable is generally, though not always completely, characterized by its variance. An unbiased estimator of the population variance σ^2 of a random variable u under extremely general conditions* is the sample variance, S^2:

---

*One general set of conditions is that u_1, u_2 ..., u_n simply be a set of independent observations from some population having a finite variance. Mood, Graybill, and Boes (1974) present a more complete theoretical discussion of this property.

$$S^2 = \sum_{i=1}^{n} \frac{(u_i - \bar{u})^2}{n - 1} .$$

The denominator, $n - 1$, is the degrees of freedom or number of independent pieces of information in S^2. If the population mean μ is known, it would be inserted in place of the sample mean $\bar{u}$ in S^2 and the denominator would be n. So the cost of estimating one parameter, μ, is one degree of freedom in the unbiased estimation of σ^2.

Details of estimating the variance of ε in a regression analysis will be given in Chapter 6. Essentially, the role of $u_i - \bar{u}$ or $u_i - \hat{\mu}$ in S^2 is performed by $r_i = Y_i - \hat{\alpha} - \hat{\beta} X_i$ in the estimation of $\text{Var}(\varepsilon) = \sigma^2$. Since two parameters are estimated in order to form the residuals, the estimator of σ^2 has $n - 2$ degrees of freedom and is given by:

$$\hat{\sigma}^2 = \frac{\sum_{i=1}^{n} (Y_i - \hat{\alpha} - \hat{\beta} X_i)^2}{n - 2} = \frac{\sum_{i=1}^{n} r_i^2}{n - 2} = \frac{\text{SSE}}{n - 2} . \tag{3.3.8}$$

The value of this estimator is now demonstrated.

Again consider fitting a single-variable prediction equation for the homicide rate in Detroit using the variables in Data Set A.3. From R^2 values for the nine single-variable prediction equations listed in Table 3.5, FTP and CLEAR are the two best candidate single-variable predictors. Figure 3.5 in Section 3.3.1 shows that both variables result in fitted models that appear to approximate the homicide rate well. The estimated standard deviations ("standard errors") of the error components of the two single-variable models allow a closer examination of the approximation.

In any population of observations, virtually 90% or more of the population lies within 3 standard deviations of the population mean. This fact can be adapted to give a better notion of how accurate the prediction equation can be expected to perform. For a fixed value of X, the response variable, Y, actually represents a population of observations with mean $\alpha + \beta X$ if we assume that ε measures purely random error and fluctuates around zero. Then it is reasonable to expect the responses to vary within three standard deviations of $\alpha + \beta X$, and any response within these limits should not generally be considered an unusual observation.

Similarly, with a fitted model one should not consider a response that is predicted for a specific value of X to be a precise estimate. The true response can crudely be expected to lie within 3 standard errors of the predicted response. Very quickly, then, by glancing at the standard errors, $\hat{\sigma}$,

**Table 3.5. Summary Measures of Fit, Single-Variable
Prediction Equations of Detroit Homicide Rate: 1961-1973**

Predictor Variable	R^2	$\hat{\sigma}^2$	$\hat{\sigma}$
ACC	.042	280.649	16.753
ASR	.680	93.641	9.677
FTP	.929	20.674	4.547
UEMP	.044	279.956	16.732
MAN	.299	205.440	14.333
LIC	.528	138.385	11.764
GR	.666	97.730	9.886
CLEAR	.938	18.184	4.264
WM	.907	27.122	5.208

for the single-variable prediction models using FTP and CLEAR it is apparent that the predicted homicide rates are not likely to be as accurate as one would like. Three standard errors for the fitted model using FTP is 13.64 while that for CLEAR is 12.79. Thus there is a 27 unit range around predicted values using FTP and a 25 unit range around those using CLEAR for which we can say that the true responses are likely to be. Despite the large R^2 values for these two prediction equations—a requirement we have labeled as necessary for sensible predictions and credible interpretations of the fitted model—the standard errors quickly cast doubt on the ability of the fitted models to predict with a high degree of accuracy.

Before leaving this example we should point out that there are several possible reasons for the large standard errors using FTP and CLEAR. One is that either model may be a correct theoretical relationship between HOM and FTP and HOM and CLEAR, respectively, but the error terms are indeed highly variable. Another more plausible explanation is that HOM is influenced by more than one predictor variable and the residuals, and hence $\hat{\sigma}$, are measuring the variability associated with both the random error component and the predictor variables erroneously left out of the specification of the model. Perhaps all the predictor variables in Data Set A.3 are needed to predict the homicide rate accurately.

APPENDIX

3.A   Derivation of Least Squares Estimators

For single-variable intercept models we desire to choose $\hat{\alpha}$ and $\hat{\beta}$ to minimize

$$\Sigma \, (Y_i - \hat{\alpha} - \hat{\beta} X_i)^2 \, . \tag{3.A.1}$$

The derivative of eqn. (3.A.1) with respect to $\hat{\alpha}$ is

$$-2 \, \Sigma (Y_i - \hat{\alpha} - \hat{\beta} X_i) = \quad -2[\, \Sigma \, Y_i - n \, \hat{\alpha} - \hat{\beta} \Sigma \, X_i],$$

Equating this expression to zero and solving for $\hat{\alpha}$ yields

$$\hat{\alpha} = \overline{Y} - \hat{\beta} \overline{X} \, .$$

The derivative of eqn. (3.A.1) with respect to $\hat{\beta}$ is

$$
\begin{aligned}
-2 \, \Sigma \, X_i (Y_i - \hat{\alpha} - \hat{\beta} X_i) \;\; &= -2[\, \Sigma \, X_i Y_i - \hat{\alpha} \, \Sigma \, X_i - \hat{\beta} \, \Sigma \, X_i^2] \\
&= -2[\, \Sigma \, X_i Y_i \;\; - \overline{Y} \, \Sigma \, X_i + \hat{\beta} \overline{X} \, \Sigma \, X_i - \hat{\beta} \, \Sigma \, X_i^2] \, ,
\end{aligned}
\tag{3.A.2}
$$

the latter expression arising by the insertion of the solution for $\hat{\alpha}$. Equating (3.A.2) to zero and solving for $\hat{\beta}$ yields

$$\hat{\beta} = \frac{\displaystyle\sum_{i=1}^{n} (X_i - \overline{X})(Y_i - \overline{Y})}{\displaystyle\sum_{i=1}^{n} (X_i - \overline{X})^2} \, .$$

An equivalent formula for $\hat{\beta}$ is obtained by noting that

$$\Sigma \, (X_i - \overline{X})(Y_i - \overline{Y}) \; = \; \Sigma \, X_i Y_i - n \overline{X} \, \overline{Y}$$

and

$$\Sigma \, (X_i - \overline{X})^2 = \; \Sigma \, X_i^2 - n \overline{X}^2 \, .$$

Hence, equating (3.A.2) to zero and making these substitutions gives

$$\hat{\beta} = \frac{\displaystyle\sum_{i=1}^{n} X_i Y_i - n \overline{X} \, \overline{Y}}{\displaystyle\sum_{i=1}^{n} X_i^2 - n \overline{X}^2} \, .$$

With no-intercept models the sum of squared residuals is

$$\Sigma\,(Y_i - \hat{\beta}X_i)^2\,. \tag{3.A.3}$$

Differentiating this expression with respect to $\hat{\beta}$ and equating the result to zero yields the estimator shown in eqn. (3.2.10).

3.B Coefficient Estimates for Standardized Variables

Relationships between coefficient estimates for original and standardized prediction equations can be obtained for any of the standardizations of Section 3.2.3. We derive only the relationships for the normal deviate standardization; other standardizations are derived in a similar fashion. Beginning with the original predictor variable X, the standardized model is obtained through the following series of transformations:

$$Y_i = \alpha + \beta X_i + \varepsilon_i$$

$$= (\alpha + \beta \overline{X}) + \beta(X_i - \overline{X}) + \varepsilon$$

$$= (\alpha + \beta \overline{X}) + \beta S_X (X_i - \overline{X})/S_X + \varepsilon_i$$

$$= \alpha^* + \beta^* Z_i + \varepsilon_i, \tag{3.B.1}$$

where $\alpha^* = \alpha + \beta \overline{X}$ and $\beta^* = \beta S_X$.

The least squares coefficient estimates for model (3.B.1) are

$$\hat{\alpha}^* = \overline{Y} - \hat{\beta}^* \overline{Z}$$

and

$$\hat{\beta}^* = \frac{\displaystyle\sum_{i=1}^{n} Z_i Y_i - n\overline{Z}\,\overline{Y}}{\displaystyle\sum_{i=1}^{n} Z_i^2 - n\overline{Z}^2}\,.$$

Recall now that $\overline{Z} = 0$ and $S_Z^2 = (\Sigma\,Z_i^2 - n\overline{Z}^2)/(n-1) = 1$, so that

$$\hat{\alpha}^* = \overline{Y} \qquad \text{and} \qquad \hat{\beta}^* = \sum_{i=1}^{n} \frac{Z_i Y_i}{n-1}\,.$$

From the relationships between the original and standardized parameters in eqn. (3.B.1), it follows that

$$\hat{\alpha} + \hat{\beta}\overline{X} = \hat{\alpha}^* \quad \text{and} \quad \hat{\beta}S_X = \hat{\beta}^* \, .$$

Consequently,

$$\hat{\alpha} = \hat{\alpha}^* - \hat{\beta}\overline{X} \quad \text{and} \quad \hat{\beta} = \frac{\hat{\beta}^*}{S_X} .$$

3.C Partitioning the Total Sum of Squares

The additive partitioning of TSS can be derived in two steps. First, by adding and subtracting $\overline{Y}$ to each Y_i one finds that

$$\text{TSS} = \Sigma (Y_i - \overline{Y} + \overline{Y})^2$$

$$= \Sigma (Y_i - \overline{Y})^2 + 2\Sigma (Y_i - \overline{Y})\overline{Y} + \Sigma \overline{Y}^2$$

$$= \Sigma (Y_i - \overline{Y})^2 + n\overline{Y}^2, \tag{3.C.1}$$

since

$$\overline{Y} \Sigma (Y_i - \overline{Y}) = \overline{Y}[\Sigma Y_i - n\overline{Y}] = \overline{Y}[\Sigma Y_i - \Sigma Y_i] = 0.$$

Thus, from eqn. (3.C.1),

$$\text{TSS} = \text{TSS(adj)} + \text{SSM}.$$

Next, add and subtract $\hat{Y}_i = \hat{\alpha} + \hat{\beta}X_i$ to each term in TSS(adj):

$$\text{TSS} = \text{SSM} + \Sigma (Y_i - \hat{Y}_i + \hat{Y}_i - \overline{Y})^2$$

$$= \text{SSM} + \Sigma (Y_i - \hat{Y}_i)^2 + 2\Sigma (Y_i - \hat{Y}_i)(\hat{Y}_i - \overline{Y}) + \Sigma (\hat{Y}_i - \overline{Y})^2$$

$$= \text{SSM} + \Sigma (Y_i - \hat{Y}_i)^2 + \Sigma (\hat{Y}_i - \overline{Y})^2 , \tag{3.C.2}$$

since

$$\Sigma (Y_i - \hat{Y_i})(\hat{Y_i} - \overline{Y}) = \Sigma [Y_i - \overline{Y} - \hat{\beta}(X_i - \overline{X})][\overline{Y} + \hat{\beta}(X_i - \overline{X}) - \overline{Y}]$$

$$= \hat{\beta} \Sigma (Y_i - \overline{Y}) (X_i - \overline{X}) - \hat{\beta}^2 \Sigma (X_i - \overline{X})^2$$

$$= \hat{\beta}[\Sigma (Y_i - \overline{Y}) (X_i - \overline{X}) - \Sigma (Y_i - \overline{Y}) (X_i - \overline{X})]$$

$$= 0.$$

From eqn. (3.C.2), noting that $\Sigma (Y_i - \hat{Y_i})^2 = \Sigma r_i^2$,

$$\text{TSS} = \text{SSM} + \text{SSR} + \text{SSE}.$$

3.D Calculation of the Coefficient of Determination

The coefficient of determination, R^2, is defined to be the square of the ordinary correlation coefficient, r, between observed and predicted responses, eqn. (3.3.6). An equivalent means of calculating R^2 is to use the formula:

$$R^2 = \frac{\text{SSR}}{\text{TSS(adj)}} .$$

To verify this calculation, recall that

$$\text{SSR} = \hat{\beta}^2 \Sigma (X_i - \overline{X})^2 \qquad (3.D.1)$$

and

$$\text{TSS(adj)} = \Sigma (Y_i - \overline{Y})^2 . \qquad (3.D.2)$$

Next observe that the average predicted response M equals $\overline{Y}$ since

$$M = n^{-1} \Sigma (\hat{\alpha} + \hat{\beta}X_i) = \hat{\alpha} + \hat{\beta}\overline{X} = \overline{Y}.$$

Consequently the numerator of eqn. (3.3.6) is

$$\Sigma (Y_i - \overline{Y})(\hat{Y_i} - M) = \Sigma (Y_i - \overline{Y})(\hat{Y_i} - \overline{Y})$$

$$= \Sigma (Y_i - \overline{Y})(\hat{\alpha} + \hat{\beta}X_i - \overline{Y})$$

$$= \Sigma (Y_i - \overline{Y}) \hat{\beta}(X_i - \overline{X})$$

$$= \hat{\beta} \Sigma (Y_i - \overline{Y})(X_i - \overline{X})$$

$$= \hat{\beta}^2 \Sigma (X_i - \overline{X})^2$$

$$= \text{SSR}. \qquad (3.D.3)$$

Finally, squaring eqn. (3.3.6) and utilizing eqns. (3.D.1) to (3.D.3) yields

$$R^2 = r^2 = \frac{SSR^2}{TTS(adj) \cdot SSR}$$

$$= \frac{SSR}{TTS(adj)} \; .$$

EXERCISES

1. Smooth the data in Table 2.1 and plot the points on a scatter diagram. Eyeball a line ignoring the first three points and another one with the last two points ignored and sketch these on the graph. Does smoothing eliminate the problems associated with eyeballing illustrated in Figure 3.3?

2. In order to specify a theoretical regression model both theoretical and empirical verifications are required. Consider eqn. (1.4.1) that describes a relationship between the weight and height of children. What would be necessary in order to establish a cause-and-effect inference about the regression model?

3. State two reasons why the fitted regression lines graphed in Figure 3.6, although offering excellent fits, do not represent a theoretical relationship between average suicide rate and age.

4. Fit a regression line using GNP as the response variable and INFD as a single predictor variable (exclude Hong Kong and Singapore). Repeat the process excluding the intercept term. Compare the resulting coefficients. Is a no-intercept model plausible in this situation?

5. Compute the residuals for the GNP vs. INFD data in Exercise 4 using an intercept model. Notice anything unusual about the resultant magnitudes? How might one correct this problem?

6. Multiply all the values of INFD by 8. Verify that the estimated coefficients determined by regressing GNP on $8 \cdot INFD$ are equal to those found in Exercise 4 divided by 8.

7. Multiply all the GNP values by 3 and regress the data on INFD. Verify that the estimated coefficients are equivalent to those found in Exercise 4 multiplied by 3.

8. Add 10 to the INFD data values and regress GNP on INFD + 10. Compare the resultant estimated coefficients to those obtained in Exercise 4.

9. Standardize the INFD data using eqns. (3.2.11) and (3.2.14). Compare the estimated coefficients from regressions with the GNP data to those obtained in Subsection 3.2.3 using DENS and AGDS. Which of the three variables has the most influence on GNP?

10. Consider the male suicide rates in Table 3.4. Fit a straight line to the data and verify eqns. (3.3.5) and (3.3.7). Why do you think the coefficient of determination is so low?

11. Calculate $\hat{\sigma}^2$ for the male suicide rate data. Why is there such a large standard error?

CHAPTER 4

MULTIPLE-VARIABLE PRELIMINARIES

Prediction equations that contain several predictor variables require closer examination and more skills from the data analyst than do single-variable ones. The methodology for single-variable analysis is generalizable to multiple-variable analyses, only in the latter situation there are more inter-relationships among variables to consider. Assessing the information contained in these interrelationships necessitates some basic knowledge of matrix algebra and a large amount of insight on the part of the investigator.

The need to be concerned about interrelationships among the predictor variables in multiple-variable models is well-illustrated by a controversy described in Blalock (1963). Blalock describes three studies that were conducted to examine the influence of anomie and authoritarianism on prejudice. One study, conducted by Srole (1956) on 401 subjects, concluded that anomie is more important as a predictor of prejudice than is authoritarianism. A second study by Roberts and Rokeach (1956) involving 86 subjects drew the opposite conclusion: authoritarianism is more important than anomie. The third study, McDill (1961) using 266 subjects, determined that both anomie and authoritarianism are equally important as predictors of prejudice.

Blalock points out that the different conclusions can be attributed to either the existence of or the lack of an association between the predictor variables, anomie (AN) and authoritarianism (AU). These associations are measurable by calculating the correlation coefficients among the variables in the three studies, correlations which are displayed in Table 4.1. Observe that the pairwise correlations between AN and AU in the first two studies are about equal and much smaller than in the third. What are the implications of these relationships?

Table 4.1. Comparison of Variable Interrelationships Between Anomie (AN), Authoritarianism (AU), and Prejudice (PR)

		Investigation	
	Srole	Roberts and Rokeach	McDill
Sample Size	401	86	266
Variables		Pairwise Correlations	
AN and AU	.45	.47	.67
AU and PR	.29	.64	.64
AN and PR	.43	.55	.63

Source: Blalock (1963). ©1963 by The University of North Carolina. Reprinted with permission.

One is first led to question whether all the investigators sampled the same population. Particularly since the sample correlation between AN and AU differs so greatly for McDill's study, the population sampled in McDill's investigation should be carefully compared to the other two. All three sampled populations could differ greatly in characteristics that would affect the three variables being examined. Generalizations from these studies, as was stressed in Chapter 1, may not be comparable if the populations that were sampled are in fact different.

The magnitude of the pairwise correlation between AN and AU in McDill's study suggests that the two variables contain much common information. The scales of measurement may differ for the two but $r = 0.67$ indicates that there is a moderately strong linear correspondence between increases (decreases) in one of the variables and increases (decreases) in the other. One would expect, therefore, that the two estimated regression coefficients would be similar and that the pairwise correlations of each with prejudice (PR) would likewise be similar. Table 4.1 reveals that the pairwise correlations are indeed close to one another (as were the estimated regression coefficients).

Smaller pairwise correlations between AN and AU in Srole's and Roberts and Rokeach's investigations suggest that the effects of these predictor variables on the response need not necessarily be similar. The pairwise correlations between PR and each predictor variable are reversed in relative magnitude for the two studies: AN and PR have a larger correlation than AU and PR in Srole's data but a smaller one in Roberts and Rokeach's. Coupled with the earlier comparison of the correlations between AN and AU for the three studies, it appears that all three sampled populations do differ and the contradictory conclusions of the studies are no longer surprising. If three different populations have been sampled, the influences of anomie and authoritarianism on prejudice can indeed vary.

This controversy illustrates that interrelationships among predictor variables in multiple-variable prediction equations are as important to the data analyst as are relationships between the response and predictor variables. Not only can strong relationships identify characteristics of sampled populations and point out differences in data bases, but as we shall see in Chapter 9 they can also distort inferences on fitted models.

Before proceeding to the analysis of data bases that contain several predictor variables, this chapter surveys several topics of importance to the preliminary specification of multiple-variable prediction equations. Since the definition of multiple-variable regression models and the detection of complicated relationships among predictor variables are facilitated by using matrix notation, we begin with a brief review of matrix algebra.

4.1 REVIEW OF MATRIX ALGEBRA

Matrix algebra provides short-cut techniques for expressing multiple-variable prediction equations and the associated computational manipulations needed for a regression analysis. Using matrix notation, multiple linear regression analysis appears as a natural extension of single-variable regression. The following sections provide a brief exposure to the main features of matrix algebra as they relate to multiple linear regression. While the more advanced reader may desire to skip this material, it is recommended that the section be scanned since the notation will be employed throughout the remainder of this book.

4.1.1 Notation

A multiple linear regression model for n response variables can be written as

$$Y_i = \alpha + \beta_1 X_{i1} + \beta_2 X_{i2} + \ldots + \beta_p X_{ip} + \varepsilon_i \qquad i = 1, 2, \ldots, n. \qquad (4.1.1)$$

In this definition, Y_i represents the response for the ith observation in a data base that contains a total of n responses. Associated with the ith response in the data base is a value on each of the p predictor variables, denoted by X_{i1},

$X_{i2}, \ldots, X_{ip}$. Also associated with the ith response is a random error component that is indicated by ε_i. As discussed in Section 1.2.1, the predictor variables in eqn. (4.1.1) enter as linear functions; hence, the name multiple linear regression model.

Collectively, the n observations on the response and each of the predictor variables can be indicated through the use of vectors. We define* $\underline{Y}, \underline{X}_1$, $\underline{X}_2, \ldots, \underline{X}_p$ to be vectors of observations on the response and predictor variables, respectively, so that

$$\underline{Y} = \begin{pmatrix} Y_1 \\ Y_2 \\ \vdots \\ Y_n \end{pmatrix}, \underline{X}_1 = \begin{pmatrix} X_{11} \\ X_{21} \\ \vdots \\ X_{n1} \end{pmatrix}, \underline{X}_2 = \begin{pmatrix} X_{12} \\ X_{22} \\ \vdots \\ X_{n2} \end{pmatrix}, \ldots, \underline{X}_p = \begin{pmatrix} X_{1p} \\ X_{2p} \\ \vdots \\ X_{np} \end{pmatrix}.$$

Symbolically, then, a vector is a shorthand technique for compactly representing n variables or numerical values.

Matrices are collections of vectors. If one desires to collectively discuss the observations on the predictor variables, one can define a matrix, say X, that contains n rows and p columns (referred to as an "n by p" or "$n \times p$" matrix), each column of which contains the n observations on a predictor variable. Notationally,

$$X = [\underline{X}_1, \quad \underline{X}_2, \quad \ldots, \quad \underline{X}_p]$$

$$= \begin{bmatrix} X_{11} & X_{12} & \ldots & X_{1p} \\ X_{21} & X_{22} & \ldots & X_{2p} \\ \vdots & \vdots & & \vdots \\ X_{n1} & X_{n2} & \ldots & X_{np} \end{bmatrix}.$$

In order to incorporate the constant term as an additional predictor variable, one can add a column vector all of whose elements are unity to X. We denote such a vector by $\underline{1}$ and the corresponding matrix by X^*:

---

*In this book vectors will be written as upper case or Greek letters that are underlined; e.g., $\underline{a}$, $\underline{b}, \underline{C}, \underline{D}, \underline{\alpha}, \underline{\beta}, \underline{\gamma}$. All matrices are written as upper case letters without underlining.

$$X^* = [\underline{1}, \underline{X}_1, \underline{X}_2, ..., \underline{X}_p]$$

$$= \begin{bmatrix} 1 & X_{11} & X_{12} & ... & X_{1p} \\ 1 & X_{21} & X_{22} & ... & X_{2p} \\ \vdots & \vdots & \vdots & & \vdots \\ 1 & X_{n1} & X_{n2} & ... & X_{np} \end{bmatrix}.$$

It will often be advantageous to express column vectors as row vectors. This is accomplished by taking the "transpose" of a vector, denoted $\underline{Y}'$. Thus,

$$\underline{Y}' = (Y_1, Y_2, ..., Y_n).$$

Similarly, transposes of matrices interchange rows with columns; the jth column of X becomes the jth row of X':

$$X' = \begin{bmatrix} X_{11} & X_{21} & ... & X_{n1} \\ X_{12} & X_{22} & ... & X_{n2} \\ \vdots & \vdots & & \vdots \\ X_{1p} & X_{2p} & ... & X_{np} \end{bmatrix}.$$

With this notation, the important elements in a multiple linear regression model are $\underline{Y}$, X (or X^*) and $\underline{\varepsilon}$ (a vector of error terms defined similarly to $\underline{Y}$). The parameters of the model can be symbolized as α and $\underline{\beta}$, where

$$\underline{\beta}' = (\beta_1, \beta_2, ..., \beta_p),$$

or by including α and the elements of $\underline{\beta}$ in a single vector, $\underline{B}$, where

$$\underline{B}' = (\alpha, \beta_1, \beta_2, ..., \beta_p).$$

After a discussion of vector and matrix operations we will see that the model (4.1.1) can now be defined in a very concise form.

4.1.2 Vector and Matrix Operations

Addition and subtraction of vectors and matrices are defined to be the addition and subtraction of each corresponding element in the vectors or matrices. Thus, if $\underline{a}$ and $\underline{b}$ are $n \times 1$ vectors,

$$\underline{a} + \underline{b} = \begin{pmatrix} a_1 + b_1 \\ a_2 + b_2 \\ \vdots \\ a_n + b_n \end{pmatrix} \quad \text{and} \quad \underline{a} - \underline{b} = \begin{pmatrix} a_1 - b_1 \\ a_2 - b_2 \\ \vdots \\ a_n - b_n \end{pmatrix}.$$

More generally, a "linear combination" of r vectors $\underline{a}_1, \underline{a}_2, \ldots, \underline{a}_r$, each of dimension n, is defined to be

$$\sum_{j=1}^{p} c_j \underline{a}_j = \begin{pmatrix} c_1 a_{11} + c_2 a_{12} + \ldots + c_r a_{1r} \\ c_1 a_{21} + c_2 a_{22} + \ldots + c_r a_{2r} \\ \vdots \\ c_1 a_{n1} + c_2 a_{n2} + \ldots + c_r a_{nr} \end{pmatrix}. \qquad (4.1.2)$$

This linear combination of the vectors $\underline{a}_1, \underline{a}_2, \ldots, \underline{a}_r$ indicates that the vector $\underline{a}_1$ is to be multiplied by the constant c_1, $\underline{a}_2$ by c_2, etc. and the results summed. Equation (4.1.2) shows that these same multiplications and summations are done on each element of the respective vectors.

An example of these computations is now given using the data in Table 4.2. For simplicity let $\underline{Y}$, $\underline{X}_1$, $\underline{X}_2$, and $\underline{X}_3$ contain only the first four observations on the response and predictor variables shown in Table 4.2. From the basic relationship in eqn. (4.1.2), all the following calculations are obtained:

$$2\underline{X}_3 = \begin{pmatrix} 2 \\ 168 \\ 1{,}096 \\ 602 \end{pmatrix}, \; 3\underline{Y} - 4\underline{X}_2 = \begin{pmatrix} 508 \\ -770 \\ -11{,}400 \\ 312 \end{pmatrix}, \; \text{and } 2\underline{X}_1 - \underline{X}_3 + \underline{1} = \begin{pmatrix} 30.0 \\ -8.0 \\ -426.2 \\ -229.2 \end{pmatrix}.$$

Similar linear combinations are valid for matrices, where again the linear combinations are taken element by element. The only requirement for performing these operations is that all matrices have the same number of rows and columns. For example, if A denotes the 4×4 matrix of values in the first four rows of Table 4.2 and B is the 8×4 matrix of values in the last eight rows, $A + B$ is meaningless since A and B do not have the same number of rows.

Multiplication of two column vectors is accomplished by multiplying the corresponding elements and summing the products. It is a generalization of

Table 4.2. First Twelve Observations on GNP, INFD, PHYS, and DENS for the Demographic Data in Data Set A.1

GNP (Y)	INFD (X_1)	PHYS (X_2)	DENS (X_3)
1,316	19.5	860	1
670	37.5	695	84
200	60.4	3,000	548
1,196	35.4	819	301
235	67.1	3,900	3
365	45.1	740	72
1,947	27.3	900	2
379	127.9	1,700	11
357	78.9	2,600	24
467	29.9	1,400	62
680	31.0	620	108
1,057	23.7	830	107

ordinary multiplication in that the definition reduces to ordinary multiplication if each vector contains only one element. Notationally, we write one of the column vectors as a row vector and follow it with the other vector in column form to indicate multiplication. Thus multiplication of $\underline{a}$ by $\underline{b}$ is defined to be

$$\underline{a}'\underline{b} = (a_1, a_2, \ldots, a_n) \begin{pmatrix} b_1 \\ b_2 \\ \vdots \\ b_n \end{pmatrix}$$

$$= \sum_{i=1}^{n} a_i b_i. \tag{4.1.3}$$

Since the order of the variables (whether a_i preceeds or follows b_i) is immaterial to the evaluation of the summation in eqn. (4.1.1), $\underline{a}'\underline{b} = \underline{b}'\underline{a} = a_1 b_1 + \ldots + a_n b_n$. Again, the only requirement for the multiplication of two vectors is that they have the same dimension.

Multiplication of matrices consists of a sequence of vector multiplications. Let U be a $q \times r$ matrix and V be an $s \times t$ matrix. If $r = s$, multiplication of U and V is defined, regardless of the values of q and t. So if $r = s$ and U and V are given by

$$U = \begin{bmatrix} U_{11} & U_{12} & \cdots & U_{1r} \\ U_{21} & U_{22} & \cdots & U_{2r} \\ \vdots & \vdots & & \vdots \\ U_{q1} & U_{q2} & \cdots & U_{qr} \end{bmatrix} \quad \text{and} \quad V = \begin{bmatrix} V_{11} & V_{12} & \cdots & V_{1t} \\ V_{21} & V_{22} & \cdots & V_{2t} \\ \vdots & \vdots & & \vdots \\ V_{r1} & V_{r2} & & V_{rt} \end{bmatrix}$$

then the product of U and V is a $q \times t$ matrix W, symbolized $W = UV$, whose (g, h)th element is given by

$$W_{gh} = \sum_{m=1}^{r} U_{gm} V_{mh} \qquad (4.1.4)$$

i.e., W_{gh} is the product of the gth row of U and the hth column of V. Notationally, if the rows of U are denoted $\underline{U}_1'$, $\underline{U}_2'$, ..., $\underline{U}_q'$ and the columns of V are $\underline{V}_1$, $\underline{V}_2$, ..., $\underline{V}_t$ then

$$W_{gh} = \underline{U}_g' \underline{V}_h \, .$$

While UV may be defined (if $r = s$), it is not necessarily true that UV is defined since q and t may differ. Even if $r = s$ and $q = t$, it is not necessarily true that UV is equal to VU since the dimension of UV is $q \times q$ and that of VU is $r \times r$. Finally, it is not even necessarily true that $UV = VU$ when $q = r = s = t$. The reader can construct simple examples using the data in Table 4.2 to verify these remarks.

Consider again the data in Table 4.2 with $\underline{Y}$, $\underline{X}_1$, $\underline{X}_2$, and $\underline{X}_3$ defined as above. From eqn. (4.1.3),

$$\underline{X}_1' \underline{X}_2 = \underline{X}_2' \underline{X}_1 = 253{,}025.1$$

and

$$\underline{X}_1' \underline{Y} = \underline{Y}' \underline{X}_1 = 105{,}250.4.$$

Note, however, that $\underline{X}_1' \underline{Y} \neq \underline{X}_1 \underline{Y}'$ since the latter symbol represents a 4×4 matrix that results from the matrix multiplication of a 4×1 matrix, $\underline{X}_1$, and a 1×4 matrix, $\underline{Y}'$. Applying eqn. (4.1.4),

$$
\underline{X}_1\underline{Y}' = \begin{bmatrix}
25{,}662.0 & 13{,}065.0 & 3{,}900.0 & 23{,}322.0 \\
49{,}350.0 & 25{,}125.0 & 7{,}500.0 & 44{,}850.0 \\
79{,}486.4 & 40{,}468.0 & 12{,}080.0 & 72{,}238.4 \\
46{,}586.4 & 23{,}718.0 & 7{,}080.0 & 42{,}338.4
\end{bmatrix}.
$$

For another example of matrix multiplication, let A represent the 4×3 matrix of the first four observations on the predictor variables in Table 4.2 and B be the 4×3 matrix of the second four observations on the predictor variables. Then

$$
A'B = \begin{bmatrix}
9{,}176.28 & 218{,}348.00 & 3{,}268.70 \\
275{,}700.68 & 7{,}960{,}600.00 & 67{,}629.00 \\
57{,}313.80 & 1{,}070{,}960.00 & 10{,}458.00
\end{bmatrix}.
$$

Division of matrices can also be viewed as a generalization of univariate operations. For two numbers, u and v, division of v by a nonzero constant u is equivalent to multiplication of v by a number u^{-1}, where u^{-1} is the unique constant satisfying the relationship

$$
u(u^{-1}) = u^{-1} u = 1.0 ;
$$

i.e., u^{-1} is the number that multiplies u to give one. Once u^{-1} is determined, v can be multiplied by u^{-1} to yield the division of v by u. We now generalize these concepts to matrices.

An $r \times s$ matrix A is called a square matrix if the number of rows of A is the same as the number of columns, $r = s$. If A is square, its diagonal elements are those that form the intersection of the rows and columns which have identical subscripts; i.e., the elements a_{ii} ($i = 1, 2, ..., r$) are the diagonal elements of A. All other elements (a_{ij}, $i \neq j$) are termed off-diagonal elements. If A is a square matrix with all off-diagonal elements zero (with or without any diagonal elements zero), it is called a diagonal matrix and can be denoted $A = \text{diag}(a_{11}, a_{22}, ..., a_{rr})$.

Univariate values of zero and one have matrix generalizations. Any matrix all of whose elements are zero is termed a null or zero matrix and symbolized by Φ. The generalization of the number 1 is a square, diagonal matrix that has the value 1 for all its diagonal elements. Thus,

$$\Phi = \begin{bmatrix} 0 & 0 & \ldots & 0 \\ 0 & 0 & \ldots & 0 \\ \vdots & \vdots & & \vdots \\ 0 & 0 & \ldots & 0 \end{bmatrix} \quad \text{and} \quad I = \begin{bmatrix} 1 & 0 & \ldots & 0 \\ 0 & 1 & \ldots & 0 \\ \vdots & \vdots & & \vdots \\ 0 & 0 & \ldots & 1 \end{bmatrix},$$

where I, the "identity" matrix, is the matrix generalization of the number one and Φ, the "null" matrix (not necessarily square), is the matrix generalization of zero. One can easily verify that

$$A + \Phi = A \qquad BI = B \qquad IC = C,$$

provided that the matrices A, B and C are such that the respective operations are defined. The corresponding univariate operations are

$$a + 0 = a \qquad b \cdot 1 = b \qquad 1 \cdot c = c.$$

An inverse of an $n \times n$ square matrix A, if it exists*, is a matrix A^{-1} that satisfies the relationship

$$AA^{-1} = A^{-1}A = I.$$

Provided that the inverse of a matrix exists, division of a matrix B by a matrix A is defined to be the multiplication of B by A^{-1}. For example, if one desires to solve the system of equations

$$A\underline{x} = \underline{b}$$

for $\underline{x}$, where A is a known $n \times n$ matrix which has an inverse and $\underline{b}$ is a known $n \times 1$ vector, premultiplication of both sides of this equation by A^{-1} yields

$$A^{-1}A\underline{x} = A^{-1}\underline{b}$$

or

$$I\underline{x} = A^{-1}\underline{b}$$

or

$$\underline{x} = A^{-1}\underline{b}.$$

---

*Square matrices that have inverses are often referred to as "nonsingular" matrices while those that do not are called "singular" matrices.

Appendix 4.A presents a technique for determining the inverse of a square matrix. Also defined in Appendix 4.A is the determinant of a square-matrix, symbolized by $|A|$. Readers who are not familiar with the calculation of the determinant or inverse of a matrix should consult the appendix. To illustrate the computations in a simple example, define the matrix Z to be the first two elements of the first two predictor variables in Table 4.2:

$$Z = \begin{bmatrix} 19.5 & 860 \\ 37.5 & 695 \end{bmatrix}.$$

Using the techniques described in Appendix 4.A, we find that

$$|Z| = -18,697.50 \quad \text{and} \quad Z^{-1} = \begin{bmatrix} -0.03717 & 0.04600 \\ 0.00201 & -0.00104 \end{bmatrix}.$$

One can verify that Z^{-1} is the inverse of Z by directly multiplying Z by Z^{-1}; the identify matrix will be the result.

It was remarked earlier that $n \times n$ matrices do not necessarily have inverses. The existence of a matrix inverse plays an important role in regression methodology. Conditions under which matrix inverses exist are therefore important to the data analyst. A square matrix will have a matrix inverse if its columns are "linearly independent."

Linear independence of a set of vectors is defined as follows. Let $\underline{a}_1$, $\underline{a}_2$, ..., $\underline{a}_n$ be a set of vectors. If the *only* constants c_1, c_2, ..., c_n that satisfy

$$\sum_{j=1}^{n} c_j \underline{a}_j = \underline{0} \tag{4.1.5}$$

are $c_1 = c_2 = ... = c_n = 0$, then the vectors are said to be linearly independent. Thus if any set of constants, some of which are nonzero, exist that allow eqn. (4.1.5) to hold for the columns of an $n \times n$ matrix A, the matrix will not have a matrix inverse. Equivalently, one can show that a matrix A will have an inverse if and only if its determinant is nonzero.

Finally, two properties of the transpose and inverse of products of matrices are important in later discussions. These properties are

$$\begin{aligned} \text{(i)} \quad & (AB)' = B'A' \\ \text{(ii)} \quad & (AB)^{-1} = B^{-1}A^{-1}. \end{aligned} \tag{4.1.6}$$

In other words, the transpose of the product of two (or more) matrices is the product of the transposes of the individual matrices, where the multiplication is in *reverse* order. Likewise, the inverse of the product of two (or more) matrices is the product of the inverses in reverse order, provided that all the inverses exist.

4.1.3   Model Definition

Model (4.1.1) for n response variables can now be written in terms of the vectors and matrices defined in Section 4.1.1 as

$$\underline{Y} = X^*\underline{B} + \underline{\varepsilon}$$

or, equivalently,

$$\underline{Y} = \alpha\underline{1} + X\beta + \underline{\varepsilon}.$$

Using this latter formulation, the ith element of $\underline{Y}$ is equated to the sum of the ith element of $\alpha\underline{1}$, α, the ith element of $X\beta$, $X_{i1}\beta_1 + X_{i2}\beta_2 + \dots + X_{ip}\beta_p$, and the ith element of $\underline{\varepsilon}$, ε_i. But this summation is precisely the right side of eqn. (4.1.1).

Once estimates $\hat{\alpha}$ and $\hat{\beta}$ (or $\hat{\underline{B}}$) have been obtained for the model parameters, the prediction equation for the response can be written in matrix notation as

$$\hat{Y} = \hat{\alpha} + \underline{\hat{u}}'\hat{\beta},$$

where $\underline{u}' = (u_1, u_2, \dots, u_p)$ can contain any values of the predictor variables for which a predicted response is desired. In particular, if one desires to predict the response for each observation in the data base, the vector of predicted responses is

$$\hat{\underline{Y}} = X^*\hat{\underline{B}} \qquad \text{or} \qquad \hat{\underline{Y}} = \hat{\alpha}\underline{1} + X\hat{\beta}.$$

A matrix that will be seen to have great importance in regression analysis is $X'X$. Observe that

$$X'X = \begin{bmatrix} \Sigma X_{i1}^2 & \Sigma X_{i1}X_{i2} & \dots & \Sigma X_{i1}X_{ip} \\ \Sigma X_{i1}X_{i2} & \Sigma X_{i2}^2 & \dots & \Sigma X_{i2}X_{ip} \\ \vdots & \vdots & & \vdots \\ \Sigma X_{ip}X_{i1} & \Sigma X_{ip}X_{i2} & \dots & \Sigma X_{ip}^2 \end{bmatrix}.$$

The diagonal elements of $X'X$ are the sums of squares of the observations on each predictor variable while the off-diagonal elements are the sums of products of the elements of pairs of predictor variables. Similarly,

$$
X^{*\prime}X^{*} = \begin{bmatrix}
n & \Sigma X_{i1} & \cdots & \Sigma X_{ip} \\
\Sigma X_{i1} & \Sigma X_{i1}^2 & \cdots & \Sigma X_{i1}X_{ip} \\
\vdots & \vdots & & \vdots \\
\Sigma X_{ip} & \Sigma X_{ip}X_{i1} & \cdots & \Sigma X_{ip}^2
\end{bmatrix}.
$$

The matrix, due to the additional column of ones in X^*, provides the sample size and sum of each predictor variable in addition to the sums of squares and products of the predictor variables. Virtually all the information about interrelationships among the predictor variables that we will make use of in this book is contained in these two matrices. For example, from the quantities contained in these matrices all pairwise correlations between predictor variables can be calculated. As we shall see later, there is much more information of relevance to regression analysis in $X'X$ and $X^{*\prime}X^{*}$.

4.1.4   Latent Roots and Vectors

Latent roots and latent vectors (also referred to as characteristic roots and vectors or eigenvalues and eigenvectors) of a square matrix contain valuable information about the properties of that matrix. In this subsection we are concerned with the definition of the latent roots and vectors of a matrix and their potential for diagnosing interrelationships among the columns of the matrix.

The only matrices for which we will be interested in finding latent roots and latent vectors are $n \times n$ "symmetric" matrices. A matrix is symmetric if it is square and the elements above the diagonal are the "mirror image" of those below it. In other words if $a_{ij} = a_{ji}$ for all off-diagonal elements of the matrix A, then A is called a symmetric matrix. Both $X'X$ and $X^{*\prime}X^{*}$ are symmetric matrices.

An $n \times n$ symmetric matrix A has n latent roots, denoted $\ell_1, \ell_2, \ldots, \ell_n$, and n associated latent vectors, $\underline{V}_1, \underline{V}_2, \ldots, \underline{V}_n$, that are defined by the algebraic relationship

$$
A\underline{V}_j = \ell_j\underline{V}_j \qquad j = 1, 2, \ldots, n. \tag{4.1.7}
$$

Equation (4.1.7) states that $\underline{V}_j$ is a latent vector of A if multiplication of A by $\underline{V}_j$ yields a multiple, ℓ_j, of the vector itself. This equation does not state how the latent roots and latent vectors can be calculated, but only that the latent roots and latent vectors must satisfy eqn. (4.1.7).

Calculation of the latent roots and vectors for matrices used in regression analyses are invariably performed by computers. Mathematical expressions needed for these calculations can be found in most linear algebra texts (e.g., Hadley (1961), Chapter 7). We will assume that the latent roots and vectors can be calculated and will concentrate attention on their properties.

The latent vectors satisfying eqn. (4.1.7) can always be chosen so that they are "orthonormal." Orthonormality consists of two properties: the latent vectors are mutually "orthogonal" and are "normalized" so that their length is one. Two vectors are orthogonal if their vector product is zero. The latent vectors are thus chosen so that

$$\underline{V}_i'\underline{V}_j = 0 \qquad \text{for all } i \neq j. \tag{4.1.8}$$

The latent vectors of a matrix are normalized if the product of each vector with itself is one:

$$\underline{V}_i'\underline{V}_i = 1 \qquad \text{for } i = 1, 2, \ldots, n. \tag{4.1.9}$$

Orthonormality of the latent vectors of a matrix A can be invoked to allow a reexpression of A in terms of its latent roots and latent vectors. To accomplish this decomposition of the matrix A, let $V = (\underline{V}_1, \underline{V}_2, \ldots, \underline{V}_n)$ be a matrix containing all the latent vectors of A and $L = \text{diag}(\ell_1, \ell_2, \ldots, \ell_n)$ be a diagonal matrix with the corresponding latent roots of A on the diagonal. Then

$$AV = (A\underline{V}_1, A\underline{V}_2, \ldots, A\underline{V}_n)$$

$$= (\ell_1\underline{V}_1, \ell_2\underline{V}_2, \ldots, \ell_n\underline{V}_n)$$

$$= VL.$$

From this expression,

$$AVV' = VLV' = \sum_{j=1}^{n} \ell_j\underline{V}_j\underline{V}_j'.$$

But from eqns. (4.1.8) and (4.1.9),

$$V'V = \begin{bmatrix} \underline{V}_1'\underline{V}_1 & \underline{V}_1'\underline{V}_2 & \cdots & \underline{V}_1'\underline{V}_n \\ \underline{V}_2'\underline{V}_1 & \underline{V}_2'\underline{V}_2 & \cdots & \underline{V}_2'\underline{V}_n \\ \vdots & \vdots & & \vdots \\ \underline{V}_n'\underline{V}_1 & \underline{V}_n'\underline{V}_2 & \cdots & \underline{V}_n'\underline{V}_n \end{bmatrix} = I.$$

By the definition of a matrix inverse, V' is the matrix inverse of V, hence $V'V = VV' = I$. Thus $AVV' = A$ and

$$A = \sum_{j=1}^{n} \ell_j \underline{V}_j \underline{V}'_j. \tag{4.1.10}$$

Interestingly, if A has an inverse the inverse can also be written in terms of the latent roots and vectors of A:

$$A^{-1} = \sum_{j=1}^{n} \ell_j^{-1} \underline{V}_j \underline{V}'_j. \tag{4.1.11}$$

The only difference between the expressions for A and A^{-1} is that the inverse of each of the latent roots is used in eqn. (4.1.11). The determinant of a matrix is also a function of the latent roots:

$$|A| = \ell_1 \cdot \ell_2 \cdot \ldots \cdot \ell_n = \prod_{j=1}^{n} \ell_j; \tag{4.1.12}$$

i.e., the determinant of a matrix equals the product of its latent roots. The proofs of these last properties are contained in Appendix 4.B.

The important properties just described for the latent roots and latent vectors of a square, symmetric matrix can be summarized as follows:

(i)   Latent roots and latent vectors are defined by eqn. (4.1.7),

(ii)   Latent vectors are orthonormal and collectively comprise an "orthogonal matrix" V such that $V'V = VV' = I$,

(iii) A matrix can be decomposed into a function of its latent roots and vectors as in eqn. (4.1.10),

(iv)   A matrix inverse, when it exists, can be expressed as in eqn. (4.1.11), and

(v)   The determinant of a matrix is the product of its latent roots, eqn. (4.1.12).

4.1.5   Rank of a Matrix

An important application of latent roots and vectors is in the determination of the rank of a matrix. The rank of a matrix is equal to the largest number of linearly independent columns it contains. If A is an $n \times m$ matrix (for simplicity assume $n \geqslant m$; all the matrices we examine in this book will have at least as many rows as they do columns) and all its column vectors are linearly independent, its rank is m. On the other hand, if some group of r of the m column vectors of A are linear independent but no set of $r + 1$ vectors is, the rank of the matrix is r.

Intuitively, the rank of a matrix indicates the true amount of information contained in its columns. For example if the rank of an $m \times n$ matrix is $n - 1$, there exists a linear dependency among the columns. From eqn. (4.1.5) there exist constants c_1, c_2, ..., c_n, not all of which are zero, such that

$$\sum_{j=1}^{n} c_j \underline{a}_j = \underline{0} .$$

Suppose now that $c_1 \neq 0$ in the above linear dependence. Then

$$\underline{a}_1 = -c_1^{-1} \sum_{j=2}^{n} c_j \underline{a}_j . \qquad (4.1.13)$$

In other words, $\underline{a}_1$ is redundant in A since the other column vectors can be used [through the linear combination on the right side of eqn. (4.1.13)] to provide precisely the same information that is contained in $\underline{a}_1$. By replacing $\underline{a}_1$ with any other vector having a nonzero c_j, a relationship similar to eqn. (4.1.13) can be derived, showing that any one of the vectors with nonzero c_j can be considered redundant. If the rank of A is $n - 2$, two linear dependencies exist in A; two vectors are redundant. Thus the rank of a matrix indicates the number of columns in the matrix that are required to provide all the information contained in the entire set of column vectors.

Now suppose that for an $n \times n$ symmetric matrix one of its latent roots, say ℓ_1, is zero. From eqn. (4.1.7), $A\underline{V}_1 = \underline{0}$. However,

$$A\underline{V}_1 = \sum_{i=1}^{n} V_{i1} \underline{a}_i = \underline{0} \qquad (4.1.14)$$

implies that the columns of A are linearly dependent. Each zero latent root of A defines a linear dependence $A\underline{V}_j = \underline{0}$ as in eqn. (4.1.14). The total number of linear dependencies is equal to the number of zero latent roots of the matrix. Another way of saying this is that the rank of a square, symmetric matrix is equal to the number of nonzero latent roots of the matrix.

Linear dependence and rank are important to the specification of predictor variables, particularly for categorical predictor variables. In Section 2.2.1 we cautioned against assigning arbitrary numerical codes to categorical variables such as defining high school designators to be 1, 2, and 3 for three different high schools. We then espoused the use of indicator variables for the high schools, only we pointed out that two variables should be used and not three. The reason for this specification is to allow X^* to have full column rank.

Suppose we had defined three high school designators:

$$X_j = \begin{cases} 1 & \text{if high school } j \quad j = 1, 2, \text{ or } 3 \\ 0 & \text{otherwise.} \end{cases}$$

Each row of X^* then contains a one in the first element and a one in the second, third, or fourth column depending on which school the student attended. The other two elements in each row are zero. With this specification X^* does not have rank four since

$$\underline{1} - \underline{X}_1 - \underline{X}_2 - \underline{X}_3 = \underline{0} \, .$$

Deletion of column three and using only two designator variables removes the redundancy and leaves the new $n \times 3$ matrix X^* of full column rank. This is because $1 - X_1 - X_2$ is zero for all students from schools 1 and 2 but equals 1 for students from school 3 since $X_1 = X_2 = 0$ for these students.

4.2   CARE IN MODEL BUILDING

Throughout Chapters 2 and 3 misspecification, whether due to the exclusion of relevant predictor variables or to the inclusion of incorrect functional forms for the predictors, has been shown to be of great concern to the data analyst. Regression models that contain many predictor variables or several specifications of each predictor variable can still suffer from the effects of misspecification. An additional problem, that of redundancy, may also be incurred in multiple-variable prediction equations. This section, using two of the data sets in Appendix A, focuses on some of the practical problems of specification in multiple linear regression models.

4.2.1   Misspecification Bias

Just because a prediction equation contains many predictor variables, one cannot be certain that the model has been adequately formulated. As an oversimplification of the problem, if there is only one predictor variable that will insure accurate prediction of a response, failure to include this predictor variable in the specification of the prediction equation will result in poor prediction regardless of how many other variables have been placed in the equation.

Suppose that the true regression model for a single response can be written as

$$Y = \alpha + \underline{u}_1' \beta_1 + \underline{u}_2' \beta_2 + \varepsilon,$$

where $\underline{u}_1$ is a vector containing p predictor variables that the data analyst feels are influential on the response and $\underline{u}_2$ contains an additional q variables that are unknown to the data analyst. Even if the data analyst can get a precise estimate of α and β_1, his predictions will be biased. The prediction equation only using the p predictors in $\underline{u}_1$ is

$$\hat{Y} = \hat{\alpha} + \underline{u}_1' \hat{\beta}_1 .$$

If $\hat{\alpha}$ and $\hat{\beta}_1$ are very close to α and β_1, respectively, then even if the random errors are ignored

$$Y - \hat{Y} \approx \underline{u}_2' \hat{\beta}_2 .$$

The bias, $\underline{u}_2' \beta_2$, in the prediction equation may not allow accurate prediction of the response for any set of nonzero predictor-variable values. If the random errors are not trivial, moreover, inaccuracies due to both bias and random error can render the prediction equation highly imprecise.

Unfortunately, it is no simple task to determine whether a poor fit to a data set is due to misspecification. Indeed, some data sets may be fit extremely well (large R^2, etc.) with a misspecified model. That is, a good fit to n specific response values in a data set does not necessarily mean that future responses will be accurately predicted. Obvious reasons for this occurrence were discussed in Chapter 1 with the comments on generalization and extrapolation.

As an illustration of these concepts, the least squares prediction equation for GNP using the 47 observations (excluding Hong Kong and Singapore) on the six predictor variables in Data Set A.1 is

$$\hat{Y} = 159.87 - 3.96X_1 + 0.02X_2 + 0.20X_3 - 0.05X_4 + 5.90X_5 + 0.68X_6 .$$

(The fit was obtained using techniques to be discussed in Chapter 5.) The coefficient of determination in multiple-variable prediction equations measures the proportion of the variability of the response variable that can be accounted for by the predictor variables just as it does for single-variable ones. For this data its value is $R^2 = 0.60$. An estimate of the standard deviation of the response variable is $\hat{\sigma} = 353.84$.

Neither R^2 nor $\hat{\sigma}$ indicate that the fit to the GNP data is exceptionally good. Fully 40% of the variability of the response remains unexplained. The estimated standard deviation, moreover, is of the same order of magnitude as most of the responses for GNP; in fact, the average GNP value is only about twice as large as $\hat{\sigma}$ ($\overline{Y} = 661.17$). Could misspecification be causing the poor fit?

Misspecification of both types could be a problem with this data. The two-variable plots of response with predictor variables (e.g., Figures 2.2 and 2.4) suggest that the logarithm or inverse of GNP should be examined as appropriate transformations of the response. Only six of many possible predictor variables are included in Data Set A.1. Measures of industrialization, employment rate, inflation rate, relative currency values, and many other economic indices might influence GNP more than those variables in the data set.

It is again beyond the scope of this book to attempt to definitely answer the questions we have just posed. Our purpose in discussing them is to reinforce their importance and stress the obligations of the data analyst in seeking the answers. By being aware of these problems the data analyst can often avoid their occurrence; but even if they are unavoidable, one's awareness of them may prevent erroneous inferences or improper uses of the fitted model.

4.2.2 Overspecification Redundancy

Data analysts who are concerned about obtaining good fits to responses have a natural tendency to include any predictor variable that might influence the response in a prediction equation. This is especially true when pilot studies are conducted or in research endeavors where there is no theoretical model or past experimental evidence that suggests how the model should be specified. In some instances, fear of bias in the predictor likewise leads investigators to collect data on as many variables as possible.

Instead of misspecification (although misspecification can still be a problem here), prediction equations formed in this manner may suffer from overspecification. Too many predictor variables in the model can create problems when the goal of the investigation is to make inferences on the model parameters. This is because there is a great danger of redundancy among the predictor variables. If several predictor variables repeat information, it is difficult to determine which of the redundant variables should be retained in the model and which should be eliminated.

The Detroit homicide data, Data Set A.3, provides a good illustration of redundancy among the predictor variables. Several of the predictor variables have large pairwise correlations: ACC and UEMP ($r = -0.88$), ASR and

FTP ($r = 0.88$), ASR and CLEAR ($r = -0.86$), FTP and CLEAR ($r = -0.97$), FTP and WM ($r = -0.88$), LIC and GR ($r = 0.90$), GR and WM ($r = -0.85$), and CLEAR and WM ($r = 0.89$). The extreme amount of repetition in these variables will make inferences on the individual predictor variables very difficult.

One of the major goals of the investigation described in Appendix A.3 was to assess the relative influences of the predictor variables on the homicide rate. But consider an attempt to determine whether the full-time police rate or the rate of clearance of crimes is the more influential determinant of the homicide rate. The large pairwise correlation between FTP and CLEAR ($r = -0.97$) reveals that as FTP has increased over these 13 years, CLEAR has dropped proportionately. In fact since r is so close to -1, the decrease in CLEAR has been almost exactly proportional to the increase in FTP. Since the correspondence between the two variables is so close to being exact, one cannot expect to differentiate their effects on the homicide rate. Any inference, for example, that FTP has a great effect on HOM is simultaneously an inference that CLEAR has a great effect on HOM.

Observational studies such as this one frequently suffer from unavoidable redundancy: too many variables are increasing or decreasing in an approximate linear fashion with time. Yet balancing the need for sufficient variables to insure that no large biases occur in the prediction equation with the danger of introducing unnecessary redundancy is not an easy task. As a minimum precaution, the data analyst should be careful not to include predictor variables just because they are available.

4.3 STANDARDIZATION

Having discussed matrix algebra and the selection of prediction variables for a multiple-variable regression analysis in the first two sections of this chapter, we now wish to concentrate attention on some preliminary examinations of the data base. Assuming that the data base has been edited for obvious outliers and two-variable plots have been scrutinized for both outliers and variable misspecifications, whether to use raw or scaled response and predictor variables is one of the next decisions to be made.

Standardization of response and predictor variables in single-variable least squares model fitting was presented in Section 3.2.3. The mechanics for this type of scaling are identical for single- and multiple-variable data bases, only there are more variables to standardize in multiple-variable models. Since these models tend to have predictors that are dissimilar in both location and variability, standardization can be very beneficial. The next two subsections extend standardization to multiple-variable regression analyses.

4.3.1 Benefits

Many of the benefits of standardization were discussed and illustrated in Section 3.2.3 of Chapter 3. Briefly recapping those comments, standardization was shown to transform widely different variables so that they were more closely related in terms of their average and variability. Scaling in either the normal deviate form or the unit length variant results in all predictor variables having the same average (zero) and the same measure of variability ($S_X = 1.0$ or $d_X = 1.0$, respectively). This type of standardization also enables individual estimated regression coefficients to be more directly comparable.

Standardization of the response variable is also an attractive transformation since all variables, response and predictor, will be scaled alike. One must be cautious in scaling the response, however, since the beta weights do not have the same theoretical probability distributions as do estimated regression coefficients when only the predictor variables are standardized.

In addition to these advantageous features, the computational complexities of multiple analyses offer another important rationale for standardization. Estimated regression coefficients can theoretically be determined from raw or standardized predictor variables. If standardized predictor variables are used, the estimated coefficients can be found by simple transformations such as those shown in Section 3.2.3. The numerical accuracy of the computations, unlike the theoretical relationships, can have a great impact on the decision to standardize.

Suppose the middle four observations on X_2 and X_3 in Table 4.2 comprise a matrix X:

$$X = \begin{bmatrix} 3,900 & 3 \\ 740 & 72 \\ 900 & 2 \\ 1,700 & 11 \end{bmatrix}.$$

The values in the first column of X are roughly two to three orders of magnitude larger than those in the second column. Such disparity is common in multiple-variable analyses as a scan of the data sets in Appendix A will reveal.

Problems with numerical accuracy in regression analyses can be illustrated by finding $|X'X|$ and $(X'X)^{-1}$ for the above matrix, X. For this matrix,

$$X'X = \begin{bmatrix} 19{,}457{,}600 & 85{,}480 \\ 85{,}480 & 5{,}318 \end{bmatrix}.$$

To eight significant figures $|X'X| = 96{,}168{,}686{,}000$. The inverse of $X'X$ is

$$(X'X)^{-1} = \begin{bmatrix} .000{,}000{,}055{,}298 & -.000{,}000{,}888{,}855 \\ -.000{,}000{,}888{,}855 & .000{,}202{,}327{,}813 \end{bmatrix}.$$

Manipulation of determinants and matrices with such extremely large and small numerical values can lead to severe roundoff error. The problem is compounded even further when many additional predictor variables of possibly even greater disparity must be included in X.

Contrast the magnitudes of the values in $|X'X|$ and $(X'X)^{-1}$ using the raw data with those of standardized predictor variables. Let $W_{ij} = (X_{ij} - \overline{X}_j)/d_j$ as in eqn. (3.2.14). The matrix of standardized values is

$$W = \begin{bmatrix} 0.82918 & -0.32671 \\ -0.42451 & 0.85977 \\ -0.36103 & -0.34391 \\ -0.04364 & -0.18915 \end{bmatrix}.$$

Then

$$W'W = \begin{bmatrix} 1.00000 & -0.50347 \\ -0.50347 & 1.00000 \end{bmatrix}$$

$$|W'W| = 0.74652$$

and

$$(W'W)^{-1} = \begin{bmatrix} 1.33955 & 0.67442 \\ 0.67442 & 1.33955 \end{bmatrix}.$$

All the numerical values in these matrix operations are of the same order of magnitude and there is far less risk of severe roundoff error.

4.3.2   Correlation Form of $X'X$

Two standardizations of the predictor variables specifically singled out in Section 3.2.3 are the normal deviate form

$$Z_{ij} = \frac{X_{ij} - \overline{X}_j}{S_j}.$$

where $\overline{X}_j = n^{-1} \Sigma X_{ij}$ and $S_j^2 = \Sigma (X_{ij} - \overline{X}_j)^2/(n-1)$, and the unit length scaling

$$W_{ij} = \frac{X_{ij} - \overline{X}_j}{d_j}.$$

where $d_j^2 = \Sigma (X_{ij} - \overline{X}_j)^2$. Both standardizations are equally effective in reducing numerical inaccuracies associated with multiple-variable regression analyses. The unit length scaling offers the additional advantage of transforming $X'X$ into "correlation form."

Again denoting the unit length standardization matrix of predictor variables by W, $W'W$ takes on the form

$$W'W = \begin{bmatrix} 1 & r_{12} & r_{13} & \cdots & r_{1p} \\ r_{21} & 1 & r_{23} & \cdots & r_{2p} \\ r_{31} & r_{32} & 1 & \cdots & r_{3p} \\ \vdots & \vdots & \vdots & & \vdots \\ r_{p1} & r_{p2} & r_{p3} & \cdots & 1 \end{bmatrix}.$$

where

$$r_{uv} = \sum_{i=1}^{n} W_{iu} W_{iv}$$

$$= \frac{\displaystyle\sum_{i=1}^{n} (X_{iu} - \overline{X}_u)(X_{iv} - \overline{X}_v)}{d_{x_u} \cdot d_{x_v}}.$$

Hence r_{uv} is in the form of a correlation coefficient between the uth and vth predictor variables. A quick glance at $W'W$ can therefore indicate whether two predictor variables are highly correlated and provide redundant information.

If standardization of the predictor variables is accomplished with the normal deviate scaling, the scaled $X'X$ matrix is closely related to $W'W$. Specifically,

$$Z'Z = (n-1)W'W.$$

The diagonal elements of $Z'Z$ are equal to $(n-1)$ and the (u, v)th off-diagonal element is $(n-1)r_{uv}$. Detecting pairwise redundancies between predictor variables is not as readily apparent as it is with $W'W$, although it is a simple matter to obtain r_{uv} from $Z'Z$.

4.4 MULTICOLLINEARITY

Linear dependence of columns of predictor variables [eqn. (4.1.5)] is a consequence of the duplication of information that is provided by the variables. When linear dependencies occur in X^* the effects of some predictor variables are not distinguishable and the matrix inverse of $X^{*'}X^*$ does not exist.

Of increasing concern to researchers in recent years is the tendency for data bases to have "approximate" linear dependencies among columns of predictor variables. Often these approximate dependencies occur among predictor variables for which there are no theoretical or physical explanations for the dependencies. Worse still, since the dependencies are inexact they can go unnoticed by the data analyst. Computer programs can invert $X^{*'}X^*$, estimated regression coefficients can be calculated, and inferences can be drawn on the parameters of the model. But just as with the approximate dependence between FTP and CLEAR in the homicide data, clear-cut decisions on the relative importance of predictor variables which are approximately linearly dependent can be impossible to make.

Economists have long referred to approximate linear dependencies as "multicollinearities." The next few subsections formally define multicollinearities, propose techniques for detecting them, and distinguish between multicollinearities that arise due to sampling deficiencies and those that are inherent to a population.

4.4.1 Definition

A multicollinearity is an approximate linear dependence. More formally, if there exist constants $c_1, c_2, \ldots, c_n$, not all of which are zero, such that

$$\sum_{j=1}^{n} c_j \underline{a}_j \approx \underline{0}, \qquad (4.4.1)$$

then the vectors $\underline{a}_1$, $\underline{a}_2$, ..., $\underline{a}_n$ are said to be multicollinear. Multicollinearity is more a question of degree than existence: the closer the linear combination in eqn. (4.4.1) is to zero, the stronger is the multicollinearity. (Note: In eqn. (4.4.1) we assume that the vectors are centered $\underline{a}'_j \underline{1} = 0$, otherwise $\underline{0}$ on the right-hand side of eqn. (4.4.1) would be replaced by a constant vector, $m\underline{1}$, for some $m \neq 0$.)

Large coefficients, c_j, in eqn. (4.4.1) identify which vectors contribute most to the multicollinearity, provided that the vectors are scaled similarly. For example, if

$$.70\underline{a}_1 + .69\underline{a}_2 + .10\underline{a}_3 - .14\underline{a}_4 \approx \underline{0},$$

$\underline{a}_1$ and $\underline{a}_2$ cannot be judged to be the dominant vectors in the multicollinearity if the elements of $\underline{a}_3$ and $\underline{a}_4$ happen to be, say, two or three orders of magnitude larger than those of $\underline{a}_1$ and $\underline{a}_2$. If all four vector are standardized, however, the coefficients do indicate which vectors are most involved in the multicollinearity.

Standardization of vectors using the unit length scaling transforms all variable values to numbers between -1 and $+1$. Table 4.3 lists the standardized values for FTP and CLEAR for the Detroit homicide data. The large negative correlation ($r = -0.97$) between these two variables reveals that $\underline{W}_1 \approx -\underline{W}_2$ or $\underline{W}_1 + \underline{W}_2 \approx \underline{0}$. To confirm this pairwise multicollinearity, the standardized predictor-variable values have been summed in the last column of the table. The largest magnitude in the last column is -0.1434 while the smallest is 0.0003.

Table 4.3. Standardized (Unit Length) Predictor-Variable Values for FTP and Clear, Detroit Homicide Data

| Year | Standardized Values | | FTP + CLEAR |
	FTP(W_1)	CLEAR(W_2)	
1961	-.2723	.2726	.0003
62	-.2141	.1609	-.0532
63	-.2002	.2954	.0952
64	-.1946	.2407	.0461
65	-.1973	.2179	.0206
66	-.2662	.1358	-.1304
67	-.2997	.1563	-.1434
68	-.0526	.1061	.0535
69	.0947	-.0558	.0389
1970	.2277	-.1721	.0556
71	.3212	-.4115	-.0939
72	.4451	-.4320	.0131
73	.5284	-.5141	.0143

For comparison purposes, Table 4.4 records the standardized values of ACC and GR. Since $r = 0.03$ for these two variables, if they formed a pairwise multicollinearity it would be $\underline{W}_1 \approx \underline{W}_2$ or $\underline{W}_1 - \underline{W}_2 \approx \underline{0}$. Differences between the standardized values of ACC and GR are shown in the last column of Table 4.4. These values are much larger than the summed values of FTP and CLEAR in Table 4.3, the largest one (in magnitude) being –0.7781 and the smallest –0.0348.

Table 4.4. Standardized (Unit Length) Predictor-Variable Values for ACC and GR, Detroit Homicide Data

Year	Standardized Values		
	ACC(W_1)	GR(W_2)	ACC - GR
1961	-.4355	-.3060	-.1295
62	-.3737	-.3389	-.0348
63	-.0906	-.3119	.2213
64	.1453	-.2914	.4367
65	.4565	-.2302	.6867
66	.3919	-.1652	.5571
67	.2076	.0658	.1418
68	.2554	.4493	-.1939
69	.1256	.2233	-.0977
1970	-.0631	.1560	-.2191
71	-.1338	.1901	-.3239
72	-.3310	.4471	-.7781
73	-.1546	.1122	-.2668

Pictorially the distinction between a pairwise multicollinearity and two variables that are not strongly multicollinear can be viewed by plotting the raw or standardized predictor variables. Plotted in Figure 4.1 are the standardized predictor-variable values listed in Tables 4.3 and 4.4. Figure 4.1(a) reveals that the values for FTP and CLEAR lie in a narrow band along the line $W_1 = -W_2$. Only a very restricted region of the possible values of FTP and CLEAR are included in the data base.

Figure 4.1(b) shows that the scatter of points for ACC and GR is much more representative of the entire set of points that could have been included in the data base. These variables are not restricted to a narrow band and thus will more clearly reflect the individual effects of the predictors.

Strictly speaking, both of these examples exhibit multicollinearities in that some information is shared by FTP and CLEAR and by ACC and GR. The pairwise multicollinearity between FTP and CLEAR is much stronger than that between ACC and GR, as was revealed by the pairwise correlation coefficients, by how close the multicollinearities were to $\underline{0}$ (the last columns of Table 4.3 and 4.4), and by the two-variable plots. It is not as

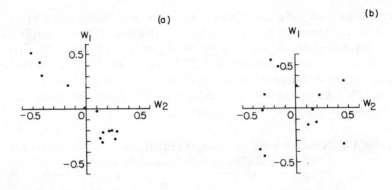

Figure 4.1   Two-variable plots of the standardized predictor variables in Tables 4.3 and 4.4: (a) FTP (W_1) versus CLEAR (W_2), (b) ACC (W_1) versus GR (W_2).

straightforward to analyze multicollinearities involving three or more predictor variables. Such multicollinearities cannot be plotted in p dimensions but have the same tendencies as illustrated in Figure 4.1(a). Rather than representing a wide range of possible combinations of the predictor variables, multicollinearities indicate that only a restricted range of predictor variables is represented in the data base. The next subsection demonstrates that the latent roots and latent vectors of $W'W$ are valuable diagnostic tools for data sets with multiple-variable multicollinearities.

4.4.2   Detection

Pairwise multicollinearities can be identified by scrutinizing the off-diagonal elements of $W'W$, the standardized matrix of sums of squares and products of the predictor variables. Just how large these off-diagonal elements must be before one should become concerned about multicollinearities is somewhat subjective. Any pairwise correlation larger in magnitude than, say, 0.70 or 0.80 should be investigated further.

Multicollinearities consisting of three or more predictor variables may not be readily identified from an examination of the pairwise correlation coefficients. A multivariable multicollinearity can occur without any of the pairwise correlations being large. To be certain that all multicollinearities are found in a data base, other diagnostic procedures are needed.

Suppose a latent root of $W'W$, while not exactly zero, is very close to zero. Since all the elements of a latent vector are less than 1.0 in magnitude, $\ell_j \underline{V}_j \approx \underline{0}$ for small latent roots. From the definition of latent roots and vectors, eqn. (4.1.7), it then follows that

$$W'W\underline{V}_j = \ell_j\underline{V}_j \approx \underline{0} \qquad (4.4.2)$$

for any latent vector whose corresponding latent root is nearly zero. Premultiplying eqn. (4.4.2) by $\underline{V}_j$ yields

$$\underline{V}_j' W' W \underline{V}_j \approx 0 .$$

Now let $\underline{U} = W\underline{V}_j$. The above equation becomes $\underline{U}'\underline{U} = \Sigma\ U_r^2 \approx 0;$ but since all terms in the summation are nonnegative, $\underline{U}'\underline{U} \approx 0$ implies $\underline{U} \approx \underline{0}$. Thus if $\ell_j \approx 0$,

$$W\underline{V}_j = \sum_{r=1}^{p} V_{rj}\underline{W}_r \approx \underline{0} . \qquad (4.4.3)$$

Implied by eqn. (4.4.3) is that the elements of latent vectors corresponding to small latent roots provide the coefficients in eqn. (4.4.1) that define multicollinearities among the predictor variables. The larger elements in $\underline{V}_j$ determine which predictor variables are most involved in the multicollinearities since the predictor variables are all standardized in W. These remarks parallel those made in Section 4.1.5 for exact multicollinearities (linear dependencies).

Returning to the two examples of the previous subsection, the correlation matrices for FTP and CLEAR and ACC and GR are, respectively,

$$W'W = \begin{bmatrix} 1.0000 & -0.9743 \\ -0.9743 & 1.0000 \end{bmatrix} \text{ and } W'W = \begin{bmatrix} 1.0000 & 0.0318 \\ 0.0318 & 1.0000 \end{bmatrix} .$$

For the first of these matrices the latent roots are $\ell_1 = 0.0257$ and $\ell_2 = 1.9743$. The corresponding latent vectors are

$$\underline{V}_1 = \begin{pmatrix} 0.7071 \\ 0.7071 \end{pmatrix} \text{ and } \underline{V}_2 = \begin{pmatrix} 0.7071 \\ -0.7071 \end{pmatrix} .$$

Applying eqn. (4.4.3), the multicollinearity between FTP and CLEAR is

$$.7071\ \underline{W}_1 + .7071\ \underline{W}_2 \approx \underline{0}$$

or

$$\underline{W}_1 + \underline{W}_2 \approx \underline{0},$$

as was discovered in the last subsection.

The latent roots for the second correlation matrix shown above are $\ell_1 =$ 0.9682 and $\ell_2 = 1.0318$. The corresponding latent vectors are

$$\underline{V}_1 = \begin{pmatrix} 0.7071 \\ -0.7071 \end{pmatrix} \quad \text{and} \quad \underline{V}_2 = \begin{pmatrix} 0.7071 \\ 0.7071 \end{pmatrix}.$$

Since ℓ_1 is so large, however, there is not a strong multicollinearity between ACC and GR.

This example illustrates how a pairwise multicollinearity affects the latent roots of $W'W$ for prediction equations that contain two predictor variables. In later chapters we will consider larger data sets and multicollinearities that involve more than two predictor variables. The essence of detecting multicollinearities* is the same in all cases: small latent roots define multicollinearities, large elements of the corresponding latent vectors identify which variables are involved in them.

4.4.3 Sample vs. Population Multicollinearities

Some multicollinearities arise in populations through the definitions of the predictor variables or because they are characteristic of the population itself. For example, using indicator variables often produces exact multicollinearities (linear dependencies). As pointed out in Section 4.1.5, using the three indicator variables

$$X_j = \begin{cases} 1 & \text{if high school } j \qquad j = 1, 2, \text{ or } 3 \\ 0 & \text{otherwise} \end{cases}$$

for students from the three high schools sampled in Data Set A.2 induces the exact multicollinearity $\underline{1} - \underline{X}_1 - \underline{X}_2 - \underline{X}_3 = \underline{0}$. Moreover, this multicollinearity exists for all students from these three high schools, not only those in Data Set A.2. It is thus a population multicollinearity caused by the definition of the predictor variables.

Inexact population multicollinearities can similarly arise because of the way the predictor variables are defined. It appears reasonable to expect that

---

*A discussion of the merits of several alternative procedures that have been advocated for identifying multicollinearities can be found in Mason, Gunst, and Webster (1975).

handgun licenses and handgun registrations will remain highly multicollinear in Detroit so long as the laws governing the requirement for licensing and registering handguns are not altered. Here again a multicollinearity belongs to the population and is not just a property of the sample values in Data Set A.3. In this case the multicollinearity is not due to the definition of the predictor variables per se; rather it occurs because of the nature of the laws governing handgun ownership.

One strategy for dealing with highly multicollinear predictor variables, especially when linear dependencies occur, is to delete one of the multicollinear variables. If the multicollinearity is strong enough (i.e., if the latent root of $W'W$ that corresponds to the multicollinearity is close enough to zero) and if the multicollinearity is a characteristic of the population of interest, this strategy is a good one. Every sample that is drawn from such a population and every individual or element of the population for which prediction is desired will have the same multicollinearities as those observed in the data base. The variable eliminated, therefore, can be compensated for by the other variables with which they are multicollinear.

Many multicollinearities are not inherent to populations but are due to sample deficiencies. The term "sample deficiency" is meant to imply unwanted concommitant variation among the predictor variables rather than ineffective sampling techniques employed by the researcher. The controversy discussed in the introduction to this chapter arose because of sample multicollinearities. In one sample anomie and authoritarianism were more strongly multicollinear than in the other two ($r = 0.67$ vs. $r = 0.45$ and 0.47). If a multicollinearity between these two predictor variables does exist in the population, the sampling schemes did not adequately reflect the multicollinearity for the two samples with smaller correlation coefficients. On the other hand, if a strong multicollinearity does not exist in the population, the sampling technique for the third study erroneously indicates that one exists. As was mentioned in the introduction, however, it is likely that three different populations, not one, were sampled.

The large negative correlation between FTP and CLEAR is a multicollinearity that is characteristic of the sample data in Data Set A.3. and not of a population consisting of all previous and future observations for Detroit. One might even expect that an increase in police manpower should be coupled with an increase, not a decrease, in the clearance rate of crimes. The decrease in the clearance rate in the sampled 13 years could be attributable to a rapidly rising population and a great increase in the overall crime rate in Detroit that the increased police manpower could not keep up with. Perhaps in future years the trend will be slowed, even reversed, with increased full-time police rates. Many other factors could also affect the clearance regardless of the trend in the full-time police rate.

Depending on one's knowledge of the sampled population, several remedial strategies can be considered for sample (and population) multicollinearities. Deletion of one or more of the multicollinear predictor variables (Chapter 8) is certainly one course of action, but this may result in biased predictions if the multicollinearities in the data base are due to sampling deficiencies. Another strategy is to employ biased regression estimators (Chapter 10) to construct prediction equations. An often overlooked possibility is to utilize the least squares prediction equation but restrict its application to predicting responses for which the predictor variables exhibit the same multicollinearities as occur in the data base. We are now ready to explore these topics in detail, beginning with least squares model-fitting techniques in the next chapter.

APPENDIX

4.A Determinants and Inverse Matrices

The determinant of a matrix is a natural extension of numbers for variables. A determinant is a number that in certain respects indicates the size of the matrix. Symbolically the determinant of an $n \times n$ matrix A is written

$$
|A| \quad \text{or} \quad
\begin{vmatrix}
a_{11} & a_{12} & \cdots & a_{1n} \\
a_{21} & a_{22} & \cdots & a_{2n} \\
\vdots & \vdots & & \vdots \\
a_{n1} & a_{n2} & \cdots & a_{nn}
\end{vmatrix} .
$$

For a 2×2 matrix the determinant of A is the difference between the product of the diagonal elements and the product of the off-diagonal elements:

$$
|A| =
\begin{vmatrix}
a_{11} & a_{12} \\
a_{21} & a_{22}
\end{vmatrix}
= a_{11}a_{22} - a_{12}a_{21} . \tag{4.A.1}
$$

For 3×3 or higher-order matrices the determinant is obtained by reexpressing the determinant of A in terms of 2×2 determinants involving the elements of A. Before showing the general form for $|A|$, we need to define a "cofactor" of an element a_{ij} of A. The cofactor of a_{ij} is $(-1)^{i+j}$ times the determinant of the $(n-1) \times (n-1)$ submatrix of A obtained by eliminating the ith row and jth column of A. For example if A is 4×4, the cofactor of a_{12} is

$$A_{12} = (-1)^3 \cdot \begin{vmatrix} a_{21} & a_{23} & a_{24} \\ a_{31} & a_{33} & a_{34} \\ a_{41} & a_{43} & a_{44} \end{vmatrix} .$$

Expanding the determinant of A by its cofactors provides a sequential method for evaluating the determinant. This expansion can be done by either rows or columns. If one desires to expand using the rth row of A, then

$$|A| = \sum_{j=1}^{n} a_{rj} A_{rj} . \tag{4.A.2}$$

Each of the cofactors A_{rj} is itself a determinant of an $(n - 1) \times (n - 1)$ matrix and can be expressed along any of its rows or columns to yield a linear combination of $(n - 2) \times (n - 2)$ cofactors. This process can be continued until only 2×2 cofactors remain, each of which can be evaluated accordingly to eqn. (4.A.1). Note that eqn. (4.A.1) is itself an expansion by cofactors.

For a 3×3 matrix A, an expansion along the first row yields

$$|A| = a_{11}A_{11} - a_{12}A_{12} + a_{13}A_{13}$$
$$= a_{11}(a_{22}a_{33} - a_{23}a_{32}) - a_{12}(a_{21}a_{33} - a_{23}a_{31}) + a_{13}(a_{21}a_{32} - a_{22}a_{31}) .$$

Equivalently, one could expand along one of the columns, say the second one:

$$|A| = - a_{12}A_{12} + a_{22}A_{22} - a_{32}A_{32}$$
$$= -a_{12}(a_{21}a_{33} - a_{23}a_{31}) + a_{22}(a_{11}a_{33} - a_{13}a_{31}) - a_{32}(a_{11}a_{23} - a_{13}a_{21}) .$$

The general expression for expansion along the cth column of an $n \times n$ matrix A is

$$|A| = \sum_{i=1}^{n} a_{ic} A_{ic} . \tag{4.A.3}$$

For a more complete discussion of determinants the reader is referred to Hadley (1961) or some other linear algebra textbook.

A matrix inverse of an $n \times n$ matrix A is defined to be an $n \times n$ matrix A^{-1} that satisfies

$$AA^{-1} = A^{-1}A = I.$$

When a matrix has an inverse it can be found by

$$A^{-1} = \frac{1}{|A|} A^+ , \qquad (4.A.4)$$

where A^+, referred to as the adjoint of A, is an $n \times n$ matrix whose (i,j)th element is the cofactor of element a_{ji} of A. Specifically,

$$A^+ = \begin{bmatrix} A_{11} & A_{21} & \dots & A_{n1} \\ A_{12} & A_{22} & \dots & A_{n2} \\ \vdots & \vdots & & \vdots \\ A_{1n} & A_{2n} & \dots & A_{nn} \end{bmatrix} .$$

The inverse of A does not exist when $|A| = 0$. When $|A| \neq 0$, A^{-1} exists and is given by eqn.(4.A.4).

4.B   Determinants and Inverses Using Latent Roots and Latent Vectors

A general property of determinants of square matrices is that the determinant of the product of square matrices equals the product of the individual determinants. For instance, the determinant of the product of three $n \times n$ matrices A, B, and C, can be found from the individual determinants:

$$|ABC| = |A| \cdot |B| \cdot |C| . \qquad (4.B.1)$$

Another important property of determinants is that the determinant of a diagonal matrix equals the product of the diagonal elements. Thus if

$$D = \begin{bmatrix} d_1 & 0 & \dots & 0 \\ 0 & d_2 & \dots & 0 \\ \vdots & \vdots & & \vdots \\ 0 & 0 & \dots & d_n \end{bmatrix} ,$$

then

$$|D| = d_1 \cdot d_2 \cdot \dots \cdot d_n = \prod_{j=1}^{n} d_j . \qquad (4.B.2)$$

This result can easily be verified by the expansion of D by cofactors using either eqns.(4.A.2) or (4.A.3).

Relationships (4.B.1) and (4.B.2) allow an important property of the latent roots and vectors of a matrix to be verified. The determinant of a square matrix equals the product of its latent roots; i.e.,

$$|A| = \prod_{j=1}^{n} \ell_j ; \qquad (4.B.3)$$

To verify this result, recall from Section 4.1.4 that $A = VLV'$. Thus

$$
\begin{aligned}
|A| \quad &= |VLV'| \\
&= |V| \cdot |L| \cdot |V'| \qquad \text{from eqn. (4.B.1)} \\
&= |V| \cdot |V'| \cdot |L| \\
&= |VV'| \cdot |L| \\
&= |I| \cdot |L| \\
&= \left(\prod_{j=1}^{n} 1 \right) \left(\prod_{j=1}^{n} \ell_j \right) \\
&= \prod_{j=1}^{n} \ell_j .
\end{aligned}
$$

From eqn. (4.B.3), if any latent root of A is zero so is its determinant, implying that the inverse of A does not exist if any of its latent roots are zero.

Provided that $|A|$ is nonzero, the inverse of A can be written as

$$A^{-1} = \sum_{j=1}^{n} \ell_j^{-1} \underline{V}_j \underline{V}'_j . \qquad (4.B.4)$$

This expression gives an important alternative method to eqn. (4.A.4) for computing inverses. Verification of eqn.(4.B.4) consists of showing that $AA^{-1} = I$. First note that if D is a diagonal matrix with all its diagonal elements nonzero, its inverse is also a diagonal matrix with diagonal elements that are the inverses of those of D, i.e.,

$$D^{-1} = \begin{bmatrix} d_1^{-1} & 0 & \ldots & 0 \\ 0 & d_2^{-1} & \ldots & 0 \\ \vdots & \vdots & & \vdots \\ 0 & 0 & \ldots & d_n^{-1} \end{bmatrix}. \tag{4.B.5}$$

Thus

$$AA^{-1} = (VLV')(VL^{-1}V')$$

$$= VLL^{-1}V' \quad \text{since } V'V = I$$

$$= VV' \quad \text{since } LL^{-1} = I$$

$$= I.$$

EXERCISES

1. Using the matrix notation of Subsection (4.1.1) express the regression model for relating GNP to INFD, PHYS, and DENS for the data in Table 4.2.

2. Consider only X_1 and X_2 in Table 4.2. Construct the 2×2 matrix, $W'W$, where W_{ij} is defined as in eqn. (3.2.4). Using the techniques described in Appendix 4.A find the determinant and inverse of $W'W$.

3. Verify by evaluating eqn. (4.1.7) that the latent roots of $W'W$ in Exercise 2 are $1 + r$ and $1 - r$ where r is the off-diagonal element of $W'W$, and the corresponding latent vectors are respectively, $\sqrt{.5}(1,1)$ and $\sqrt{.5}(1,-1)$.

4. Verify equations (4.1.10), (4.1.11), and (4.1.12) with the data from Exercise 2. What is the rank of $W'W$ in Exercise 2? Explain your answer.

5. Compare the fits of GNP to INFD and GNP to $(INFD)^{-2}$ by examining the resultant R^2 values and $\hat{\sigma}^2$ values. Do the results indicate that the prediction equation using INFD is misspecified? Suppose both INFD and $(INFD)^{-2}$ had been included in a single regression equation for GNP. Are these variables redundant?

6. Standardize the $X'X$ matrix of Exercise 2 using both the normal devi-
    ate and unit length scaling methods. Compare the magnitudes of
    $|X'X|$, $|Z'Z|$, and $|W'W|$ as well as $(X'X)^{-1}$, $(Z'Z)^{-1}$, and $(W'W)^{-1}$.
    Numerically verify that $Z'Z$ equals $(n-1)W'W$.

7. Using the data in Data Set A.3 standardize (using the length scaling)
    the values for LIC and GR and verify that the large correlation
    between these two variables ($r = 0.90$) implies that $\underline{W}_1 \approx \underline{W}_2$.

8. Using the results of Exercise 3 find the latent roots and corresponding
    latent vectors of $W'W$ in Exercise 7. Do these values indicate the exis-
    tence of a pairwise multicollinearity between LIC and GR? Explain.

9. Is the large correlation between LIC and GR indicative of a sample or
    population multicollinearity? Explain.

CHAPTER 5

MULTIPLE-VARIABLE LEAST SQUARES

Regression analyses that utilize several predictor variables in the specification of a prediction equation are simultaneously more flexible and more complicated than those that incorporate a single predictor variable. Flexibility is introduced into the analysis by the investigator's ability to assess the influence on the response variable of several predictor variables, several different specifications of one predictor variable, or a combination of both. Contrast this flexibility with a single-predictor-variable analysis which forces a functional specification of one predictor variable to adequately fit a response variable.

Complexity is the added price that must be paid for the flexibility of a multiple-variable regression analysis. Any analysis involving more than two or three predictor variables generally requires the use of a computer. To those who are trained in the use of computers, this complication is not a major one; to others, it may appear to be so. Most computer installations, however, contain (or have access to) libraries of computer programs that perform many of the analyses that are suggested in this book. In fact, once the data is keypunched and input to the computer, much time that would be expended in plotting and computing is saved by letting the computer perform these functions.

With several predictor variables in a prediction equation, interpretation of the information that is output by the computer becomes more difficult than in a single-variable analysis. Even if one has correctly specified a theoretical model and properly formed a data base, one cannot merely examine the magnitudes of the estimated coefficients to determine which predictor variables most affect the response. Likewise one cannot examine the influence of one of the predictor variables on the response without regard to the concurrent influence of the other predictor variables. The complexity of a multiple-variable regression analysis arises, then, not as much from the computing problems as from correctly discerning the implications of the computer output.

Data Set A.4 can be used to numerically illustrate some of the major differences in the analysis of single- and multiple-variable prediction equations. Sewell et al. (1970) set out to investigate causal models for educational and occupational attainment. Theoretical arguments and previous empirical studies enabled the authors to define a causal sequence among the variables in their data base, culminating in prediction equations for both educational and occupational achievement. Data Set A.4 concentrates attention on a causal model for educational attainment.

A complete analysis of the causal relationships suggested by the authors requires the application of material covered in later chapters; however, let us merely attempt to fit a prediction equation to educational attainment (EDATT). The six predictor variables to be used are: a measure of the occupational goals of the subject (LOA); a measure of the educational goals of the subject (LEA); an indicator of the degree of influence of close friends, parents, and teachers on an individual's educational plans (SOI); an academic performance rating (AP); a socioeconomic index of the subject's family (SES); and a measure of mental ability (MA). The predictor variables were measured on male high school seniors in Wisconsin in 1957, the response variable on a random sample of one-third of these individuals in 1964-65.

Immediately the computational difficulties of multiple-variable prediction equations become apparent. First, the authors proposed that EDATT is influenced not by a single predictor variable but by six predictors. As we shall see in the next section, calculation of least squares parameter estimates for the full prediction equation necessitates the inversion of a 6×6 or 7×7 matrix—a formidable task if attempted by hand. Second, Data Set A.4 contains summary information on only the subjects in the original data base who resided in small cities in 1957. Just summary statistics are exhibited in Data Set A.4 since the original data base consists of 1,094 subjects from small cities. Thus the large number of observations and the formulation of a prediction equation with six predictor variables strongly supports the utilization of a computer for computational purposes.

The summary information provided in Data Set A.4 will allow the computation of statistics that are needed to analyze the data base, although some information (such as residuals) that require individual data values cannot be calculated. In particular, estimates of regression coefficients can be calculated from the summary statistics in Data Set A.4 (details of the computations will be pointed out in Section 5.1.4).

If we let X_1 through X_6, respectively, denote the predictor variables LOA, LEA, SOI, AP, SES, and MA, the least squares prediction equation for EDATT(Y) is

$$\hat{Y} = 0.08X_1 + 0.39X_2 + 0.15X_3 + 0.20X_4 + 0.16X_5 + 0.05X_6.$$

In this formulation of the prediction equation both response and predictor variables are standardized. The beta weights appear to indicate that level of educational aspiration (X_2) is the most important determinant of educational attainment, followed by academic performance (X_4), the influence of significant others (X_3), and socioeconomic status (X_5), all with about the same influence. Level of occupational aspiration (X_1) and mental ability (X_6) appear to have relatively less influence on educational attainment.

Conclusions similar to those just rendered generate confusion and controversy in a multiple regression analysis. Despite the inclusion of all six predictor variables in the prediction equation and the simultaneous calculation of the estimated coefficients, we have interpreted the magnitudes of the coefficients individually; i.e., without regard to the fact that other variables are in the prediction equation and, more importantly, to the potential for differing interpretations if some of the variables were eliminated from the predictor. For example, we observed that LEA has the largest estimated coefficient and MA one of the smallest. Would these conclusions remain the same if, say, AP was removed from the model?

The present chapter examines least squares estimation of multiple-variable prediction equations. Estimation of model parameters and summarization of the adequacy of the fit are discussed. Particular attention is focussed on the interpretation of fitted models and the difficulties that arise when two or more predictor variables provide redundant information about the response.

Although computers are generally used to perform the calculations that are needed for multiple-variable regression analyses, it is instructive to examine the algebraic derivation and properties of the statistics that are used in the analyses. The benefits are twofold: (i) the data analyst must understand the operations that the computer is undertaking and the output that it provides; and (ii) theoretical properties of the estimators and prediction equations are easily described with algebraic expressions. Again we stress that the computer can facilitate the computation of statistics needed for a regression analysis but it is the obligation of the data analyst to properly interpret and utilize the results.

5.1 PARAMETER ESTIMATION

We begin our discussion of multiple-variable least squares methodology with parameter estimation, although we could again stress appropriateness of the model or model specification. Whether multiple linear regression models are appropriate for a particular regression analysis is just as important a consideration as it was for single-variable models in Chapter 3. Those admonitions (see Section 3.1) should be borne in mind just as concerns over correct model specification should lead to appropriate investigations (two-variable plots, etc.) prior to attempts to fit a multiple-variable prediction equation.

In this chapter we consider least squares parameter estimation. The basic principle in deriving the least squares estimator is the same as that enunciated in Section 3.2.1. If we write the multiple linear regression model as

$$Y = \alpha + \beta_1 X_1 + \beta_2 X_2 + \ldots + \beta_p X_p + \varepsilon, \qquad (5.1.1)$$

where Y represents the response variable, $X_1, X_2, \ldots, X_p$ are the p predictor variables, and ε is a random error term, we seek a prediction equation of the form

$$\hat{Y} = \hat{\alpha} + \hat{\beta}_1 X_1 + \hat{\beta}_2 X_2 + \ldots + \hat{\beta}_p X_p, \qquad (5.1.2)$$

where $\hat{\alpha}, \hat{\beta}_1, \ldots, \hat{\beta}_p$ are suitable estimates of the unknown regression coefficients $\alpha, \beta_1, \ldots, \beta_p$. Least squares parameter estimation chooses numerical estimates of the regression coefficients that minimize the sum (average) of the squared residuals for a data base. Specifically, if n observations are available from model (5.1.1), the least squares parameter estimates minimize

$$\sum_{i=1}^{n} r_i^2 = \sum_{i=1}^{n} (Y_i - \hat{Y}_i)^2$$

$$= \sum_{i=1}^{n} (Y_i - \hat{\alpha} - \hat{\beta}_1 X_{i1} - \hat{\beta}_2 X_{i2} - \ldots - \hat{\beta}_p X_{ip})^2 . \qquad (5.1.3)$$

There are several methods for deriving and justifying least squares parameter estimates. In this section we present two approaches, both of which produce identical estimates but provide different viewpoints of the rationale for their use. The remaining subsections of this section treat the special case of orthogonal predictor variables, standardization, and no-intercept models.

5.1.1 Matrix Algebra Formulation

Matrix algebra can be used to concisely represent the least squares estimators for multiple-variable prediction equations. The multiple linear regression model

$$Y_i = \alpha + \beta_1 X_{i1} + \beta_2 X_{i2} + \dots + \beta_p X_{ip} + \varepsilon_i \qquad i = 1, 2, \dots, n$$

can be written, as in Section 4.1.1, as

$$\underline{Y} = X^* \underline{B} + \underline{\varepsilon} \tag{5.1.4}$$

where $\underline{Y}$ is an n-dimensional vector of the response variables, $X^* = [\underline{1}, \underline{X}_1, \underline{X}_2, \dots, \underline{X}_p]$ contains a column of n ones and a column of n observations on each of the p predictor variables, $\underline{B}$ is a $(p + 1)$-dimensional vector of model parameters $(\alpha, \beta_1, \beta_2, \dots, \beta_p)$, and $\underline{\varepsilon}$ is a vector of the n random error terms. Equivalently, the constant term can be partitioned from X^* and $\underline{B}$ to give the following representation of the model:

$$\underline{Y} = \alpha \underline{1} + X\beta + \underline{\varepsilon}, \tag{5.1.5}$$

where $X = (\underline{X}_1, \underline{X}_2, \dots, \underline{X}_p)$ and $\underline{\beta}' = (\beta_1, \beta_2, \dots, \beta_p)$.

We seek a prediction equation of the form

$$\hat{\underline{Y}} = X^* \hat{\underline{B}} \qquad \text{or} \qquad \hat{\underline{Y}} = \hat{\alpha} \underline{1} + X\hat{\beta}$$

where, using either formulation of the model, the estimates of the parameters are chosen to minimize the sum of squared residuals. The residual vector, $\underline{r}$, is defined to be

$$\begin{aligned} \underline{r} &= \underline{Y} - \hat{\underline{Y}} \\ &= \underline{Y} - X^* \hat{\underline{B}} \\ &= \underline{Y} - \hat{\alpha} \underline{1} - X\hat{\beta} \end{aligned}$$

so that the sum of squared residuals is expressible as

$$\sum_{i=1}^{n} r_i^2 = \underline{r}'\underline{r} = (\underline{Y} - X^*\hat{\underline{B}})'(\underline{Y} - X^*\hat{\underline{B}}) \tag{5.1.6}$$

$$= (\underline{Y} - \hat{\alpha}\underline{1} - X\hat{\underline{\beta}})'(\underline{Y} - \hat{\alpha}\underline{1} - X\hat{\underline{\beta}}).$$

Appendix 5.A demonstrates that minimization of eqn. (5.1.6) with respect to $\hat{\alpha}$, $\hat{\beta}_1$, $\hat{\beta}_2$, ..., $\hat{\beta}_p$ in either of the above model formulations yields the least squares estimator

$$\hat{\underline{B}} = (X^{*\prime}X^*)^{-1}X^{*\prime}\underline{Y}. \tag{5.1.7}$$

Computer programs essentially invert $X^{*\prime}X^*$ and perform the matrix multiplications that are indicated in eqn. (5.1.7). The programs take advantage of known properties of the matrices and vectors in eqn. (5.1.7) to reduce computational difficulties and roundoff error and of course are much more rapid than hand calculations.

Simplification of eqn. (5.1.7) is possible if, instead of working with the original values of the predictor variables, centered predictor variable values are used. Let

$$M_{ij} = X_{ij} - \overline{X}_j \tag{5.1.8}$$

be centered predictor variable values; i.e., deviations of the predictor variables from their respective means. Denote the matrix of centered predictor variables by $M = (\underline{M}_1, \underline{M}_2, ..., \underline{M}_p)$, where $\underline{M}_j$ is the n-dimensional vector of centered values of the jth predictor variable. The least squares estimator (5.1.7) can now be partitioned (see Section 5.1.4) to yield separate estimates of α and β:

$$\hat{\underline{\beta}} = (M'M)^{-1}M'\underline{Y} \quad \text{and} \quad \hat{\alpha} = \overline{Y} - \sum_{j=1}^{p} \overline{X}_j\hat{\beta}_j. \tag{5.1.9}$$

It is instructive to note that if $p = 1$, i.e., there is a single predictor variable in the model, eqn. (5.1.9) reduces to

$$\hat{\beta}_1 = \frac{\sum_{i=1}^{n} X_{i1} Y_i - n\overline{X}_1\overline{Y}}{\sum_{i=1}^{n} X_{i1}^2 - n\overline{X}_1^2} \quad \text{and} \quad \hat{\alpha} = \overline{Y} - \hat{\beta}_1\overline{X}_1,$$

just as in eqn. (3.2.6).

Computationally, for hand calculations, one can show algebraically that

$$M'M = \begin{bmatrix} \Sigma X_{i1}^2 - n\bar{X}_1^2 & \Sigma X_{i1}X_{i2} - n\bar{X}_1\bar{X}_2 & \dots & \Sigma X_{i1}X_{ip} - n\bar{X}_1\bar{X}_p \\ \Sigma X_{i2}X_{i1} - n\bar{X}_2\bar{X}_1 & \Sigma X_{i2}^2 - n\bar{X}_2^2 & \dots & \Sigma X_{i2}X_{ip} - n\bar{X}_2\bar{X}_p \\ \vdots & \vdots & & \vdots \\ \Sigma X_{ip}X_{i1} - n\bar{X}_p\bar{X}_1 & \Sigma X_{ip}X_{i2} - n\bar{X}_p\bar{X}_2 & \dots & \Sigma X_{ip}^2 - n\bar{X}_p^2 \end{bmatrix}$$

and

$$M'\underline{Y} = \begin{pmatrix} \Sigma X_{i1}Y_i - n\bar{X}_1\bar{Y} \\ \Sigma X_{i2}Y_i - n\bar{X}_2\bar{Y} \\ \vdots \\ X_{ip}Y_i - n\bar{X}_p\bar{Y} \end{pmatrix}.$$

To illustrate these computations consider the regression of HOM on FTP, LIC, and GR for the Detroit Homicide data (the variable values are listed in Data Set A.3 and in Table 5.1 of the next subsection):

$$M'M = \begin{bmatrix} 26{,}295.98 & 101{,}052.65 & 122{,}716.73 \\ 101{,}052.65 & 1{,}201{,}422.49 & 1{,}067{,}367.92 \\ 122{,}716.73 & 1{,}067{,}367.92 & 1{,}160{,}887.96 \end{bmatrix}$$

$$(M'M)^{-1} = \begin{bmatrix} .0000787822 & .0000042170 & -.0000122053 \\ .0000042170 & .0000047703 & -.0000048318 \\ -.0000122053 & -.0000048318 & .0000065942 \end{bmatrix}$$

and

$$M'\underline{Y} = \begin{pmatrix} 8{,}873.56 \\ 45{,}187.23 \\ 49{,}921.40 \end{pmatrix}.$$

Performing the multiplications indicated in eqn. (5.1.9) gives the estimates

$$\hat{\underline{\beta}} = \begin{pmatrix} .28033 \\ .01177 \\ .00255 \end{pmatrix} \quad \text{and} \quad \hat{\alpha} = -67.95468.$$

The estimates of β_1, β_2, and β_3 are exactly the same (to the five decimal places shown) as the values that would be obtained from computer output. The estimate of α agrees with the computer output to two decimal places; the discrepancy is due to a loss of accuracy during the hand calculations just performed. This latter discrepancy would be reduced, if not eliminated, by standardizing the predictor variable prior to the calculations.

5.1.2 Fitting By Stages

The second technique for finding least squares estimates of the parameters in model (5.1.1) employs a step-by-step technique of fitting p single-variable prediction equations. It requires only the use of a desk calculator but is tedious and has a great potential for roundoff and calculation errors. We illustrate this technique because it provides valuable insight into the interpretation of estimated regression coefficients, a topic explored later in this chapter.

Fitting least squares prediction equations by stages will be illustrated with a portion of the homicide data in Data Set A.3. Table 5.1 lists $n = 13$ observations on homicide rate (Y), full-time police rate (X_1), handgun license rate (X_2), and handgun registration rate (X_3) in Detroit over the years 1961-1973. Figures 3.4 and 3.5 in Chapter 3 contain two-variable plots of the response versus each of the three predictor variables. No obvious nonlinear trends are indicated in the two-variable plots so no transformations of the variables will be attempted at this point in the analysis.

Stage one in the process of fitting a multiple-variable prediction equation for HOM is to regress the response on one of the predictor variables, say FTP, using single-variable least squares estimation. The fitted prediction equation is symbolized by

$$\hat{Y}(1) = \hat{\alpha}(1) + \hat{\beta}(1)X_1,$$

where the numbers in parentheses are used throughout this subsection to indicate at which stage in the analysis the estimates are being calculated (here, stage 1). Numerically, as is indicated at the bottom of Table 5.1,

$$\hat{Y}(1) = -77.63026 + 0.33745X_1.$$

The fit using only FTP is quite good as indicated by $R^2 = 0.929$ (recall that one would also check residuals and the estimate of the error standard deviation before reaching a conclusive statement on the fit).

Since each predictor variable contains some information on both the response variable and the other predictor variables, one method of accounting for the redundant information between FTP and the other predictor variables is to regress the other predictor variables on FTP. The *residuals* of the response variable with FTP can then be regressed on the *residuals* of one of the other predictor variables with FTP. A linear relationship between the residuals of HOM and, say, the residuals of LIC would indicate that information on the homicide rate is provided by LIC *in addition to* the information already provided by FTP.

Completion of stage one of this analysis therefore requires that both LIC (X_2) and GR(X_3) be regressed on FTP(X_1), yielding (see Table 5.1)

$$\hat{X}_2 = -632.69877 + 3.84289X_1$$

and

$$\hat{X}_3 = -875.41989 + 4.66675X_1 .$$

Note that either LIC or GR could be used in place of FTP in the stage one analysis; we are merely inserting the variables as they are listed in Table 5.1.

Stage two of the analysis consists first of the calculation of residuals from the stage one prediction equations:

$$r_{i1}(Y) = Y_i + 77.63026 - 0.33745X_{i1}$$

$$r_{i1}(X_2) = X_{i2} + 632.69877 - 3.84289X_{i1}$$

and

$$r_{i1}(X_3) = X_{i3} + 875.41989 - 4.66675X_{i1}.$$

With this notation, the subscript 1 on the residuals denotes that the prediction equations are from the stage one analysis and the symbol in parentheses shows the variable whose residuals are being computed. The respective residuals are listed in the second through the fourth columns of Table 5.2.

Next, the response residuals are regressed on the residuals of a second predictor variable, say LIC; i.e., use single-variable least squares estimators to find $\hat{\alpha}(2)$ and $\hat{\beta}(2)$ for the prediction equation

$$\hat{r}_{i1}(Y) = \hat{\alpha}(2) + \hat{\beta}(2)\, r_{i1}(X_2)$$

Table 5.1. First Stage in the Fit of a Three-Variable Prediction Equation for the Detroit Homicide Rate

Year	Homicide Rate (Y)	Full-Time Police (X_1)	Handgun Licenses (X_2)	Handgun Registrations (X_3)
1961	8.60	260.35	178.15	215.98
62	8.90	269.80	156.41	180.48
63	8.52	272.04	198.02	209.57
64	8.89	272.96	222.10	231.67
65	13.07	272.51	301.92	297.65
66	14.57	261.34	391.22	367.62
67	21.36	268.89	665.56	616.54
68	28.03	295.99	1131.21	1029.75
69	31.49	319.87	837.80	786.23
1970	37.39	341.43	794.90	713.77
71	46.26	356.59	817.74	750.43
72	47.24	376.69	583.17	1027.38
73	52.33	390.19	709.59	666.50

Regression	$\hat{\alpha}$	$\hat{\beta}$	R^2
HOM on FTP	-77.63026	0.33745	0.929
LIC on FTP	-632.69877	3.84289	0.323
GR on FTP	-875.41989	4.66675	0.493

Table 5.2. Second and Third Stages of the Least Squares Analysis of the Homicide Data in Table 5.1

	Stage 2			Stage 3	
Year	Homicide Rate	Handgun Licenses	Handgun Registrations	Homicide Rate	Handgun Registrations
1961	-1.62485	-189.68607	-123,58847	0.96247	15.40020
62	-4.51375	-247.70295	-203.18926	1.13508	-21.68988
63	-5.64964	-214.70103	-184.55278	-2.72112	-27.23489
64	-5.59009	-194.15648	-166.74619	-2.94180	-24.48191
65	-1.25824	-112.60718	-98.66615	0.27772	-16.15549
66	4.01108	19.61790	23.43145	3.74349	9.05683
67	8.25333	264.94408	237.11748	4.63949	42.98500
68	5.77843	626.45176	523.85856	-2.76637	64.83856
69	1.18013	241.07355	168.89657	-2.10811	-7.74525
1970	-0.19529	115.52084	-4.17856	-1.77099	-88.82415
71	3.55896	80.10262	-38.26649	2.46636	-96.96008
72	-2.24378	-231.70946	144.88183	0.91674	314.66230
73	-1.70936	-157.16848	-278.99929	0.43442	-163.83723

Regression	$\hat{\alpha}$	$\hat{\beta}$	R^2	Regression	$\hat{\alpha}$	$\hat{\beta}$	R^2
HOM on LIC	0.0	0.01364	0.665	HOM on GR	0.0	0.00255	0.013
GR on LIC	0.0	0.73273	0.742				

Since $r_{i1}(Y)$ and $\bar{r}_{i1}(X_2)$ are residuals from least squares prediction equations, the average of each set of residuals is zero.* Thus fitting the response residuals can be accomplished using the estimation equation for no-intercept models, eqn. (3.2.10),

$$\hat{\alpha}(2) = 0 \qquad \text{and} \qquad \hat{\beta}(2) = \frac{\Sigma r_{i1}(X_2)r_{i1}(Y)}{\Sigma[r_{i1}(X_2)]^2} \ ,$$

resulting in the prediction equation

$$\hat{r}_{i1}(Y) = 0.01364 r_{i1}(X_2).$$

Likewise, the regression of $r_{i1}(X_3)$ on $r_{i1}(X_2)$ results in the following prediction equation

$$\hat{r}_{i1}(X_3) = 0.73273 r_{i1}(X_2).$$

As indicated in Table 5.2, the residuals of the handgun license rate account for 67% of the variability remaining in the response residuals; i.e., 67% of the variability of the response residuals, not 67% of the variability of the original responses. This is an indication that the first predictor variable, full-time police rate, was not able to completely eliminate or account for trends in the homicide rate.

Combined, the joint effect of both predictor variables can be calculated from the two stages of the analysis thus far. Since

$$\hat{r}_1(Y) = 0.01364 r_{i1}(X_2)$$

and

$$\hat{r}_{i1}(X_2) = X_{i2} + 632.69877 - 3.84289 X_{i1},$$

insertion of this second equation into the first one yields

$$\hat{r}_1(Y) = 8.63001 - 0.05242 X_{i1} + 0.01364 X_{i2}.$$

Using the algebraic results of Appendix 5.B one can show that $\hat{r}_1(Y)$ can also be written as

---

*Even though the estimated regression coefficients are carried through the analyses in Tables 5.1 and 5.2 to five decimal places, the average residuals are zero only to three or four decimal places. Zero is still used as the value of the average of the residuals since it would be so if enough decimal places could be carried throughout (see Appendix 5.B).

$$\hat{r}_{i1}(Y) = \hat{Y}_i(2) - \hat{Y}_i(1)$$

$$= \hat{Y}_i(2) + 77.63026 - 0.33745X_{i1}.$$

Equating the above two expressions for $\hat{r}_{i1}(Y)$ and solving for $\hat{Y}_i(2)$ produces the two-variable least squares prediction equation:

$$\hat{Y}_i(2) = -69.00025 + 0.28503X_{i1} + 0.01364X_{i2}. \qquad (5.1.10)$$

Apart from roundoff error, this is the least squares prediction equation that results from the regression of homicide rate on both the full-time police rate and the handgun license rate. It is particularly important to observe that the estimated coefficient $\hat{\beta}(2)$ obtained from the second stage analysis is the least squares estimator of β_2 in a prediction equation consisting of both X_1 and X_2. The estimates of both α and β_1, however, are altered from the stage one results. As we shall now verify, the third stage produces the least squares estimate of β_3 for the prediction equation containing all three predictor variables but the estimates of α, β_1, and β_2 are all modified by the third stage results.

Stage three of the analysis consists of calculating a prediction equation for the stage two homicide rate residuals from the stage two handgun registration rate residuals. The last two columns of Table 5.2 display the response and third predictor variable residuals resulting from the second stage analysis. Both sets of residuals again (when enough accuracy is maintained) sum to zero so the prediction equation is of the form

$$\hat{r}_{i2}(Y) = \hat{\alpha}(3) + \hat{\beta}(3) \, r_{i2}(X_3)$$

where $\hat{\alpha}(3) = 0$. As indicated in Table 5.2, $\hat{\beta}(3) = 0.00255$.

Beginning with this last prediction equation one can make a series of substitutions similar to those used in deriving eqn. (5.1.10) in order to derive the least squares prediction equation for the joint influence of all three predictor variables. For this example, the numerical calculations produce the least squares prediction equation

$$\hat{Y} = -67.95010 + 0.28031X_1 + 0.01177\,X_2 + 0.00255X_3,$$

while the prediction equation using a computer program to calculate all the estimates is (refer also to the hand calculations in the previous subsection)

$$\hat{Y} = -67.95217 + 0.28033X_1 + 0.01176\,X_2 + 0.00255X_3. \qquad (5.1.11)$$

The differences between the two sets of coefficients are due to roundoff error.

As anticipated earlier, eqn. (5.1.11) reveals that this third stage of the analysis produces the least squares coefficient estimate of β_3 for a prediction equation containing all three predictor variables. The estimates of α, β_1, and β_2 are adjusted from the stage 2 analysis to account for the new predictor variable that has entered the prediction equation.

Let us now summarize the sequential process that is undertaken in this fitting procedure. The first stage fits the response Y to one of the predictor variables, say X_1. Both the response and the other predictor variables are then adjusted to remove the effect of X_1. The second stage then fits the residuals of the response to the residuals of a second predictor variable, say X_2. The residuals of the response and those of the remaining predictor variables are then adjusted for the residuals of the second predictor variable so that, in effect, all remaining response and predictor variables are adjusted for both of the first two predictor variables. This process continues until the pth stage when the residuals of the response are fit to the residuals of the last predictor variable, both of which have been adjusted for all $(p - 1)$ other predictor variables. The coefficients of the predictors at the p stages are then algebraically combined to produce the fitted model of the response on all p predictor variables.

We do not recommend this technique for fitting multiple-variable least squares prediction equations since the tedious nature of the calculations (if performed on a desk calculator) can easily lead to mistakes and large round-off error. Understanding the process just discussed will, however, enable one to more readily comprehend the interpretation of the estimated coefficients. In particular, note again that only the last variable entering the prediction equation yields a least squares estimate that is identical to its value in the complete prediction equation; i.e., when the response adjusted for all $(p - 1)$ predictor variables except the last one is regressed on the last predictor variable, itself adjusted for the other $(p - 1)$ predictor variables, the least squares estimate of β_p results.

5.1.3 Orthogonal Predictor Variables

Orthogonal predictor variables result in several simplifications for multiple variable regression models. Recall that a matrix A is said to be orthogonal if each of its columns is orthogonal to all the other columns of the matrix; i.e., if $\underline{a}_i'\underline{a}_j = \Sigma a_{ki}a_{kj} = 0$. If the columns of X^* are mutually orthogonal, we refer to the corresponding variables as being orthogonal.

Suppose now that X^* contains orthogonal columns. Then both $X^{*\prime}X^*$ and $(X^{*\prime}X^*)^{-1}$ are diagonal matrices:

$$X^{*\prime}X^* = \mathrm{diag}(n,\ \Sigma X_{i1}^2,\ \Sigma X_{i2}^2, ...,\ \Sigma X_{ip}^2),$$

$$(X^{*\prime}X^*)^{-1} = \mathrm{diag}[n^{-1}, (\ \Sigma X_{i1}^2)^{-1},\ (\Sigma X_{i2}^2)^{-1}, ...,\ (\Sigma X_{ip}^2)^{-1}].$$

The least squares estimators are then simply

$$\hat{\alpha} = \overline{Y} \quad \text{and} \quad \hat{\beta}_j = \frac{\Sigma X_{ij} Y_i}{\Sigma X_{ij}^2}.$$

These are the same estimators that one would obtain if p single-variable prediction equations were estimated using eqn. (3.2.6) with $\overline{X}_j = 0$. Thus when X^* is an orthogonal matrix the least squares estimators of the parameters of a multiple-variable model can be estimated as though each predictor variable acted on the response independently of all the other predictor variables.

Simplification of the least squares estimators occurs precisely because none of the predictor variables contains any (linear) information about any of the others. To appreciate this fact, consider regressing one predictor variable, say X_s, on another one, say X_t. The least squares parameter estimates of the regression are

$$\hat{\alpha} = \overline{X}_s = 0 \quad \text{and} \quad \hat{\beta} = \frac{\Sigma X_{is} X_{it}}{\Sigma X_{it}^2} = 0$$

since $\underline{1}'\underline{X}_s = 0$ and $\underline{X}_s'\underline{X}_t = 0$ for an orthogonal X^*. Orthogonality among predictor variables is the opposite extreme of linear dependence: not even approximate linear dependencies can exist in an orthogonal X^* matrix.

A simplification of least squares estimation is also possible if only some of the columns of X^* are orthogonal. As one example, suppose the first r columns of X^* are denoted X_1^* and the remaining $(p + 1) - r$ columns are denoted X_2^*. If the columns of X_1^* are orthogonal to those of X_2^*, i.e., $X_1^{*\prime}X_2^*$ = Φ, the corresponding sets of parameters in $\underline{B}$, say $\underline{B}_1$ and $\underline{B}_2$, are separably estimable:

$$\hat{\underline{B}}_1 = (X_1^{*\prime}X_1^*)^{-1}X_1^{*\prime}\underline{Y} \quad \text{and} \quad \hat{\underline{B}}_2 = (X_2^{*\prime}X_2^*)^{-1}X_2^{*\prime}\underline{Y}.$$

The next subsection illustrates this property with centered and standardized predictor variables.

Apart from the obvious benefits due to simplified calculation of regression estimates, many other attractive properties accompany regression data bases that contain orthogonal predictor variables. Intuitively, the interpretation of estimated regression coefficients should be easier since there are no interrelationships among the predictor variables. Due to these and other desirable features of orthogonal predictor variables, it is important for the investigator who can control or specify the values of the predictor variables in a data base to attempt to make X^* as orthogonal as possible. While it is beyond the scope of this book to comprehensively treat design aspects of regression studies, there are many experimental design books which do so, including Davies (1967).

5.1.4 Standardization

Four basic types of predictor variables have been mentioned in the last three chapters: raw, centered, normal deviate, and unit length. The benefits of standardizing the predictor variables were argued in Sections 3.2.3 and 4.3.1. We now wish to show that, although the estimates for the parameters associated with the four types of predictor variables may differ,

(i)   the least squares parameter estimates for raw predictor variables can always be obtained from those for the centered and standardized estimates, and

(ii) predicted responses for all four types of variables are identical.

When the response variable is standardized in addition to the predictor variables, (i) remains valid and predicted (raw) responses can easily be obtained from the standardized predictions. We verify these properties first by algebraic relationships and then illustrate them with a numerical example.

Equations (5.1.4) or (5.1.5) define the multiple linear regression model for any collection of raw predictor variables:

$$\underline{Y} = X^*\underline{B} + \underline{\varepsilon} \qquad \text{or} \qquad \underline{Y} = \alpha\underline{1} + X\underline{\beta} + \underline{\varepsilon}.$$

Least squares parameter estimates are given, respectively, by eqns. (5.1.7) or (5.1.9):

$$\underline{\hat{B}} = (X^{*\prime}X^*)^{-1}X^{*\prime}\underline{Y}$$

or

$$\underline{\hat{\beta}} = (M'M)^{-1}M'\underline{Y} \qquad \text{and} \qquad \hat{\alpha} = \overline{Y} - \sum_{j=1}^{p} \overline{X}_j\hat{\beta}_j.$$

Formulating the model in the second manner, eqn. (5.1.5), results in the least squares estimator of $\underline{\beta}$ being expressed in terms of the centered predictor variables in M.

Suppose now one desires to express the original model in terms of the centered predictor variables, eqn. (5.1.8). In order to make the original and centered models equivalent, the intercept term in the centered model must be altered as follows:

$$Y_{ij} = \alpha + \beta_1 X_{i1} + \beta_2 X_{i2} + \ldots + \beta_p X_{ip} + \varepsilon_i$$

$$= (\alpha + \beta_1 \overline{X}_1 + \beta_2 \overline{X}_2 + \ldots + \beta_p \overline{X}_p) + \beta_1 (X_{i1} - \overline{X}_1)$$

$$+ \beta_2 (X_{i2} - \overline{X}_2) + \ldots + \beta_p (X_{ip} - \overline{X}_p) + \varepsilon_i$$

$$= \alpha^* + \beta_1 M_{i1} + \beta_2 M_{i2} + \ldots + \beta_p M_{ip} + \varepsilon_i,$$

where $\alpha^* = \alpha + \Sigma \overline{X}_j \beta_j$. In matrix notation the centered model is

$$\underline{Y} = \alpha^* \underline{1} + M\beta + \underline{\varepsilon}. \tag{5.1.12}$$

Because $\underline{1}$ is orthogonal to each of the columns of M, α^* and β can be estimated separately so that the least squares estimators reduce to

$$\hat{\beta} = (M'M)^{-1} M' \underline{Y} \qquad \text{and} \qquad \hat{\alpha}^* = \overline{Y}. \tag{5.1.13}$$

From these estimators we see that $\hat{\beta}$ is the same for both raw and centered predictor variables. The intercept, α^*, for the centered model is estimated by the mean response. The intercept for the raw predictor variables, α, can be estimated from eqn. (5.1.13) by resubstituting the expression for $\hat{\alpha}^*$:

$$\hat{\alpha}^* = \hat{\alpha} + \sum_{j=1}^{p} \overline{X}_j \beta_j = \overline{Y} \Rightarrow \hat{\alpha} = \overline{Y} - \sum_{j=1}^{p} \overline{X}_j \hat{\beta}_j,$$

just as in eqn. (5.1.9).

To see that the prediction equation based on eqn. (5.1.13) gives the same predicted responses as the one using the raw predictor variables, let u_1, u_2, ..., u_p be any values of the raw predictor variables for which a predicted response is desired. Then using the centered coefficient estimates,

$$\hat{Y} = \hat{\alpha}^* + \hat{\beta}_1 (u_1 - \overline{X}_1) + \hat{\beta}_2 (u_2 - \overline{X}_2) + \ldots + \hat{\beta}_p (u_p - \overline{X}_p)$$

$$= \left(\hat{\alpha} + \sum_{j=1}^{p} \overline{X}_j \hat{\beta}_j \right) + \left(\sum_{j=1}^{p} u_j \hat{\beta}_j \right) - \left(\sum_{j=1}^{p} \overline{X}_j \hat{\beta}_j \right)$$

$$= \hat{\alpha} + \hat{\beta}_1 u_1 + \hat{\beta}_2 u_2 + \ldots + \hat{\beta}_p u_p.$$

The last expression is the predicted response for a prediction equation based on the raw predictor variables.

Next suppose one wishes to standardize the predictor variables using the normal deviate form, eqn. (3.2.11). Denoting by Z the matrix of standardized predictor variables, the normal deviate form of the regression model is

$$\underline{Y} = \alpha^*\underline{1} + Z\underline{\beta}^* + \underline{\varepsilon}, \qquad (5.1.14)$$

where $\alpha^* = \alpha + \Sigma \bar{X}_j\beta_j$ and $\beta_j^* = \beta_j S_j$. Again since $\underline{1}$ is orthogonal to each column of Z, the least squares estimators are

$$\hat{\underline{\beta}}^* = (Z'Z)^{-1}Z'\underline{Y} \qquad \text{and} \qquad \hat{\alpha}^* = \bar{Y}. \qquad (5.1.15)$$

To obtain the parameter estimators for the original predictor variables, make the inverse transformations:

$$\hat{\beta}_j = \frac{\hat{\beta}_j^*}{S_j} \qquad \text{and} \qquad \hat{\alpha} = \bar{Y} - \sum_{j=1}^{p} \bar{X}_j\hat{\beta}_j.$$

Finally, to see that the standardized prediction equation gives the same numerical results as the original one, straightforward algebraic transformations show that

$$\hat{Y} = \hat{\alpha}^* + \hat{\beta}_1^* Z_1 + \hat{\beta}_2^* Z_2 + \dots + \hat{\beta}_p^* Z_p$$

$$= \hat{\alpha}^* + \frac{\hat{\beta}_1^*(u_1 - \bar{X}_1)}{S_1} + \frac{\hat{\beta}_2^*(u_2 - \bar{X}_2)}{S_2} + \dots + \frac{\hat{\beta}_p^*(u_p - \bar{X}_p)}{S_p}$$

$$= \hat{\alpha}^* + \hat{\beta}_1(u_1 - \bar{X}_1) + \hat{\beta}_2(u_2 - \bar{X}_2) + \dots + \hat{\beta}_p(u_p - \bar{X}_p)$$

$$= \hat{\alpha} + \hat{\beta}_1 u_1 + \hat{\beta}_2 u_2 + \dots + \hat{\beta}_p u_p.$$

Repeating this technique for the unit length standardization, the transformed model becomes

$$\underline{Y} = \alpha^*\underline{1} + W\underline{\beta}^* + \underline{\varepsilon}, \qquad (5.1.16)$$

where $\alpha^* = \alpha + \Sigma \bar{X}_j\beta_j$ and $\beta_j^* = \beta_j d_j$, and the least squares parameter estimators are

$$\hat{\underline{\beta}}^* = (W'W)^{-1}W'\underline{Y} \qquad \text{and} \qquad \hat{\alpha}^* = \bar{Y}. \qquad (5.1.17)$$

The original parameter estimators are obtainable as

$$\hat{\beta} = \frac{\hat{\beta}_j^*}{d_j} \qquad \text{and} \qquad \hat{\alpha} = \overline{Y} - \sum_{j=1}^{p} \overline{X}_j \hat{\beta}_j .$$

Verification that the predicted responses using this standardization are identical to those using the raw predictor variables is exactly analogous to the steps employed for the normal deviate standardization.

From the form of the least squares estimators in eqn. (5.1.17) one can see that it is only necessary to have summary information on the response and predictor variables in order to calculate the parameter estimates. Recall that Data Set A.4 contains only pairwise correlations among the variables and the means and standard deviations of each variable. The estimate of the constant term, $\hat{\alpha}^*$, is immediately obtainable from $\overline{Y}$. The correlations among the predictor variables are the elements of $W'W$. The vector $W'\underline{Y}$ is calculable by multiplying the correlation coefficient of each predictor variable with the response by the standard deviation of the response, S_Y:

$$W'\underline{Y} = \begin{pmatrix} S_Y r_{Y,X_1} \\ \cdot \\ \cdot \\ \cdot \\ S_Y r_{Y,X_p} \end{pmatrix} .$$

where r_{Y,X_j} denotes the correlation coefficient between Y and X_j. The least squares estimates for the raw predictor variables or any of the other standardizations can then be found by suitable transformations of the estimates for the unit length standardization.

Standardization of both response and predictor variables only slightly alters these results. Let the normal deviate standardization of the response be denoted by $Y_i^0 = (Y_i - \overline{Y})/S_Y$. Then the regression model can be written as

$$\underline{Y}^0 = \alpha^0 \underline{1} + Z\underline{\beta}^0 + \underline{\varepsilon}^0, \qquad (5.1.18)$$

where

$$\alpha^0 = \frac{(\alpha^* - \overline{Y})}{S_Y} = \frac{(\alpha + \sum \overline{X}_j \beta_j - \overline{Y})}{S_Y} ,$$

$\beta_j^0 = \beta_j S_j/S_Y = \beta_j d_j/d_Y$ and $\varepsilon_j^0 = \varepsilon_j/S_Y$. One could also standardize the response using the unit length scaling by paralleling this derivation using $(Y_i - \overline{Y})/d_Y$.

Since both response and predictor variables are standardized,

$$\hat{\alpha}^0 = \frac{\underline{1}'\underline{Y}^0}{\underline{1}'\underline{1}} = 0$$

and the least squares estimators are

$$\hat{\underline{\beta}}^0 = (Z'Z)^{-1} Z' \underline{Y}^0 \quad \text{and} \quad \hat{\alpha}^0 = 0. \qquad (5.1.19)$$

By noting that $Y^0 = S_Y^{-1}(\underline{Y} - \overline{Y}\underline{1})$ and $Z'\underline{1} = \underline{0}$, $\hat{\underline{\beta}}^0$ simplifies to

$$\hat{\underline{\beta}}^0 = (Z'Z)^{-1}Z'[S_Y^{-1}(\underline{Y} - \overline{Y}\underline{1})]$$

$$= S_Y^{-1}(Z'Z)^{-1}Z'\underline{Y}$$

$$= S_Y^{-1}\hat{\underline{\beta}}*.$$

Thus by setting $\hat{\alpha}^0 = 0$ and dividing the least squares parameter estimates, $\hat{\underline{\beta}}*$, from the normal deviate standardization of only the predictor variables by S_Y, one can immediately obtain the beta weights for a prediction equation in which all variables are in normal deviate form. The least squares estimators of the original model can be found from eqn. (5.1.19) by

$$\hat{\beta}_j = \frac{\hat{\beta}_j^0 S_Y}{S_j} \quad \text{and} \quad \hat{\alpha} = \overline{Y} - \sum_{j=1}^{p} \overline{X}_j \hat{\beta}_j.$$

Predicted responses for the beta weight standardization are not identical to those of the original prediction equation but can easily be transformed to the equivalent values. Since $Y^0 = (Y - \overline{Y})/S_Y$, $\hat{Y}^0$ is equal to $(\hat{Y} - \overline{Y})/S_Y$ and so $\hat{Y} = \overline{Y} + S_Y\hat{Y}^0$, where $\hat{Y}^0$ is the predicted normal deviate response using eqn. (5.1.28):

$$\hat{Y}^0 = \hat{\beta}_1^0 Z_1 + \hat{\beta}_2^0 Z_2 + \ldots + \hat{\beta}_p^0 Z_p.$$

As an example of the relationships among the various response and predictor variable scalings just described, consider again the high school grade data of Data Set A.2. For illustrative purposes let us examine a prediction equation for only the first two schools in the data set. The proposed model is

$$Y_i = \alpha + \beta_1 X_{i1} + \beta_2 X_{i2} + \varepsilon_i,$$

where

$$X_{i1} = \begin{cases} 0 & \text{if student } i \text{ is from school 1} \\ 1 & \text{if student } i \text{ is from school 2} \end{cases}$$

and X_{i2} is the grade 13 average for student i. For this example centering or standardizing the predictor variables only necessitates the inversion of a 2×2 matrix while using the raw predictors requires the inversion of a 3×3 matrix. Let us therefore compute the prediction equation for the unit length scaling first. The model is now

$$Y_i = \alpha^* + \beta_1^* W_{i1} + \beta_2^* W_{i2} + \varepsilon_i,$$

where W_{ij} is the standardized score for the ith student on the jth predictor variable. From the data on the 37 subjects in the first two high schools in Data Set A.2,

$$\bar{Y} = 61.359 \qquad \bar{X}_1 = 0.514 \qquad \bar{X}_2 = 71.686$$

$$d_Y = 57.415 \qquad d_1 = 3.040 \qquad d_2 = 41.116$$

$$S_Y = 9.569 \qquad S_1 = 0.507 \qquad S_2 = 6.853$$

$$W'W = \begin{bmatrix} 1.0000 & 0.1116 \\ 0.1116 & 1.0000 \end{bmatrix}, \qquad (W'W)^{-1} = \begin{bmatrix} 1.0126 & -0.1131 \\ -0.1131 & 1.0126 \end{bmatrix}$$

$$W'\underline{Y} = \begin{pmatrix} 1.4704 \\ 42.6999 \end{pmatrix}.$$

The standardized least squares estimates are

$$\underline{\hat{\beta}}^* = (W'W)^{-1} \, W'\underline{Y} = \begin{pmatrix} -3.34 \\ 43.07 \end{pmatrix} \quad \text{and} \quad \hat{\alpha}^* = \bar{Y} = 61.36.$$

From these estimates all the numerical estimates listed in Table 5.3 can be found by the algebraic manipulations that were derived earlier.

The estimates in Table 5.3 again point out the importance of standardization. Coefficient estimates for the raw or the centered predictor variables might be erroneously interpreted as indicating that the two predictor variables are of equal value in predicting the response. When differences in the variability and scaling of the two predictors are accounted for in the last three reexpressions, the coefficient estimate for the grade 13 averages turns out to be much larger than that for the high school designator. Despite the differences in the numerical values for the three standardizations, the ratios of the estimates for the two regression coefficients are identical: the first predictor variable's coefficient estimate is only about 7.7% of the magnitude of the second one for each of the standardizations. The second predictor variable now appears much more important as a predictor of the response, as one might expect.

**Table 5.3. Comparison of Regression Estimates for
Two High Schools in Data Set A.2**

| | Coefficient Estimate | | |
Transformation	Constant	Predictor #1	Predictor #2
Raw Data	-13.17	-1.10	1.05
Centered Predictors	61.36	-1.10	1.05
Normal Deviate Predictors	61.36	-0.56	7.18
Unit Length Predictors	61.36	-3.34	43.07
Beta Weights	0	-0.058	0.75

Let us now compute predicted first year scores for students from high
school 1 ($X_1 = 0$) who have an average score in grade 13 of 70 ($X_2 = 70$).
Using the summary values given above we find that

$$M_1 = 0 - 0.514 = -0.514 \qquad M_2 = 70 - 71.686 = -1.686$$

$$Z_1 = -0.514/0.507 = -1.014 \qquad Z_2 = -1.686/6.853 = -0.246$$

$$W_1 = -0.514/3.040 = -0.169 \qquad W_2 = -1.686/41.116 = -0.041.$$

Predictions are then obtained as follows:

Raw Data:	$\hat{Y} = -13.17 - 1.10(0) + 1.05(70)$	$= 60.33$
Centered Predictors:	$\hat{Y} = 61.36 - 1.10(-0.514) + 1.05(-1.686)$	$= 60.16$
Normal Deviate:	$\hat{Y} = 61.36 - 0.56(-1.014) + 7.18(-0.246)$	$= 60.16$
Unit Length:	$\hat{Y} = 61.36 - 3.34(-0.169) + 43.07(-0.041)$	$= 60.16$
Beta Weight:	$\hat{Y}0 = -0.058(-1.014) + 0.75(-0.246)$	$= -0.13$
	$\hat{Y} = 61.36 + 9.569(-0.13)$	$= 60.12.$

Differences in these predictions are all ascribable to roundoff error. These
calculations emphasize the importance of carrying a sufficient number of
digits in the computations to insure adequate accuracy. All the raw predicted values are 60 units when rounded to the nearest whole number but
some differ considerably in the first two decimal places. The original data
only contained scores to the nearest tenth of a unit and consequently one
cannot hope to predict average grades to more accuracy than the first decimal place; nevertheless, these predicted grades differ in the first decimal
place. Carrying more significant figures in the intermediate calculations will
eliminate the discrepancies in the predicted grades.

5.1.5   No-Intercept Models

Multiple-variable models that do not contain intercepts can be evaluated for
raw predictor variables with virtually no modification from the intercept
models. The theoretical model is now written

$$\underline{Y} = X\underline{\beta} + \underline{\varepsilon},$$

where $X = [\underline{X}_1, \underline{X}_2, ..., \underline{X}_p]$ only contains p columns of predictor variables and no column of ones. The least squares estimator of $\underline{\beta}$ is

$$\hat{\underline{\beta}} = (X'X)^{-1}X'\underline{Y}.$$

Standardized models are altered when no intercept term occurs in the original model. For example, if one desires to center the predictor variables, the original model becomes

$$\underline{Y} = \alpha^*\underline{1} + M\underline{\beta} + \underline{\varepsilon},$$

where $\alpha^* = \sum \bar{X}_j\beta_j$. The centered model now appears to contain $(p + 1)$ parameters whereas the original one only contains p. This is not the case, however, since α^* is actually only a function of the original p parameters and does not need to be separately estimated. Once estimates of $\beta_1, \beta_2, ..., \beta_p$ are obtained the estimate of α^* is automatically fixed at $\sum \bar{X}_j\hat{\beta}_j$.

No computational advantage acrues by centering predictor variables for no-intercept models. To avoid confusion, moreover, from the appearance of an intercept in standardized models when none occurs in the original model, normal deviate and unit length standardizations of predictor variables do not center the predictor variables before scaling them. For no-intercept models, then,

$$W_{ij} = \frac{X_{ij}}{d_j^*} \quad \text{and} \quad Z_{ij} = \frac{X_{ij}}{S_j^*}$$

where

$$d_j^* = \left[\sum_{i=1}^{n} X_{ij}^2\right]^{1/2} \quad \text{and} \quad S_j^* = \left[\sum_{i=1}^{n} \frac{X_{ij}^2}{n}\right]^{1/2}$$

Parameter estimators using these standardizations are as in eqns. (5.1.15) and (5.1.17) only without an estimate of α^*. Beta weights can be calculated as in eqn. (5.1.19).

5.2 INTERPRETATION OF FITTED MODELS

Central to most applications of multiple linear regression methodology is the interpretation of the final fitted model. Even if one only desires to construct a prediction equation in order to predict the response variable, questions relating to which variables most influence the predicted response invariably arise. For several reasons already mentioned (different scales for the predictor variables, covariation, etc.), this is often not a simple task. In this section we concentrate attention on the meaning attached to estimated regression coefficients.

5.2.1   General Interpretation

As was stressed in Chapters 1 and 2, inferences on model parameters and their interpretations must always be conditioned on the data base. The broader, more representative a data base is of the population or phenomenon under study, the more global and generalizable are the conclusions that can be drawn from the resulting fit. Assuming then that a representative data base has been constructed for the purpose of the investigation, how does one interpret the magnitudes of the estimated regression coefficients?

Often the estimated coefficients in a multiple-variable prediction equation are said to measure "The change in the estimated response that is due to increasing one predictor variable by a unit amount while all other predictor variables are held constant." While, strictly speaking, this interpretation is correct, it overlooks two important factors, both of which can be illustrated from the analysis of the homicide data in Section 5.1.2.

The three stage-wise prediction equations for homicide rate (Y) using, sequentially, full-time police rate (X_1), handgun license rate (X_2), and handgun registration rate (X_3) are:

$$\hat{Y} = -77.63 + 0.3375X_1,$$

$$\hat{Y} = -67.00 + 0.2850X_1 + 0.0136X_2,$$

and

$$\hat{Y} = -67.95 + 0.2803X_1 + 0.0118X_2 + 0.0026X_3.$$

Obviously the magnitudes of the estimated coefficients depend on which predictor variables are included in the fitted model. In extreme cases, estimated coefficients can change signs and dramatically fluctuate in magnitude as predictor variables are inserted or removed from prediction equations.

Extreme covariation among the predictor variables in the homicide data precludes the condition that some predictor variables can be increased by one unit while others are held fixed. Even if separate variation is possible theoretically, the data from which the coefficient estimates must be calculated include variables such as FTP and CLEAR which have large pairwise correlations. Interpretations of estimated coeficients should, therefore, take into account possible covariation among the predictor variables.

A more adequate interpretation of the estimated regression coefficients which takes into account the difficulties just discussed is suggested by the derivation of least squares estimates by stages. We pointed out in Section 5.1.2 that when multiple-variable prediction equations are fit by stages, only the last variable entering the predictor has its coefficient estimated identically to its numerical value in the full prediction equation. We can thus view each parameter estimate as having arisen from a fitting-by-stages process in which its corresponding predictor variable was the last one to enter: $\hat{\beta}_j$ can be calculated by regressing $r_{p-1}(Y)$ [i.e., the residuals of Y from its regression on the $(p - 1)$ predictor variables other than X_j] on $r_{p-1}(X_j)$ (i.e., the residuals of X_j from its regression on the other $(p - 1)$ predictor variables). As will be demonstrated in Appendix 5.B, the estimator of β_j can also be calculated by regressing the unadjusted response, Y, on the adjusted jth predictor variable, $r_{p-1}(X_j)$.

From this discussion and the algebraic derivation in Appendix 5.B, the estimated regression coefficients can be interpreted as follows:

$\hat{\beta}_j$ measures the change in the predicted response variable which results from a unit change in the *adjusted* value of X_j.

The adjustment referred to in this interpretation is an adjustment of X_j for its covariation with the other predictor variables; i.e., the residual of X_j after adjusting for the $(p - 1)$ other predictor variables. This interpretation pointedly reflects the fact that X_j may not be capable of change independently of some of the other predictor variables and that the numerical value of $\hat{\beta}_j$ may depend on which predictor variables are included in the fitted model. Only if X_j is orthogonal (or nearly so) to the other predictor variables can "adjusted" be eliminated from the above interpretation.

The three-predictor-variable prediction equation for homicide rate given above has coefficient estimates $\hat{\beta}_1 = 0.2803$, $\hat{\beta}_2 = 0.0118$, and $\hat{\beta}_3 = 0.0026$. The estimated coefficient $\hat{\beta}_1$ indicates that the predicted homicide rate increases by 0.2803 homicides per 100,000 population for each unit increase in the adjusted full-time police rate. Similarly, the homicide rate increases by 0.0118 and 0.0026 units for each unit increase in the adjusted handgun license and handgun registration rates, respectively. One would anticipate that the adjustments on the handgun license and handgun registration rates would be more severe than that of the full-time police rate since the former two variables are so highly correlated (multicollinear) in this data set. To see whether this is so, let us examine the calculations needed for the adjustment.

5.2.2 Adjustment Computations

In order to assess the numerical effects of multicollinearities on predicted responses and to illustrate the need for the interpretation of estimated regression coefficients given in the previous subsection, let us once again consider the three-variable homicide data example. If one ignores the multicollinearities in the data, one would conclude that $\hat{Y}$ increases by 0.2803 units for a unit increase in X_1 (holding X_2 and X_3 fixed), by 0.0118 units for a unit increase in X_2 (holding X_1 and X_3 fixed), and by 0.0026 units for a unit increase in X_3 (holding X_1 and X_2 fixed). For example if one considers the effect of increasing GR from 600 to 700, the net effect on $\hat{Y}$ is to increase the estimated homicide rate by

$$(0.0026)(700 - 600) = 0.26$$

units.

Actually, the true effect of increasing X_3 from 600 depends on the values of X_1 and X_2. The net effect on $\hat{Y}$ due to a one unit change in X_3 when X_1 and X_2 are fixed is $\hat{\beta}_3 r_2(X_3)$ and not $\hat{\beta}_3$ (Appendix 5.B). This fact underlies the rationale for the interpretation of $\hat{\beta}_3$ given earlier: all three predictor variables must be assigned numerical values before the true effect of X_3 on the response can be ascertained. In assigning values to the predictor variables, moreoever, care must be exercised to insure that the numerical values retain the same multicollinear properties as in the data base; to do otherwise would be extrapolation.

Now suppose that when $X_3 = 600$, $X_1 = 400$ and $X_2 = 605$ and when $X_3 = 700$, $X_1 = 400$ and $X_2 = 710$ (note that X_2 and X_3 retain the high correlation exhibited in Data Set A.3). From Section 5.1.2

$$r_2(X_3) = r_1(X_3) - 0.73273 r_1(X_2)$$

$$= [X_3 + 875.41989 - 4.66675 X_1]$$

$$- 0.73273 [X_2 + 632.69877 - 3.84289 X_1]$$

so that changing X_3 from 600 to 700 (accompanied by the values of X_1 and X_2 shown above) results in a change in $r_2(X_3)$ from -171.86 to -148.80. Thus the net effect on the estimated homicide rate is

$$(0.0026)(-148.80 + 171.86) = 0.06.$$

The decrease from 0.26 to 0.06 in the change in $\hat{Y}$ for the two techniques indicates that once the change in X_2 has been accounted for in the latter scheme—and not ignored as in the former one—the effect due to X_3 is reduced. The correlation between X_2 and X_3 reduces the separate influence of each predictor variable on $\hat{Y}$.

To summarize, $\hat{\beta}_j$ measures the effect on the estimated response, $\hat{Y}$, attributable to unit changes in X_j after adjusting X_j for covariation with other predictor variables; i.e., $\hat{\beta}_j$ measures the effect due to unit changes in $r_{p-1}(X_j)$. If one wishes to assess the effect of changes in X_j on Y, predictor variable values for all variables for which X_j is multicollinear must also be given. The net changes in $\hat{Y}$ due to the changes in X_j are then conditional (i.e., dependent) on the specified changes in the other predictor variables.

5.2.3 Special Case: Indicator Variables

Following the suggestions in Sections 2.2.1 and 4.1.5 for the definition of indicator variables leads to interpretation problems that are similar to those mentioned in the last subsection, only more extreme. Consider again the high school grade data in Data Set A.2. If X_1 is an indicator variable designating high schools ($X_1 = 0$ if school 1, 1 if school 2) and X_2 denotes the grade 13 average score, the least squares prediction equation is

$$\hat{Y} = -13.17 - 1.10X_1 + 1.05X_2.$$

The estimated coefficient for the first predictor variable measures the increase (here, a decrease since $\hat{\beta}_1$ is negative) in the estimated first year of college average grade of high school 2 *over* that for high school 1, *after* adjusting the school designator for the grade 13 average. The important point to realize with the indicator variable is that it actually represents two predictor variables that would be linearly dependent if both were included in the prediction equation and, therefore, the effect of $\hat{\beta}_1$ is the net effect of high school 1 compared to high school 2.

With two predictor variables designating all three schools in Data Set A.2, the interpretation of the estimated coefficients depends on the manner in which the indicator variables are defined but the estimates still represent comparisons. If

$$X_1 = \begin{cases} 1 & \text{if school 1} \\ 0 & \text{otherwise} \end{cases}, \quad X_2 = \begin{cases} 1 & \text{if school 2} \\ 0 & \text{otherwise} \end{cases}$$

and X_3 = grade 13 average, then $\hat{\beta}_1$ and $\hat{\beta}_2$ measure, respectively, the increase (or decrease) in the estimated first year of college average grade of high school 1 over high school 3 and high school 2 over high school 3 after adjusting the school designator for the grade 13 averages. Note that there is no need to adjust high school 1 and 2 designators for one another since X_1 and X_2 are orthogonal.

If one specifies the high school designators as

$$X_1 = \begin{cases} 1 & \text{if school 2 or 3} \\ 0 & \text{otherwise} \end{cases} \quad \text{and} \quad X_2 = \begin{cases} 1 & \text{if school 3}, \\ 0 & \text{otherwise} \end{cases}$$

then X_1 and X_2 are not orthogonal and one must interpret each estimated coefficient in terms of all the other predictor variables. For this specification $\hat{\beta}_1$ estimates the increase (decrease) in the estimated first year of college average grade of high school 2 over high school 1 and $\hat{\beta}_2$ measures the *additional* increase attributable to high school 3 over high school 2 when the designators are adjusted for the grade 13 scores. Once again the coefficient estimates measure comparisons among the high schools and not just the effect due to a particular high school.

5.3 INITIAL ASSESSMENT OF FIT

Before interpretations of fitted model parameters can be judged meaningful, the fitted model must adequately predict the response variable. While the ultimate decision of whether the fitted model does adequately predict the response variable often must be based on the purpose of the investigation and the degree of accuracy desired by the experimenter or data analyst, statistical measures of the fit are also available to aid in the assessment. In this section we discuss several measures that can be used in an initial assessment of the fit of a prediction equation. The material in this section parallels that of Section 3.3 for single-variable models.

5.3.1 Analysis of Variance Table

Algebraically the total sum of squares for multiple-variable models can be partitioned into three components. The steps in the derivation are exactly analogous to those for single-variable models:

$$\sum_{i=1}^{n} Y_i^2 = n\overline{Y}^2 + \sum_{i=1}^{n} (\hat{Y}_i - \overline{Y})^2 + \sum_{i=1}^{n} (Y_i - \hat{Y}_i)^2 . \tag{5.3.1}$$

Notationally we write eqn. (5.3.1) as

$$TSS = SSM + SSR + SSE.$$

A better appreciation of the relationship between these sums of squares and those for single-variable prediction equations can be obtained by reexpressing the sums of squares in terms of the estimators of the model parameters. First note that the total sum of squares can be written in vector notation as

$$TSS = \underline{Y}'\underline{Y},$$

the sum of squares attributable to the mean as

$$SSM = n^{-1}(\underline{1}'\underline{Y})^2,$$

and the residual sum of squares as

$$SSE = \underline{r}'\underline{r} = (\underline{Y} - \hat{\underline{Y}})'(\underline{Y} - \hat{\underline{Y}}),$$

where $\hat{\underline{Y}}$ can be written in many equivalent forms, including the following:

$$\hat{\underline{Y}} = X^*\hat{\underline{B}} = \hat{\alpha}\underline{1} + X\hat{\underline{\beta}} = \hat{\alpha}^*\underline{1} + M\hat{\underline{\beta}}.$$

Next let us express SSM and SSR in terms of the estimated model parameters. The sum of squares attributable to the mean is readily seen to be

$$SSM = n\hat{\alpha}^{*2}$$

since $\hat{\alpha}^* = \overline{Y}$. The sum of squares attributable to the regression coefficients on the predictor variables is (see Appendix 5.C)

$$SSR = \hat{\underline{\beta}}'M'\underline{Y} = \hat{\underline{\beta}}'M'M\hat{\underline{\beta}},$$

the latter equality resulting since $\hat{\underline{\beta}} = (M'M)^{-1}M'\underline{Y}$. The partitioning of the total sum of squares can thus be written as

$$TSS = \Sigma Y_i^2, \quad SSM = n\hat{\alpha}^{*2}, \quad SSR = \hat{\underline{\beta}}'M'\hat{\underline{Y}} \qquad (5.3.2)$$

$$SSE = \Sigma r_i^2 = TSS - SSM - SSR.$$

Associated with the partitioning of the total sum of squares in eqn. (5.3.2) is a similar partitioning of the sample size or degrees of freedom, n. There are n independent random variables, each of the Y_i, in the total sum of squares. Only when each of these n random variables have been given numerical values can TSS be calculated. We thus say that the responses have n degrees of freedom associated with them.

The sum of squares attributable to the mean response, SSM, has one degree of freedom associated with it since only one random variable, $\overline{Y}$, needs to be assigned a numerical value before SSM can be calculated. Despite the fact that $\overline{Y}$ utilizes all n responses for its calculation, it is $\overline{Y}$ and not the individual Y_i that determines SSM. This is because many sets of Y_i can generate the same $\overline{Y}$ and subsequently SSM. For example, the following sets of values all generate the same mean:

Values (Y_i)					$\overline{Y}$
10,	10,	10,	10,	10	10
0,	10,	10,	10,	20	10
-10,	0,	10,	20,	30	10
-43,	27,	-4,	16,	54	10

Consequently once $\overline{Y}$ is specified, regardless of the magnitudes of the individual Y_i, SSM is fixed. Associated with SSM, therefore, is the single degree of freedom for $\overline{Y}$.

Nonzero values of the $\hat{\beta}_j$ contribute to the numerical value of the regression sum of squares and so there are p degrees of freedom for SSR. The individual Y_i affect SSR only through $\hat{\beta}$ just as they affected SSM only through $\overline{Y}$. Although as we shall see in the next chapter the $\hat{\beta}_j$ are correlated, if $X^{*'}X^*$ is nonsingular (has an inverse) there are p independent pieces of information contributing to the value of SSR. This will be demonstrated in Section 10.1 when $\hat{\beta}$ is transformed to p independent random variables and SSR is written as a function of these independent variables.

One other property of SSR is important to the partitioning of the total sum of squares. The sum of squares attributable to the regression of the response on the predictor variables, SSR, is invariant (numerically unchanged) if one uses the centered, normal deviate, or unit length transformations of the predictor variables. As an illustration of this property let us examine using the unit length scaling. Consider defining SSR for the unit length standardization as

$$\text{SSR} = \hat{\underline{\beta}}^{*'} W' \underline{Y} = \hat{\underline{\beta}}^{*'} W' W \hat{\underline{\beta}}^*,$$

again the latter equality holding since $\hat{\underline{\beta}}^* = (W'W)^{-1} W'\underline{Y}$. But $\hat{\beta}_j^* = \hat{\beta}_j d_j$ and $W_{ij} = M_{ij}/d_j$ so that

$$\hat{\underline{\beta}}^* = D\hat{\underline{\beta}} \text{ and } W = MD^{-1},$$

where

$$D = \text{diag}(d_1, \ldots, d_p).$$

Thus

$$\hat{\beta}^{*\prime} W' \underline{Y} = \hat{\beta}' DD^{-1} M' \underline{Y}$$

$$= \hat{\beta}' M' \underline{Y} .$$

Hence, from eqn. (5.3.2) it is true that SSR $= \hat{\beta}^{*\prime} W' \underline{Y}$ and one can calculate SSR using either the centered predictor variables or the unit length standardization. This property also holds for the normal deviate standardization and for the beta weights with a slight alteration for the latter scaling:

$$SSR = S_Y^2 \hat{\beta}^{0\prime} Z' \underline{Y}^0 .$$

In all the standardizations, SSM $= n\overline{Y}^2$.

Since $\hat{\underline{B}}$ contains both $\hat{\alpha}$ and $\hat{\beta}$, SSR for the raw predictor variables must be obtained by subtraction:

$$SSR = \hat{\underline{B}}' X^{*\prime} \underline{Y} - SSM$$

$$= \hat{\underline{B}}' X^{*\prime} \underline{Y} - n\overline{Y}^2 .$$

The numerical value of SSR that is obtained from the raw predictor variables in this fashion is identical to that in eqn. (5.3.2).

A total of $n - p - 1$ degrees of freedom remain for estimating the error variance, σ^2. The error sum of squares, SSE, contains n residuals only $n - p - 1$ of which are independent of one another. In other words, if a particular group of $n - p - 1$ residuals are specified, SSE can be calculated from them. This property of the residuals will become clearer in Chapter 6.

Table 5.4 summarizes the algebraic partitioning of the total sum of squares in an *Analysis of Variance* table. The first three columns list, respectively, a description of the partitioning, the number of degrees of freedom (d.f.) associated with each source of variability, and the corresponding

Table 5.4. Symbolic Analysis of Variance (ANOVA) Table for Multiple-Variable Prediction Equations

			ANOVA		
Source	d.f.	S.S.	M.S.	F	R
Mean	1	SSM	MSM	MSM/MSE	SSR/(SSR + SSE)
Regression	p	SSR	MSR	MSR/MSE	
Error	n-p-1	SSE	MSE		
Total	n	TSS			

sums of squares (S.S.). Individual mean squares (M.S.) in the fourth column are the respective sums of squares divided by their degrees of freedom. The F statistics, described more fully in the next chapter, can be used to test certain hypotheses about the model parameters and are the ratios of the mean squares due to the mean and regression, respectively, divided by the mean squared error. Lastly, the coefficient of determination, R^2, is shown in the rightmost column.

5.3.2 Error Variance

Under certain assumptions about the model errors, including the assumption that they all have the same variance, σ^2, the residuals of the fitted model can be used to estimate σ^2. The estimator is similar to that for single-variable predictors:

$$\hat{\sigma}^2 = \frac{\sum_{i=1}^{n} r_i^2}{n - p - 1} \ . \tag{5.3.3}$$

The denominator of eqn. (5.3.3), $n - p - 1$, is the sample size less the number of parameters in the model, p regression coefficients and an intercept. For a single-variable prediction equation, $p + 1 = 2$ as in eqn. (3.3.8).

Instead of computing $\hat{\sigma}^2$ as eqn. (5.3.3), a simpler technique that poses less danger of roundoff error is available. The numerator of eqn. (5.3.3) can be expressed as (see Appendix 5.D):

$$\Sigma r_i^2 = \underline{Y}'\underline{Y} - \underline{\hat{B}}'X^{*'}\underline{Y}$$

$$= \Sigma Y_i^2 - n\overline{Y}\hat{\alpha} - \hat{\beta}'X'\underline{Y} \tag{5.3.4}$$

$$= \Sigma Y_i^2 - n\overline{Y}^2 - \hat{\beta}^{*'}W'\underline{Y} \ .$$

Any of the three representations in eqn. (5.3.4) will save one a great amount of time if the calculations are performed on a desk calculator. The first two expressions are identical expressions using the parameter estimates for the raw predictor variables, eqn. (5.1.7). The third expression shows how any one of the centered or scaled predictor variables can be utilized to find the residual sum of squares: for any scaling except the beta weights, just replace $\hat{\beta}^*$ and $W'\underline{Y}$ by the corresponding values for the particular standardization desired.

Computation of $\hat{\sigma}^2$ is illustrated with the high school grade example. We will estimate σ^2 using the raw and unit length predictors. The other predictor variable forms will yield identical results, as the reader can verify.

For the 37 students from high schools 1 and 2 in Data Set A.2,

$$\Sigma Y_i^2 = 142{,}600.8100 \quad \text{and} \quad \bar{Y} = 61.3595 \;.$$

Also,

$$X'\underline{Y} = \begin{pmatrix} 1{,}170.3000 \\ 164{,}505.5025 \end{pmatrix} \quad \text{and} \quad W'\underline{Y} = \begin{pmatrix} 1.4704 \\ 42.6999 \end{pmatrix} \;.$$

From Table 5.3, $\hat{\alpha} = -13.1738$

$$\hat{\underline{\beta}} = \begin{pmatrix} -1.0982 \\ 1.0476 \end{pmatrix} \quad \text{and} \quad \hat{\underline{\beta}}^* = \begin{pmatrix} -3.3387 \\ 43.0727 \end{pmatrix} \;.$$

Thus, from the raw data

$$\hat{\sigma}^2 = \frac{1{,}458.57}{34} = 42.90$$

while from the unit length standardization

$$\hat{\sigma}^2 = \frac{1{,}461.95}{34} = 43.00 \;.$$

It was necessary to carry additional digits in these calculations or the estimates would have suffered greatly from roundoff error. Even so, the raw predictors yield an estimate of σ^2 that is in error in the first decimal place. The value of $\hat{\sigma}^2$ obtained with the standardized predictors is numerically correct to the two decimal places that are shown. Beta weight standardizations do not allow computation of the residual sum of squares as in eqn. (5.3.4) but again the modification is simple. For the beta weight standardization (see Appendix 5.D)

$$\sum_{i=1}^{n} r_i^2 = S_y^2 \hat{\underline{\beta}}^{0\prime} Z'\underline{Y}^0 \;. \tag{5.3.5}$$

Estimation of σ^2 still utilizes eqn. (5.3.3) with this computation of the residual sum of squares.

APPENDIX

5.A   Derivation of Least Squares Estimators

Whether one expresses a multiple-variable linear regression model as

$$\underline{Y} = X^*\underline{B} + \underline{\varepsilon} \quad \text{or} \quad \underline{Y} = \alpha\underline{1} + X\underline{\beta} + \underline{\varepsilon},$$

the residual sum of squares (5.1.6) can be written in summation form as

$$\Sigma(Y_i - \hat{\alpha} - \hat{\beta}_1 X_{i1} - \hat{\beta}_2 X_{i2} - \ldots - \hat{\beta}_p X_{ip})^2. \tag{5.A.1}$$

The derivative of eqn. (5.A.1) with respect to $\hat{\alpha}$ is

$$-2 \Sigma (Y_i - \hat{\alpha} - \hat{\beta}_1 X_{i1} - \hat{\beta}_2 X_{i2} - \ldots - \hat{\beta}_p X_{ip}).$$

Equating this expression to zero and simplifying yields

$$n\hat{\alpha} + \hat{\beta}_1 \Sigma X_{i1} + \hat{\beta}_2 \Sigma X_{i2} + \ldots + \hat{\beta}_p \Sigma X_{ip} = \Sigma Y_i. \tag{5.A.2}$$

The derivative of eqn. (5.A.1) with respect to $\hat{\beta}_j$ is

$$-2\Sigma(Y_i - \hat{\alpha} - \hat{\beta}_1 X_{i1} - \hat{\beta}_2 X_{i2} - \ldots - \hat{\beta}_p X_{ip})X_{ij}.$$

Equating this expression to zero yields

$$\Sigma X_{ij}\hat{\alpha} + \hat{\beta}_1 \Sigma X_{i1}X_{ij} + \ldots + \hat{\beta}_j \Sigma X_{ij}^2 + \ldots + \hat{\beta}_p \Sigma X_{ij}X_{ip} = \Sigma X_{ij}Y_i. \tag{5.A.3}$$

Together eqns. (5.A.2) and (5.A.3) comprise $(p + 1)$ equations which must be simultaneously solved to produce the least squares parameter estimators.
In matrix notation, eqns. (5.A.2) and (5.A.3) are

$$X^{*\prime}X^*\underline{\hat{B}} = X^{*\prime}\underline{Y}.$$

Premultiplication of both sides of this equation by $(X^{*\prime}X^*)^{-1}$ yields

$$(X^{*\prime}X^*)^{-1}X^{*\prime}X^*\underline{\hat{B}} = (X^{*\prime}X^*)^{-1}X^{*\prime}\underline{Y}$$

or

$$\underline{\hat{B}} = (X^{*\prime}X^{*})^{-1} X^{*\prime}\underline{Y},$$

which is the least squares estimator (5.1.7).

To verify that the solutions to the normal equations actually minimize the sum of squared residuals, suppose that $\underline{\tilde{B}}$ is any solution to the normal equations other than $\underline{\hat{B}}$. The sum of squared residuals using $\underline{\tilde{B}}$ can be written as

$$(\underline{Y} - X^*\underline{\tilde{B}})'(\underline{Y} - X^*\underline{\tilde{B}}) = (\underline{Y} - X^*\underline{\hat{B}} + X^*\underline{\hat{B}} - X^*\underline{\tilde{B}})' (\underline{Y} - X^*\underline{\hat{B}} + X^*\underline{\hat{B}} - X^*\underline{\tilde{B}})$$

$$= (\underline{Y} - X^*\underline{\hat{B}})'(\underline{Y} - X^*\underline{\hat{B}}) + 2(\underline{\hat{B}} - \underline{\tilde{B}})' X^{*\prime}(\underline{Y} - X^*\underline{\hat{B}})$$

$$+ (\underline{\hat{B}} - \underline{\tilde{B}})' X^{*\prime}X^{*}(\underline{\hat{B}} - \underline{\tilde{B}}).$$

But $X^{*\prime}(\underline{Y} - X^*\underline{\hat{B}}) = X^{*\prime}\underline{Y} - X^{*\prime}X^{*}\underline{\hat{B}} = X^{*\prime}\underline{Y} - X^{*\prime}\underline{Y} = \underline{0}$. Thus

$$(\underline{Y} - X^*\underline{\tilde{B}})' (\underline{Y} - X^*\underline{\tilde{B}}) = \underline{r}'\underline{r} + \underline{s}'\underline{s},$$

where $\underline{r} = \underline{Y} - X^*\underline{\hat{B}}$ is the vector of least squares residuals and $\underline{s} = X^*(\underline{\hat{B}} - \underline{\tilde{B}})$. Since $\underline{s}'\underline{s} = \Sigma s_i^2 \geqslant 0$ with equality holding if and only if $\underline{s} = \underline{0}$, (i.e., $\underline{\hat{B}} = \underline{\tilde{B}}$), the least squares estimator, $\underline{\hat{B}}$, minimizes the sum of squared residuals.

5.B Algebraic Derivation of Fitting by Stages

Algebraically, $\hat{\beta}_j$ can be obtained by regressing the adjusted response variable on the adjusted jth predictor variable, where each of these variables is adjusted for the other $(p - 1)$ predictor variables (and the constant term of the model). For notational convenience let us operate on model (5.1.4):

$$\underline{Y} = X^*\underline{B} + \underline{\varepsilon},$$

where $X^* = [\underline{1}, \underline{X}_1, ..., \underline{X}_j, ..., \underline{X}_p]$ and $\underline{B}' = (\alpha, \beta_1, ..., \beta_j, ..., \beta_p)$. Now let D_j denote the $n \times p$ matrix consisting of all the columns of X^* except for $\underline{X}_j$. Thus D_j is X^* with the $(j + 1)$-th column deleted. Similarly, let $\underline{d}_j$ contain all the elements of $\underline{B}$ except β_j.

Regressing $\underline{Y}$ on D_j results in the least squares estimator

$$\underline{\hat{d}}_j = (D_j'D_j)^{-1}D_j'\underline{Y}.$$

The residuals for the corresponding prediction equation, i.e., the adjusted response variables, are

$$\underline{r}_{p-1}(Y) = \underline{Y} - \hat{\underline{Y}}$$

$$= \underline{Y} - D_j\hat{\underline{\delta}}_j$$

$$= \underline{Y} - D_j(D_jD_j)^{-1}D_j\underline{Y}$$

$$= [I - D_j(D_jD_j)^{-1}D_j]\underline{Y} .$$

In a similar fashion the residuals of $\underline{X}_j$, i.e., the adjusted values of the jth predictor variable, are

$$\underline{r}_{p-1}(X_j) = [I - D_j(D_jD_j)^{-1}D_j]\underline{X}_j .$$

Both sets of residuals (adjusted values), $\underline{r}_{p-1}(Y)$ and $\underline{r}_{p-1}(X_j)$, involve the matrix $[I - D_j(D_jD_j)^{-1}D_j']$. Two important properties of this matrix will simplify the calculation of $\hat{\beta}_j$:

(i) $[I - D_j(D_j'D_j)^{-1}D_j']\underline{1} = \underline{0}$

(ii) $[I - D_j(D_j'D_j)^{-1}D_j']D_j = D_j - D_j(D_j'D_j)^{-1} D_j'D_j$

$$= D_j - D_j = \Phi,$$

i.e., $[I - D_j(D_j'D_j)^{-1}D_j']$ is orthogonal to a vector of ones and to the matrix D_j. Property (i) actually follows from property (ii) since $\underline{1}$ is the first column of D_j.

Regressing $\underline{r}_{p-1}(Y)$ on $\underline{r}_{p-1}(X_j)$ using the no-intercept formula (3.2.10) [no intercept since from (i), $\underline{r}_{p-1}(Y)'\underline{1} = 0$] yields

$$\hat{\beta}_j = \frac{\underline{r}_{p-1}(X_j)'\underline{r}_{p-1}(Y)}{\underline{r}_{p-1}(X_j)'\underline{r}_{p-1}(X_j)}$$

But $\underline{r}_{p-1}(X_j)'\underline{r}_{p-1}(Y_j) = \underline{X}_j'[I - D_j(D_j'D_j)^{-1}D_j'] [I - D_j(D_j'D_j)^{-1}D_j')] \underline{Y}$

$$= \underline{X}_j'[I - D_j(D_j'D_j)^{-1} D_j'] \underline{Y}   \quad \text{from (ii)}$$

$$= \underline{r}_{p-1}(X_j)' \underline{Y} .$$

Thus $\hat{\beta}_j$ can be obtained from the *original* response variables and the *adjusted* jth predictor variable:

$$\hat{\beta}_j = \frac{\underline{r}_{p-1}(X_j)' \underline{Y}}{\underline{r}_{p-1}(X_j)'\underline{r}_{p-1}(X_j)} \tag{5.B.1}$$

Through this algebraic derivation and the arguments offered in Section 5.1.2, we see that the estimator in eqn. (5.B.1) is identical to that obtained through a fitting-by-stages process and that calculated from the matrix eqns. (5.1.7) or (5.1.9).

We are now able to show how the interpretation given in Section 5.2.1 is justified. The prediction equation for the observed response variables can be written as

$$\hat{\underline{Y}} = X^* \hat{\underline{B}}$$

$$= D_j \hat{\underline{\delta}}_j + \underline{X}_j \hat{\beta}_j .$$

This prediction equation is unchanged if one adds and subtracts the same algebraic quantity, so let us add and subtract $\hat{\beta}_j D_j (D_j' D_j)^{-1} D_j' \underline{X}_j$ and rearrange the terms as follows:

$$\hat{\underline{Y}} = D_j \hat{\underline{\delta}}_j + \hat{\beta}_j D_j (D_j' D_j)^{-1} D_j' \underline{X}_j - \hat{\beta}_j D_j (D_j' D_j)^{-1} D_j' \underline{X}_j + \underline{X}_j \hat{\beta}_j$$

$$= D_j [\hat{\underline{\delta}}_j + (D_j' D_j)^{-1} D_j' \underline{X}_j \hat{\beta}_j] + [\underline{X}_j - D_j (D_j' D_j)^{-1} D_j' \underline{X}_j] \hat{\beta}_j$$

$$= D_j \hat{\underline{B}}_j + \underline{r}_{p-1}(X_j) \hat{\beta}_j \qquad (5.B.2)$$

Since $\underline{r}_{p-1}(X_j)$ and D_j are orthogonal in eqn. (5.B.2), $\hat{\beta}_j$ can be estimated and interpreted separately from $\hat{\underline{B}}_j$. If one uses the full set of equations in (5.A.2) and (5.A.3) to simultaneously estimate $\hat{\beta}_j$ and $\hat{\underline{B}}_j$, the resulting estimator of β_j will be identical with that obtained from eqn. (5.B.1). It also follows from the orthogonality of D_j and $\underline{r}_{p-1}(X_j)$ that there is no covariation between $\underline{r}_{p-1}(X_j)$ and the variables in D_j (this follows because $\underline{r}_{p-1}(X_j)$ contains the adjusted X_j values, adjusted for the variables in D_j). Hence, one can indeed interpret $\hat{\beta}_j$ as the increase in the predicted response attributable to a unit increase in the adjusted value of X_j.

The notational change from $\hat{\underline{\delta}}_j$ to $\hat{\underline{B}}_j$ in the expression

$$\hat{\underline{B}}_j = \hat{\underline{\delta}}_j + (D_j' D_j)^{-1} D_j' \underline{X}_j \hat{\beta}_j \qquad (5.B.3)$$

is meant to point out that $\hat{\underline{B}}_j$ does contain the least squares estimates of α, $\beta_1, ..., \beta_{j-1}, \beta_{j+1}, ..., \beta_p$ from the full prediction equation. The estimates in $\hat{\underline{\delta}}_j$ are the least squares estimates from a model containing all the predictor variables except X_j (or, equivalently, the estimates from the $(p - 1)$-th stage of a fitting-by-stages process). The last term in eqn. (5.B.3) is the adjustment needed to change the $(p - 1)$-th stage estimates after $\hat{\beta}_j$ has been calculated during the last stage of the fitting process. These are the adjustments that were made in deriving eqns. (5.1.10) and (5.1.11) in Section 5.1.2.

5.C   Calculating SSR

To show that

$$\text{SSR} = \hat{\beta}' M' M \hat{\beta},$$

observe that

$$M\hat{\beta} = \begin{pmatrix} \sum\limits_{j=1}^{p} M_{1j}\hat{\beta}_j \\ \vdots \\ \sum\limits_{j=1}^{p} M_{nj}\hat{\beta}_j \end{pmatrix}$$

so that

$$\hat{\beta}' M' M \hat{\beta} = (M\hat{\beta})'(M\hat{\beta})$$

$$= \sum_{i=1}^{n} \left(\sum_{j=1}^{p} M_{ij}\hat{\beta}_j \right)^2 . \qquad (5.C.1)$$

By expressing M_{ij} as $X_{ij} - \overline{X}_j$, eqn. (5.C.1) can be written as

$$\hat{\beta}' M' M \hat{\beta} = \sum_{i=1}^{n} \left(\sum_{j=1}^{p} X_{ij}\hat{\beta}_j - \sum_{j=1}^{p} \overline{X}_j\hat{\beta}_j \right)^2$$

$$= \sum_{i=1}^{n} \left(\overline{Y} - \sum_{j=1}^{p} \overline{X}_j\hat{\beta}_j + \sum_{j=1}^{p} X_{ij}\hat{\beta}_j - \overline{Y} \right)^2$$

$$= \sum_{i=1}^{n} \left(\hat{\alpha} + \sum_{j=1}^{p} X_{ij}\hat{\beta}_j - \overline{Y} \right)^2$$

$$= \sum_{i=1}^{n} (\hat{Y}_i - \overline{Y})^2$$

which is SSR.

5.D   Equivalence of Residual Expressions

In order to show that all three expressions for the sum of squared residuals, eqn. (5.3.4), are identical, first recall that

$$\hat{\underline{B}}' = (\hat{\alpha}, \hat{\beta}') \quad \text{and} \quad (X^{*\prime}\underline{Y})' = (n\overline{Y}, X'\underline{Y}).$$

Then

$$\hat{\underline{B}}'X^{*\prime}\underline{Y} = n\overline{Y}\hat{\alpha} + \hat{\beta}'X'\underline{Y},$$

verifying the equivalence of the first two lines of eqn. (5.3.4).
   Now let us rearrange $\hat{\underline{B}}'X^{*\prime}\underline{Y}$ as follows:

$$\hat{\underline{B}}'X^{*\prime}\underline{Y} = n\overline{Y}\left(\overline{Y} - \sum_{j=1}^{p}\hat{\beta}_j\overline{X}_j\right) + \sum_{j=1}^{p}\hat{\beta}_j\left(\sum_{i=1}^{n}X_{ij}Y_i\right)$$

$$= n\overline{Y}^2 + \sum_{j=1}^{p}\hat{\beta}_j\left(\sum_{i=1}^{n}X_{ij}Y_i - n\overline{X}_j\overline{Y}\right)$$

$$= n\overline{Y}^2 + \sum_{j=1}^{p}\hat{\beta}_j\sum_{i=1}^{n}(X_{ij} - \overline{X}_j)Y_i$$

since $n\overline{X}_j\overline{Y} = \overline{X}_j\Sigma Y_i$. Multiplying and dividing $\hat{\beta}_j$ by d_j yields the desired result,

$$\hat{\underline{B}}'X^{*\prime}\underline{Y} = n\overline{Y}^2 + \sum_{j=1}^{p}\hat{\beta}_j d_j \sum_{i=1}^{n}\frac{(X_{ij} - \overline{X}_j)Y_i}{d_j}$$

$$= n\overline{Y}^2 + \sum_{j=1}^{p}\hat{\beta}_j^* \sum_{i=1}^{n}W_{ij}Y_i$$

$$= n\overline{Y}^2 + \hat{\underline{\beta}}^{*\prime}W'\underline{Y}.$$

EXERCISES

1. Using the Detroit homicide data in Table 5.1 and eqn. (5.1.9) find the coefficient estimators for the regression of HOM on FTP and LIC and, hence, verify eqn. (5.1.10).

2. Repeat the fitting-by-stages analysis described in Section 5.1.2 for the data in Table 5.1, except regress HOM on LIC in stage one, on GR in stage two, and on FTP in stage three. Compare the final prediction equation with eqn. (5.1.11).

3. Construct a prediction equation using the high school grade data of Data Set A.1 and the first specification of the two indicator variables, X_1 and X_2, in Section 5.3.2 (in addition to X_3 = grade 13 average). Verify that X_1 and X_2 are orthogonal predictor variables. Compare the coefficient estimates for X_1 and X_2 from this equation with those obtained by regressing high school averages separately on X_1 and X_3, and X_2 and X_3.

4. Using the data in Table 5.3 verify that the raw data coefficient estimates can be derived from the centered, normal deviate, unit length, and beta weight estimators by making appropriate transformations of the coefficients.

5. Recalculate the coefficient estimates in Table 5.3 using a no-intercept model. Are there any changes in the beta weight estimators? Why (not)?

6. Compare the coefficients in the three stage-wise prediction equations obtained in Exercise 2. Are there any dramatic fluctuations as predictor variables are inserted? Explain.

7. Verify the results in Section 5.2.2 for the homicide data using the stage-wise fit of Exercise 2. What is the effect of changing X_3 from 1,000 to 1,500 when X_1 and X_2 change, respectively, from 550 and 1,100 to 600 and 1,000? Comment on the numerical results.

8. Obtain an ANOVA table for the regression of HOM on FTP, LIC, and GR.

9. Compute $\hat{\sigma}^2$ for the high school grade example of Table 5.3 using eqn. (5.3.3) and the centered, normal deviate, and beta weight estimators. Compare the numerical accuracy with the values given in Section 5.3.2.

CHAPTER 6

INFERENCE

Statistical inference refers, in general, to drawing logical conclusions from a data base. In many research studies, the goals of the investigations can be achieved by drawing inferences on regression models. These inferences could involve the true regression coefficients of the model, the error variance, the probability distribution of the errors, or many other important characteristics of the assumed model. In this chapter we discuss statistical procedures for making proper inferences on the parameters of linear regression models. The need for these procedures can be appreciated from an analysis of Data Set A.5.

Concern over environmental effects of automobile and truck exhaust emissions in recent years has led to concentrated efforts to measure and limit potentially dangerous emissions. Engineering studies have shown, however, that the measurement of automobile emissions can be affected by local environmental conditions such as temperature and humidity. Testing procedures and emissions standards must, therefore, account for these environmental factors.

A laboratory study was conducted to measure the effect, if any, of temperature (TEMP), humidity (HUMID), and barometric pressure (BPRES) on several types of diesel emissions, including nitrous oxides (NOX). A total of 174 tests were run on four light-duty diesel trucks. Data Set A.5 records the 44 observations made on one of the vehicles, a Datsun 220C. The ultimate purpose of the investigation was to specify correction factors (possibly as a function of TEMP, HUMID, and BPRES) for nitrous oxide so that nitrous oxide emission readings could be converted to a standard scale. These standardized nitrous oxide readings could then be compared with legal limits or readings from other studies conducted under different local environmental conditions.

Once the experimental portion of the study was concluded, the first task facing the researchers was to obtain adequate prediction equations for NOX. Prior to the analysis the relative influence of the three environmental factors was not completely known. Correct functional specification of the individual predictor variables was also unknown. Due to a desire to produce correction factors having as simple a form as possible (provided, of course, that adequate prediction could be obtained), linear, quadratic, and interaction terms for the predictor variables were evaluated.

Several of the possible prediction equations for NOX constructed from Data Set A.5 are summarized in Table 6.1. In all these prediction equations a constant term is included along with the predictor variables listed in the first column of the table. From an examination of the R^2 and $\hat{\sigma}$ values in Table 6.1 it seems apparent that BPRES is the best single variable to use in a prediction equation for NOX. It also seems clear that HUMID and BPRES are the best pair of predictor variables. Moreover, the inclusion of quadratic terms for HUMID and TEMP appears to substantially improve the fit over the prediction equation containing only the linear and interaction terms.

Table 6.1. Several Fits to Emissions Data (Data Set A.5)

Predictor Variables*	R^2	$\hat{\sigma}$
H	.572	.0711
P	.667	.0628
T	.055	.1057
H,P	.691	.0612
H,T	.601	.0695
P,T	.668	.0635
H,T,HT	.717	.0593
H,T,HT,HH,TT	.798	.0514

*H = HUMID, P = BPRES, T = TEMP, HT = HUMID × TEMP, HH = $(HUMID)^2$, TT = $(TEMP)^2$

All the preceding conclusions were based on simple numerical comparisons of R^2 and $\hat{\sigma}$ values. Like all statistics, R^2 and $\hat{\sigma}$ are random variables which are subject to variability from sample to sample. Perhaps if another experiment is run on the same Datsun 220C under the same test conditions, HUMID would be the best single predictor variable and many of the other conclusions that were drawn earlier would be altered. It is also true (and can be shown theoretically) that the addition of any predictor variables to a regression model generally increases (perhaps only trivially) and never decreases R^2. Perhaps the increase in R^2 associated with the addition of quadratic terms for HUMID and TEMP is a result of this latter property and does not indicate a substantial improvement to the fit of the response variable.

Measures are needed, therefore, which will allow one not only to assess the overall fit of the prediction equation but also the contributions of individual predictor variables to the fit. In order to construct such inferential statistics, further assumptions about the true regression model must be made.

6.1 MODEL DEFINITION

On casual reflection one might think that good prediction requires good estimates of model parameters and that if the residuals, R^2, and $\hat{\sigma}$ values are found to be acceptable the coefficient estimates must be close to the true parameter values. The discussion of the fit to the suicide rates in Section 3.1 is just one illustration of the fallacy of such a belief. Over the restricted age range in Figure 3.6 the fitted model predicts the suicide rates well, but it is clear from Figure 3.7 that the one-variable prediction equation is badly misspecified and that the model parameters are poorly estimated due to the model misspecification.

Inference procedures for regression models are dependent on assumptions about the true definition of the model. As we shall see, some techniques remain valid under a broad range of model assumptions while others critically depend on the validity of model assumptions.

6.1.1   Four Key Assumptions

Four key assumptions concerning the true regression model relating response and predictor variables are listed in Table 6.2, along with two alternative assumptions. We shall describe these assumptions more fully in the remainder of this subsection and examine their impact on least squares estimation in subsequent portions of this chapter. Techniques for validating model assumptions are discussed in Chapter 7.

Assumption 1 is too often automatically taken for granted in regression analyses. Strictly speaking this assumption asserts that the model is functionally correct. This implies, for multiple linear regression models, that all relevant predictor variables are included in the model and that they appropriately enter and affect the response as linear terms (i.e., linear in the

Table 6.2. Model Assumptions

Assumption	Description
1	The model is correctly specified.
2	Predictor variables are nonstochastic and are measured without error.
3	Model error terms have zero means, are uncorrelated, and have constant variances.
4	Model error terms follow a normal probability distribution.
	Alternatives
3'	Conditioned on the observed predictor variable values, Assumption 3 holds.
4'	Conditioned on the observed predictor variable values, Assumption 4 holds.

unknown regression coefficients). Apart from random error, then, Assumption 1 states that the exact form of the relationship between response and predictor variables is known to the researcher or data analyst.

Assumption 2 prescribes that the predictor variables (i) are under the control of the data analyst and (ii) that he or she can exactly record the values of all predictor variables included in the data base. If the predictor variables are merely observed and not controlled by the researcher, they are stochastic (random) variables since the researcher does not know what values will be included in the data base prior to the data collection. Also, the predictor variables are required by this assumption to be recorded with sufficient exactness to insure that neither bias nor randomness is incurred due to the measurement process itself.

Three distinct characteristics of the error term ε of the model comprise Assumption 3. For most practical purposes uncorrelated errors (zero correlation between any pair of error terms) can be thought of as implying that all the errors are independent of one another. A large or small error for one response, therefore, does not affect the magnitude of the error of any other response. Data bases compiled from carefully controlled laboratory experiments or random samples of populations can often be assumed to have uncorrelated errors while correlated errors are frequently exhibited by time series data (e.g., economic data that is recorded each week, month, or year).

The second characteristic of the error terms that is embodied in Assumption 3 is that their probability distribution has expectation (population mean) equal to zero. For regression models containing intercept terms this assumption is not at all restrictive since models of the form

$$Y = \alpha + \beta_1 X_1 + \ldots + \beta_p X_p + \varepsilon,$$

where ε has population mean $\delta \neq 0$, can always be rewritten as

$$Y = (\alpha + \delta) + \beta_1 X_1 + \ldots + \beta_p X_p + (\varepsilon - \delta)$$
$$= \gamma + \beta_1 X_1 + \ldots + \beta_p X_p + \eta.$$

In this last formulation of the model, the error term, $\eta = \varepsilon - \delta$, has mean zero. This model is indistinguishable from the usual linear regression model in which Assumption 3 is valid. Hence the two models are equivalent and one can always assume the errors have mean zero when regression models are specified with constant terms.

When no-intercept models are employed the assumption that the error terms have mean zero cannot be taken for granted. This is because if ε has mean $\delta \neq 0$, the same transformation performed above to produce errors with zero expectation has the following result:

$$Y = \beta_1 X_1 + \ldots + \beta_p X_p + \varepsilon$$

$$= \delta + \beta_1 X_1 + \ldots + \beta_p X_p + \eta,$$

where $\eta = \varepsilon - \delta$. Thus the transformation $\eta = \varepsilon - \delta$ changes a no-intercept model into an intercept model. One cannot, therefore, routinely assume that no-intercept models have error terms with zero means.

The final ingredient in Assumption 3 is that all error terms have the same variance. This will automatically be true if the errors are independent observations from the same population but for many types of response variables a common variance assumption is invalid. For example, the variability of psychological measurements is sometimes a function of age, social status, or ethnicity of the individuals in the data base. If a prediction equation is constructed for such variables the variances of the model errors might not all be the same.

Assumption 4 asserts that the probability distribution of the errors is the normal distribution. Combined with Assumption 3, this assumption declares that

$$\varepsilon \sim NID(0, \sigma^2) ;$$

i.e., that the errors are Normally, Independently, Distributed with mean 0 and constant variance σ^2. Assumption 4 can be used without Assumption 3, as with time series responses, when one might wish to assume that the errors are correlated or have unequal variances while retaining the assumed normal probability distribution.

More attention will be devoted to all these assumptions as examples serve to raise questions concerning their validity. One point should be mentioned at this time, however, regarding the general topic of model assumptions. Just as it was argued in Section 3.1 that all models are approximate, so too these assumptions are at best only approximately true in practice. Rarely can one be certain that a model is correctly specified. One hopes that no important predictor variable has been inadvertently left out of the model, that no large discrepancies between true and fitted responses will arise due to the model specification, and that the estimation techniques employed will be adequate even if the error terms do not satisfy the assumptions exactly.

Our goal in studying model assumptions is twofold: (i) to develop properties of regression estimators, and (ii) to become aware of the consequences of not satisfying the assumptions. If one is cognizant of model assumptions and the consequences of violating them, effective use of regression analysis is still possible when some assumptions are not met. We will soon see that some regression techniques based on least squares estimators can be employed when certain assumptions in Table 6.2 are violated and that alternative techniques can be utilized when other problems arise with data bases.

6.1.2 Alternative Assumptions

It is common in regression analysis for Assumption 2 to be violated. When this occurs are all the optimal estimator and distributional properties which depend on Assumption 2 destroyed? Strictly speaking yes; practically speaking, perhaps not.

Suppose one's interest lies specifically in the predictor variables obtained in a data set regardless of how those variables were obtained; i.e., regardless of whether they are random variables or known fixed constants. One might examine the data base and decide, for example, that the observations on the predictor variables represent "typical" values for the predictors. In some studies one might just select observations for analysis that have predictor variable values within predetermined ranges or only those that have specified values. For whatever the reason, if one is willing to draw inferences on responses from regression models with predictor variable values equal to those in the data base, then all the optimal properties of least squares estimators may not be forfeited even if the predictors were randomly obtained. Inferences on the regression model are then said to be "conditional" on the observed predictor variables.

Assumptions 3' and 4' in Table 6.2 correspond to Assumptions 3 and 4, except that the former are used when conditional inferences are desired. The use of Assumption 3' and 4' removes the need for Assumption 2, but the implications for using Assumptions 3' and 4' rather than adopting Assumptions 2-4 must be clearly understood. They center, again, on the difference between interpolation and extrapolation; i.e., conditioning inferences on the observed data base insures that one is not extrapolating to data points that are not represented or characterized by similar values in X^*.

The observations on the predictor variables in the GNP data, for example, must be considered random both because they are not under the control of the experimenter and because of measurement error. One is often tempted to ignore this fact when assessing the magnitudes of estimated regression coefficients. One would like to claim that, although the predictor variables are random, they can be treated as fixed constants since these countries are representative of all the countries of the world and the predictor variable values therefore represent the population of values for which inferences are desired. In other words, one would like to argue that if one *could* control the values of the predictor variables these are precisely the values that would be selected for inclusion in the data base.

The inherent danger in making an argument similar to this one was pointed out in Figure 2.2 where a scatter plot of GNP and LIT was displayed. These countries are not representative of all the countries of the world with respect to literacy and perhaps for other predictor variables as well. If inferences on regression model parameters are desired they must be made conditional on the observed values of the predictor variables; in this case, for the more literate countries of the world. Interpretation of the regression coefficients for the less literate countries cannot be justified when only the data in Data Set A.1 is included in a regression analysis.

A second study which points out the potential danger of drawing inferences conditional on the observed data when the data base might not be representative of all possible predictor variable values is the emissions study described in the introduction to this chapter. Atmospheric pressure (BPRES) was ultimately eliminated from the final prediction equations because it was strongly negatively correlated with humidity readings. It was noted, however, that the negative correlation was stronger than expected and due to "a local situation, namely that decreases in atmospheric pressure are frequently followed by southerly winds carrying humid air from the Gulf of Mexico."

In this experiment HUMID could not be controlled by the researchers, although it fluctuated between limits that were acceptable to the research goals. If one conditions inferences on the observed HUMID values, the correlation between HUMID and BPRES could inadvertently be ignored. Yet if similar experiments are conducted in other regions of the U. S. where humidity and atmospheric pressure are not so highly correlated, the fitted model might be altered substantially.

This discussion points out the need to carefully define the conditions under which Assumptions 3' and 4' are used to replace Assumptions 2-4.

Not only must care be taken to insure that the predictor variable values in the data base are acceptable for the purposes of the research study, but special conditions such as those described in the above examples which could affect the fitting or usage of prediction equations must be clearly stated.

6.2 ESTIMATOR PROPERTIES

Theoretical properties of least squares regression estimators are inextricably tied to the assumptions in Table 6.2. The performance and interpretation of the estimators and statistics derived from them depend on the correctness of the assumptions one makes about the underlying theoretical model that generates the observed responses. In this section we describe several theoretical properties of least squares estimators and show the dependence of each on the model assumptions.

6.2.1 Geometrical Representation

Even when none of the assumptions in Table 6.2 are valid, least squares parameter estimation offers an attractive rationale for its use. Algebraically we have seen that least squares estimates minimize the sum (or average) of the squared response residuals for the model that is being fit by the data; i.e., regardless of whether the assumed model is the correct one, least squares estimates minimize the sum of squared residuals for the specified model. An equivalent rationale follows from a geometrical discussion of least squares.

Data Set A.6 describes anthropometric and physical fitness measurements that were taken on 50 white male applicants to the police department of a major metropolitan city. Of the many objectives of the investigation of the original data base, one concerns the determination of which, if any, of these variables influence the reaction times of the applicants to a visual stimulus (equivalent to the time needed to draw a pistol). The researcher is interested in discovering which of these variables influence the reaction times most and desires that unimportant predictor variables be eliminated from the final prediction equation.

As part of the overall investigation of Data Set A.6, suppose the researcher wishes to assess the relationship of several of the anthropometric measurements to the applicants' weights. For ease of presentation let us first examine the use of shoulder width (SHLDR) as a predictor of weight (WEIGHT) in a single-variable, no-intercept model using only the observations on subjects 10 and 20. The reduced data set is:

Subject No.	WEIGHT(Y)	SHLDR(X)
10	54.28	37.2
20	77.66	41.4

Straightforward calculations produce the fit

$$\hat{\underline{Y}} = 1.690\underline{X}$$

and predicted weights of 62.86 and 69.95, respectively.

Geometrically, the least squares fit to the reduced data set is displayed in Figure 6.1. Note that unlike previous graphs in which variable values were labelled on the two axes, the axes in Figure 6.1 are scales for the two observations. The points on the graph represent the observations on SHLDR [$\underline{X}'$ = (37, 41)] and WEIGHT [$\underline{Y}'$ = (54, 78)]. Each vector is indicated by a solid line connecting the origin with the respective plotted points.

Specification of a single-variable, no-intercept prediction equation relating WEIGHT and SHLDR requires that the vector of predicted responses be a multiple ($\hat{\beta}$) of the vector of predictor variable values; i.e., $\hat{\underline{Y}} = \hat{\beta}\underline{X}$. Geometrically, the predicted responses must be a point on the line connecting the origin and $\underline{X}$ or its extension in either direction. In this example, $\hat{\underline{Y}}$ must be a point on the solid line in Figure 6.1 between the origin and the point $\underline{X}$, or it must be on the dashed extension of this line. The principle of least squares insures that the precise location of $\hat{\underline{Y}}$ on this line will minimize the distance between $\underline{Y}$ and $\hat{\underline{Y}}$, as we shall now see.

The residual vector, $\underline{r} = \underline{Y} - \hat{\underline{Y}}$, geometrically represents the line connecting $\underline{Y}$ with its predicted value, $\hat{\underline{Y}}$. The length of this vector, or the distance between $\underline{Y}$ and $\hat{\underline{Y}}$, is $(r_1^2 + r_2^2)^{1/2}$. This length is minimized when $r_1^2 + r_2^2$ is as small as possible; i.e., when the sum of squared residuals is minimized. Least squares parameter estimation chooses the value of $\hat{\beta}$ that makes $\underline{Y}$ as close as possible geometrically to $\hat{\underline{Y}} = \hat{\beta}\underline{X}$. Since the shortest path between a point and a line occurs along a vector that is perpendicular to the line, $\underline{r}$ is perpendicular to the dashed line in Figure 6.1.

Let us now add pelvic width (PELVIC) to the reduced data set as a second predictor variable and include subject 30 as well:

Subject No.	WEIGHT(Y)	SHLDR(X_1)	PELVIC(X_2)
10	54.28	37.2	24.2
20	77.66	41.4	31.6
30	73.78	42.8	29.7

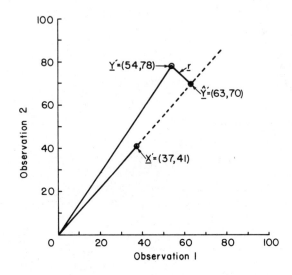

Figure 6.1 Geometrical interpretation of least squares parameter
estimation for single-variable no-intercept models.

Fitting a two-variable, no-intercept prediction equation to this reduced data
set produces the following fit:

$$\hat{\underline{Y}} = -0.747\underline{X}_1 + 3.471\underline{X}_2.$$

The corresponding predicted responses are 56.18, 78.73, and 71.09.

Figure 6.2 exhibits a plot of the three observed vectors $\underline{Y}$, $\underline{X}_1$, and $\underline{X}_2$ as
points in three dimensional space. The locus of points that are potential pre-
dicted response vectors, $\hat{\underline{Y}}$, is a plane formed by all possible linear combinations of $\underline{X}_1$ and $\underline{X}_2$. Note that $\underline{Y}$ does not fall on this plane just as it did not
fall on the line $\beta\underline{X}$ in Figure 6.1. Least squares parameter estimates choose
the precise linear combination, $\hat{\beta}_1\underline{X}_1 + \hat{\beta}_2\underline{X}_2$, that makes the distance
between $\underline{Y}$ and $\hat{\underline{Y}}$ (a point in the plane) as small as possible. Geometrically,
$\hat{\underline{Y}}$ is the intersection of the plane with a line from $\underline{Y}$ that is perpendicular to
the plane; this perpendicular line from $\underline{Y}$ to the plane is the residual vector,
$\underline{r}$.

The general geometric interpretation of least squares estimation is shown
in Figure 6.3. Schematically represented in Figure 6.3 are p points, one for
each of the n-dimensional vectors of values of predictor variables (with
$\underline{X}_1 = \underline{1}$ if an intercept is desired in the prediction equation), and a point for
the response vector. All linear combinations $\hat{\beta}_1\underline{X}_1 + \hat{\beta}_2\underline{X}_2 + \ldots + \hat{\beta}_p\underline{X}_p =$
$X\hat{\beta}$ of the p vectors of predictor variables form a "hyperplane" (a concep-
tual generalization of a two-dimensional plane) from which the predicted

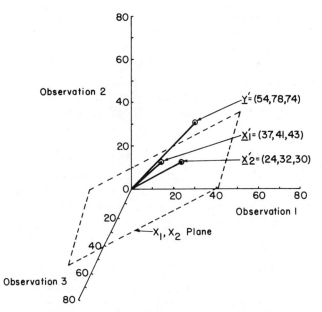

Figure 6.2   Geometrical interpretation of least squares parameter
estimator for two-variable no-intercept models.

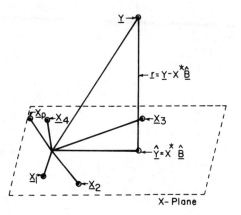

Figure 6.3   Geometrical representation of least
squares parameter estimation.

responses will be obtained. Least squares estimators select valuesfor $\hat{\beta}_1$, $\hat{\beta}_2$,
..., $\hat{\beta}_p$ (and thereby fix the point $\underline{\hat{Y}}$ in the X-plane) that minimize the distance between $\underline{Y}$ and $\underline{\hat{Y}}$. The predicted responses lie at the intersection of

the plane and a perpendicular residual vector that connects $\underline{Y}$ and the X plane.

Recall that no assumptions need be made in order to compute the least squares estimates. The previous examples and interpretations are likewise valid regardless of whether any of the specified models are correct. All that is needed is that the least squares estimates be calculable. This in turn implies that $X^{*\prime}X^*$ in eqn. (5.1.7) has a matrix inverse. Other than this requirement, no conditions need be imposed on the model nor the predictor variable values.

6.2.2 Expectation

Least squares estimators are functions of the response variable, Y; therefore, they are random variables. If many data bases are formed for a particular collection of response and predictor variables, each set of least squares parameter estimates will differ from every other set of estimates.

Knowing that the calculated values of least squares coefficients will vary from sample to sample, we would like to insure that most of these estimates are in some sense close to the true coefficients. One property of an estimator that often provides "close" estimators is that of unbiasedness. If, on the average, an estimator neither tends to overestimate nor underestimate the true parameter value it is said to be an unbiased estimator. Another way of stating that an estimator is unbiased is to declare that its expectation (expected value, population mean) equals the true parameter. Note that this does not guarantee that any single estimated value equals the true parameter value, only that if many data sets are analyzed and estimates are calculated for each data set, the average estimated value will tend to equal the true parameter value. Symbolically, if $\hat{\alpha}$ and $\hat{\beta}$ are unbiased estimators then we write $E[\hat{\alpha}] = \alpha$ and $E[\hat{\beta}] = \beta$, where E stands for the expectation or expected value of the estimator contained in brackets.

Notationally, the expectation of a random vector, $\underline{Y}$, is itself a vector containing the expectations of each of the elements of $\underline{Y}$. For example, if $\underline{Y}' = (Y_1, Y_2, \ldots, Y_n)$ and $E[Y_i] = \mu_i$, $i = 1, 2, \ldots, n$, then we write

$$E[\underline{Y}] = \underline{\mu},$$

where $\underline{\mu}' = (\mu_1, \mu_2, \ldots, \mu_n)$. An important property of the expectation of a random vector is that if A is any $r \times n$ matrix of nonrandom constants and $\underline{C}$ is a vector of constants,

$$E[A\underline{Y} + \underline{C}] = AE[\underline{Y}] + \underline{C} = A\underline{\mu} + \underline{C}. \qquad (6.2.1)$$

Also, if $\underline{Y}_1$ and $\underline{Y}_2$ are random vectors with expectations μ_1 and μ_2, respectively, then

$$E[A_1 \underline{Y}_1 + A_2 \underline{Y}_2] = A_1 \mu_1 + A_2 \mu_2 \qquad (6.2.2)$$

for any two matrices of constants, A_1 and A_2, of dimension $r \times n$.

Least squares estimators of $\underline{B}$ or any of the centered or standardized coefficients except the beta weights are unbiased under Assumptions 1 and 2 of Table 6.2 and an assumption of zero means for the error terms (see Appendix 6.A for a proof of this result). With these assumptions observe that [invoking eqn. (6.2.1)]

$$E[\underline{Y}] = X^* \underline{B} + E[\underline{\varepsilon}]$$

$$= X^* \underline{B}$$

since the errors each have mean zero. It then follows (see Appendix 6.A) that

$$E[\hat{\underline{B}}] = (X^{*\prime} X^*)^{-1} X^{*\prime} E[\underline{Y}]$$

since X^* is a matrix of constants (and, hence, nonrandom) under Assumption 2. Inserting the expectation of the response vector into this expression yields

$$E[\hat{\underline{B}}] = (X^{*\prime} X^*)^{-1} X^{*\prime} X^* \underline{B}$$

$$= \underline{B}.$$

So if Assumptions 1 and 2 are correct and the errors all have zero means, each of the least squares estimators is unbiased for its corresponding parametric value.

Expectations of the beta weights [$\hat{\beta}^0$ in eqn. (5.1.19)] are unknown and quite complex. To appreciate this fact note that

$$E[\hat{\beta}^0] = (Z'Z)^{-1} Z' E[S_Y^{-1} \underline{Y}] .$$

Both S_Y^{-1} and $\underline{Y}$ are random and the expectation of $S_Y^{-1} \underline{Y}$ is a complicated function of all the model parameters. For this reason, inferences on the model parameters should be made with the least squares estimators of the original predictor variables or one of the other standardizations.

A pertinent question that can now be addressed concerns model misspecification. If one *assumes* that the correct regression model is

$$\underline{Y} = X_1^* \underline{B}_1 + \underline{\varepsilon} \qquad (6.2.3)$$

when in reality the *true* model is

$$\underline{Y} = X_1^* \underline{B}_1 + X_2^* \underline{B}_2 + \underline{\varepsilon} \qquad (6.2.4)$$

where X_2^* contains predictor variables that should be included in the model, what are the effects on the least squares estimators of $\underline{B}_1$? In general, the estimators are biased when Assumption 1 is violated since under the erroneous assumption that eqn. (6.2.3) is the true model, $\hat{\underline{B}}_1 = (X_1^{*\prime} \, X_1^{*\prime})^{-1} X_1^* \, \underline{Y}$ and

$$E[\hat{\underline{B}}_1] = (X_1^{*\prime} X_1^*)^{-1} X_1^{*\prime} \, [X_1^* \underline{B}_1 + X_2^* \underline{B}_2]$$

$$= \underline{B}_1 + A\underline{B}_2, \qquad (6.2.5)$$

where $A = (X_1^{*\prime} X_1^*)^{-1} X_1^{*\prime} X_2^*$ is referred to as the "alias" matrix. If X_1^* and X_2^* happen to be orthogonal to one another (i.e., every vector in X_1^* is orthogonal to all the vectors in X_2^* and vice-versa), the alias matrix is Φ and the least squares estimators for $\underline{B}_1$ using the incorrect model (6.2.3) are unbiased. Even so, the prediction equation may not adequately represent the response function due to the erroneous exclusion of the variables in X_2^*.

In fitting the high school grade data in Chapters 2 and 3, concern was raised over whether a linear or a quadratic prediction equation should be fit to the data. These were not idle concerns since eqn. (6.2.5) points out that if we erroneously fit a linear predictor the estimators of the coefficients of both the constant and linear terms will be biased by the quadratic one. The nature of the bias can be derived as indicated in eqn. (6.2.5):

$$X_1^* = \begin{bmatrix} 1 & X_1 \\ 1 & X_2 \\ \vdots & \vdots \\ 1 & X_n \end{bmatrix}, \qquad X_2^* = \begin{bmatrix} X_1^2 \\ X_2^2 \\ \vdots \\ X_n^2 \end{bmatrix}$$

$$X_1^{*\prime} X_1^* = \begin{bmatrix} n & \Sigma X_i \\ \Sigma X_i & \Sigma X_i^2 \end{bmatrix}, \qquad X_1^{*\prime} X_2^* = \begin{bmatrix} \Sigma X_i^2 \\ \Sigma X_i^3 \end{bmatrix}$$

$$(X_1^{*\prime} X_1^*)^{-1} X_1^{*\prime} X_2^* = \begin{bmatrix} [(\Sigma X_i^2)^2 - (\Sigma X_i)(\Sigma X_i^3)]/\theta \\ [n(\Sigma X_i^3) - (\Sigma X_i)(\Sigma X_i^2)]/\theta \end{bmatrix}$$

$$\theta = n \sum X_i^2 - (\sum X_i)^2 ,$$

so that if the true model is

$$\underline{Y} = \alpha \underline{1} + \beta_1 \underline{X}_1 + \beta_2 \underline{X}_2 + \underline{\varepsilon} ,$$

where $X_{i1} = X_i$ and $X_{i2} = X_i^2$, then

$$E[\underline{\hat{B}}_1] = \begin{bmatrix} \alpha + \beta_2 \ [(\sum X_i^2)^2 - (\sum X_i)(\sum X_i^3)] \ / \theta \\ \beta_1 + \beta_2 [n(\sum X_i^3) - (\sum X_i)(\sum X_i^2)]/\theta \end{bmatrix} \qquad (6.2.6)$$

Numerically, if one fits a model that is linear in the grade 13 scores to the 18 observations from high school 1 in Data Set A.2 when the true model is quadratic, calculation of eqn. (6.2.6) reveals that $\underline{\hat{B}}_1$ has the following expectation:

$$E[\underline{\hat{B}}_1] = \begin{bmatrix} \alpha - 4890.97 \ \beta_2 \\ \beta_1 + 140.38 \ \beta_2 \end{bmatrix} .$$

A similar calculation for high school 2 yields

$$E[\underline{\hat{B}}_1] = \begin{bmatrix} \alpha - 5224.68 \ \beta_2 \\ \beta_1 + 149.46 \ \beta_2 \end{bmatrix} .$$

Both estimators have bias terms with large coefficients multiplying β_2. From the plots in Figure 2.7, it appears that $\underline{\hat{B}}_1$ might not be too badly biased for high school 1 since the plot looks linear (i.e., $\beta_2 \approx 0$). The estimator of β_1 for high school 2 on the other hand, could be badly biased since curvature is apparent in Figure 2.7(d).

Now consider the effect of violating Assumption 2 on the estimation of regression coefficients. Suppose that in a regression model having a single predictor variable, the predictor variable is measured with error. In particular, suppose that only Y and X can be observed and that

$$X = W + u ,$$

where W is the exact, constant value that we desire to measure and u is a random measurement error. For simplicity we assume that the errors u_i are uncorrelated with one another and are uncorrelated with the ε_i, and that

$E[u_i] = 0$, $\text{Var}[u_i] = \omega^2$ for $i = 1, 2, .., n$. The true model can now be written as:

$$Y_i = \alpha + \beta W_i + \varepsilon_i.$$

Then

$$E[Y_i] = \alpha + \beta W_i$$

and the observed responses are unbiased estimators of the true model, the one we desire to measure, despite the measurement error in the predictor variables. This is not true of the estimators of α and β.

The expectations of $\hat{\alpha}$ and $\hat{\beta}$ cannot be obtained without further specification of the probability distribution of u and ε. To appreciate this fact, observe that $\hat{\beta}$ can be reexpressed as follows:

$$\hat{\beta} = \frac{\displaystyle\sum_{i=1}^{n} (X_i - \bar{X})(Y_i - \bar{Y})}{\displaystyle\sum_{i=1}^{n} (X_i - \bar{X})^2} = \beta + \frac{\displaystyle\sum_{i=1}^{n} (X_i - \bar{X})[\,(\varepsilon_i - \bar{\varepsilon}) - \beta(u_i - \bar{u})]}{\displaystyle\sum_{i=1}^{n} (X_i - \bar{X})^2},$$

since Y_i can be written as $Y_i = \alpha + \beta X_i + (\varepsilon_i - \beta u_i)$. Consequently, it follows that (recall that X_i is a function of u_i and thus a random variable here)

$$E[\hat{\beta}] = \beta + E\left[\frac{\displaystyle\sum_{i=1}^{n} (X_i - \bar{X})[(\varepsilon_i - \bar{\varepsilon}) - \beta(u_i - \bar{u})]}{\displaystyle\sum_{i=1}^{n} (X_i - \bar{X})^2}\right]. \qquad (6.2.7)$$

The last term in eqn. (6.2.7) is zero under Assumptions 1-3, but its value is unknown (but generally nonzero) when X is stochastic (random). Thus, eqn. (6.2.7) indicates that $\hat{\beta}$ is generally a biased estimator when Assumption 2 is violated, although one can show that $E[\hat{\beta}] = \beta$ in eqn. (6.2.7) if the u_i and ε_j are all uncorrelated. The least squares estimator of α is similarly affected.

The bias in the least squares estimators when X is random does not disappear even for large samples. Under Assumptions 1 and 3 and with the above conditions on X, it can be shown* that as n becomes infinite $\hat{\beta}$ tends to underestimate β. Specifically, if Var $[u_i] = \omega^2$ and $\omega^2 > 0$, $\hat{\beta}$ tends to estimate

$$\beta \cdot \frac{\eta^2}{\eta^2 + \omega^2}$$

in large data sets, where η^2 is the limiting value of $\Sigma (W_i - W)^2/(n - 1)$ as n becomes infinite. Thus $\hat{\beta}$ tends to underestimate β in large samples when $\omega^2 > 0$ since $\eta^2/(\eta^2 + \omega^2) < 1$.

6.2.3   Variances and Correlations

The mean vector, $\underline{\mu}$, of a random vector, $\underline{Y}$, characterizes typical values for the elements of $\underline{Y}$. Variability and covariation of the elements of $\underline{Y}$ are characterized by its covariance matrix, Σ_Y. The covariance matrix of a random vector is defined to be a matrix containing the variances and covariances of the elements of $\underline{Y}$. Notationally,

$$\text{Var}[\underline{Y}] = \Sigma_Y = \begin{bmatrix} \text{Var}[Y_1] & \text{Cov}[Y_1, Y_2] & \cdots & \text{Cov}[Y_1, Y_n] \\ \text{Cov}[Y_2, Y_1] & \text{Var}[Y_2] & \cdots & \text{Cov}[Y_2, Y_n] \\ \vdots & \vdots & & \vdots \\ \text{Cov}[Y_n, Y_1] & \text{Cov}[Y_n, Y_2] & \cdots & \text{Var}[Y_n] \end{bmatrix}.$$

The diagonal elements of Σ_Y are the variances of the individual random variables in $\underline{Y}$ and the (i,j)th off-diagonal element is the covariance between Y_i and Y_j. If we let $\sigma_{ij} = \text{Cov}[Y_i, Y_j]$ and recall that $\text{Var}[Y_i] = \text{Cov}[Y_i, Y_i]$, the covariance matrix associated with $\underline{Y}$ can also be written

$$\Sigma_Y = \begin{bmatrix} \sigma_{11} & \sigma_{12} & \cdots & \sigma_{1n} \\ \sigma_{21} & \sigma_{22} & \cdots & \sigma_{2n} \\ \vdots & & & \\ \sigma_{n1} & \sigma_{n2} & & \sigma_{nn} \end{bmatrix}.$$

Finally, observe that the covariance matrix is symmetric: $\sigma_{ij} = \sigma_{ji}$.

Under Assumption 3 of Table 6.2, the covariance matrix of both the error vector, $\underline{\varepsilon}$, and the response vector, $\underline{Y}$, is

---

*A more comprehensive treatment of stochastic predictor variables can be found in many econometrics texts; e.g., Schmidt (1976) and Theil (1971).

$$\Sigma_\varepsilon = \Sigma_Y = \sigma^2 I,$$

where σ^2 is the common value of the error variances. Since the errors are uncorrelated, the off-diagonal elements of the covariance matrix are all zero. As is shown in Appendix 6.B, the covariance matrix of the least squares estimators, eqn. (5.1.7), is

$$\Sigma_B = (X^{*\prime}X^*)^{-1}\sigma^2 . \tag{6.2.8}$$

The diagonal elements of eqn. (6.2.8) contain the variances of the elements of $\underline{\hat{B}}$ and the off-diagonals contain the covariances between pairs of elements.

Correlations between pairs of regression coefficient estimators can also be found from eqn. (6.2.8) by taking the ratio of the appropriate covariance element to the square root of the product of the two corresponding diagonal elements. Let

$$C = \begin{bmatrix} C_{00} & C_{01} & C_{02} & \cdots & C_{0p} \\ C_{10} & C_{11} & C_{12} & \cdots & C_{1p} \\ C_{20} & C_{21} & C_{22} & \cdots & C_{2p} \\ \vdots & \vdots & \vdots & & \vdots \\ C_{p0} & C_{p1} & C_{p2} & \cdots & C_{pp} \end{bmatrix} = (X^{*\prime}X^*)^{-1} . \tag{6.2.9}$$

Then

$$\Sigma_B = \sigma^2 C,$$

$$\text{Corr}(\hat{\alpha}, \hat{\beta}_j) = \sigma^2 C_{0j}/(\sigma^2 C_{00} \cdot \sigma^2 C_{jj})^{1/2}$$

$$= C_{0j}/(C_{00} \cdot C_{jj})^{1/2} \qquad j = 1, 2, ..., p$$

and similarly

$$\text{Corr}(\hat{\beta}_j, \hat{\beta}_k) = C_{jk}/(C_{jj} \cdot C_{kk})^{1/2} \qquad j \neq k .$$

Note in particular that while the variances and covariances of the coefficient estimators involve an unknown model parameter, σ^2, the correlations are completely known from the elements of $(X^{*\prime}X^*)^{-1}$.

For single-variable regression models, one can verify from eqn. (6.2.8) that

$$\text{Var}[\hat{\alpha}] = \sigma^2 (n^{-1} + \overline{X}^2 / d_X^2) \qquad \text{and} \qquad \text{Var}[\hat{\beta}] = \sigma^2 / d_X^2,$$

where $d_X^2 = \Sigma (X_i - \overline{X})^2$. The correlation between $\hat{\alpha}$ and $\hat{\beta}$ is given by

$$\text{Corr}[\hat{\alpha}, \hat{\beta}] = -\overline{X} / (n^{-1} \Sigma X_i^2)^{1/2}.$$

It is particularly important to observe that the correlation between $\hat{\alpha}$ and $\hat{\beta}$ is nonzero unless the predictor variable values have a zero mean. Use of centered or standardized models, therefore, allow estimates of the (transformed) constant term α^* and the regression coefficient (β or β^*) to be uncorrelated. This property is true for multiple-variable regression models as well; for example, with centered predictor variables

$$\Sigma_B = \begin{bmatrix} n^{-1} & \underline{0}' \\ \\ \underline{0} & (M'M)^{-1} \end{bmatrix} \sigma^2 .$$

Since all the covariances of $\hat{\alpha}^*$ with each $\hat{\beta}_j$ (the first row or column of Σ_B) are zero, $\hat{\alpha}^*$ is uncorrelated with each of the coefficient estimates in $\hat{\beta}$. Note also from the above expression that $\text{Var}[\hat{\alpha}^*] = \sigma^2 / n$ (as should be expected since $\hat{\alpha}^* = \overline{Y}$, the mean of the observed responses) and $\text{Var}[\hat{\beta}] = (M'M)^{-1} \sigma^2$, which is similar to eqn. (6.2.8).

Using the techniques discussed in Appendix 6.B, the covariance matrix for each of the regression estimators mentioned thus far (except for beta weights) can be derived. Expectations and covariance matrices for each are summarized in Table 6.3. Also indicated in the table are the transformations needed to obtain estimates for the parameters of the original model from estimates for the centered and standardized ones.

One additional result can be derived under Assumptions 1-3. The variances of $\hat{\alpha}$ and $\hat{\beta}$ are unknown since eqn. (6.2.8) involves the unknown error variance, σ^2. It is shown in many linear models texts [e.g., Graybill (1976), Searle (1971)] that under these same assumptions an unbiased estimator of σ^2 is given by eqn. (5.3.3):

$$\hat{\sigma}^2 = \frac{\text{SSE}}{n-p-1} = \frac{\Sigma r_i^2}{n-p-1} .$$

Inserting $\hat{\sigma}^2$ into eqn. (6.2.8) provides unbiased estimators of the variances of $\hat{\alpha}$ and $\hat{\beta}$. In Section 6.3 these variance estimators will be used to measure the accuracy of the estimates of α and β. Before we can do so, however, we must examine Assumption 4.

Table 6.3. Moments of Least Squares Parameter Estimators Under Assumptions 1-3

	Raw	Centered†	Normal Deviate†	Unit Length†
Estimator	$\hat{\underline{B}} = (X^{*'}X^*)^{-1}X^{*'}\underline{Y}$	$\hat{\underline{\beta}} = (M'M)^{-1}M'\underline{Y}$ $\hat{\alpha}^* = \bar{Y}$	$\hat{\underline{\beta}}^* = (Z'Z)^{-1}Z'\underline{Y}$ $\hat{\alpha}^* = \bar{Y}$	$\hat{\underline{\beta}}^* = (W'W)^{-1}W'\underline{Y}$ $\hat{\alpha}^* = \bar{Y}$
Expectation	$E[\hat{\underline{B}}] = \underline{B}$	$E[\hat{\underline{\beta}}] = \underline{\beta}$ $E[\hat{\alpha}^*] = \alpha^*$	$E[\hat{\underline{\beta}}^*] = \underline{\beta}^*$ $\beta_j^* = \beta_j S_j$ $E[\hat{\alpha}^*] = \alpha^*$	$E[\hat{\underline{\beta}}^*] = \underline{\beta}^*$ $\beta_j^* = \beta_j d_j$ $E[\hat{\alpha}^*] = \alpha^*$
Variability	$\Sigma_B = \sigma^2(X^{*'}X^*)^{-1}$	$\Sigma_\beta = \sigma^2(M'M)^{-1}$ $Var[\hat{\alpha}^*] = \sigma^2/n$ $Cov[\hat{\alpha}^*, \hat{\underline{\beta}}] = \underline{0}'$	$\Sigma_{\beta^*} = \sigma^2(Z'Z)^{-1}$ $Var[\hat{\alpha}^*] = \sigma^2/n$ $Cov[\hat{\alpha}^*, \hat{\underline{\beta}}^*] = \underline{0}'$	$\Sigma_{\beta^*} = \sigma^2(W'W)^{-1}$ $Var[\hat{\alpha}^*] = \sigma^2/n$ $Cov[\hat{\alpha}^*, \hat{\underline{\beta}}^*] = \underline{0}'$

†In the centered, normal deviate, and unit length transformations, $\alpha^* = \alpha + \sum_{j=1}^{p} x_j \beta_j$.

6.2.4 Probability Distributions

Statistical inference on a population parameter is generally most reliable when an estimator of the parameter is available whose probability distribution is known. For many statistical analyses, estimators must be used whose distributions are unknown. For these problems, researchers often use distribution-free (nonparametric) techniques or approximate the distribution of the estimator. The goal in all these situations is to draw inferences based on the probability of observing various values of the parameter estimators. Tests of hypotheses and the construction of confidence intervals require the calculation of such probabilities.

Assumption 4 provides the necessary distributional properties of least squares parameter estimators. It states that the error variables follow a normal probability distribution. This is a difficult assumption to guarantee and violations of this assumption may not be readily detectable. Fortunately, such violations often do not seriously affect the probability distributions of the least squares estimators nor inferences made from these probability distributions.* We will, nevertheless, be concerned with the detection of non-normal errors and will present a graphical procedure for doing so in Chapter 7. Throughout the remainder of this subsection we assume that Assumptions 1-4 of Table 6.2 hold for both single- and multiple-variable regression models under consideration.

Response variables generated by the single-variable model $Y = \alpha + \beta X + \varepsilon$ are themselves normally distributed under Assumptions 1-4. For a fixed value of the predictor variable, X, the mean of this normal distribution is $\alpha + \beta X$ while its variance is σ^2. The mean of the normal population of response variables is thus seen to depend on the value of the predictor variable as is illustrated schematically in Figure 6.4. In Figure 6.4(a), a three-dimensional graph of the density functions of the response variable is depicted with two values of $\alpha + \beta X$, the mean of the distribution. This graph stresses the point that each value of the predictor variable results in a different normal population of responses. In most regression models a single observation on the response is taken from n normal populations, one observation for each value of X. The least squares estimators, then, are attempting to fit the expected responses: $E[Y] = \alpha + \beta X$. Figure 6.4(b) is a side view of Figure 6.4(a) obtained by looking parallel to the X-axis. It again stresses the changes in the normal distributions of the responses for different values of the predictor variables.

---

*Comprehensive discussions of the normality assumption can be found in Cochran (1947) and Box (1953). A less theoretical commentary is available in Bohrnstedt and Carter (1971).

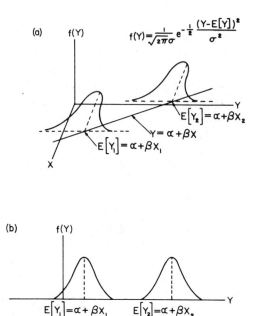

Figure 6.4   Schematic depictions of probability distributions
of the response variable for single variable models: (a) Three-
dimensional graph of a normal probability density,
(b) Side view.

Multiple linear regression models possess these same properties. The
multiple-variable model

$$\underline{Y} = \alpha \underline{1} + X\underline{\beta} + \underline{\varepsilon}$$

has expectation $E[\underline{Y}] = \underline{\mu} = \alpha\underline{1} + X\beta$, $\text{Var}[\underline{Y}] = \sigma^2 I$, and $Y_i \sim \text{NID}(\mu_i, \sigma^2)$
(see Appendix 6.C). Observe that each of the response variables are inde-
pendently distributed since the correlations of all pairs of response variables
are zero (since $\Sigma_Y = \sigma^2 I$).

A remarkable property of independent normal random variables is that
any linear transformation of the form $\underline{u} = A\underline{Y}$, where A is an $(r \times n)$ matrix
of constants with $r \leqslant n$, retains a normal probability distribution (see
Appendix 6.C). The variables in $\underline{u}$ are generally correlated so that $\underline{u}$ follows
what is referred to as a "multivariate normal" probability distribution.
Fortunately we need only concern ourselves with the distribution of individ-
ual elements of $\underline{u}$ or univariate (single-variable) transformations of $\underline{u}$ whose
probability distributions are also known.

The foregoing discussion is of relevance to least squares estimation since all the least squares estimators (except the beta weights) are linear transformations of the response vector, $\underline{Y}$. It then follows that each of the individual parameter estimates follows a normal probability distribution with mean and variance given by the appropriate quantities in Table 6.3. For example, using the raw predictor variables

$$\hat{\alpha} \sim N(\alpha, \sigma^2 C_{00}) \quad \text{and} \quad \hat{\beta}_j \sim N(\beta_j, \sigma^2 C_{jj}),$$

where $C = (X^{*\prime}X^*)^{-1}$ is notationally defined in eqn. (6.2.9).

In practice the normal probability distributions for individual least squares estimators often cannot be directly utilized since the error variance σ^2 is generally unknown. However, it can be shown that

$$\frac{\text{SSE}}{\sigma^2} \sim \chi^2\,(n-p-1)\,. \tag{6.2.10}$$

That is, $(n-p-1)\hat{\sigma}^2/\sigma^2 = \text{SSE}/\sigma^2$ has a chi-square probability distribution with $(n-p-1)$ degrees of freedom. Further, the distribution of SSE/σ^2 is independent of the normal distributions of the coefficient estimators. It then follows (see Appendix 6.C) that if $u_j \sim N(\mu_j, \sigma^2 C_{jj}^2)$,

$$\frac{u-\mu}{\hat{\sigma}C^{1/2}} \sim t\,(n-p-1)\,, \tag{6.2.11}$$

where $t\,(n-p-1)$ is a Student t random variable with $(n-p-1)$ degrees of freedom. Since all the least squares estimators listed in Table 6.3 follow normal probability distributions of the above type, all have t statistics similar to eqn. (6.2.11). Thus the least squares estimators using raw predictor variables have the following t distributions:

$$\frac{\hat{\alpha} - \alpha}{\hat{\sigma}C_{00}^{1/2}} \sim t(n-p-1) \quad \text{and} \quad \frac{\hat{\beta}_j - \beta_j}{\hat{\sigma}C_{jj}^{1/2}} \sim t(n-p-1)\,. \tag{6.2.12}$$

These t statistics will be employed in the next section to make inferences on model parameters. Tables for the normal, chi-square, and Student t probability distributions are provided in Appendix C.

6.3 TESTS OF HYPOTHESES

A wide range of test procedures are available for examining hypotheses about multiple linear regression models, only a few of which are covered in this section. Techniques that are useful for checking model assumptions are presented in the next chapter. Chapter 8 covers many of the popular variable selection procedures. In this section we introduce tests of hypotheses that are valuable in assessing the effect of individual predictor variables on the response and in evaluating the overall fit of the prediction equation. Several of the statistics discussed in this section are also used later in conjunction with the variable selection procedures.

6.3.1 Tests on Individual Parameters

One often desires to know whether an estimated regression coefficient differs from zero by a magnitude larger than can be reasonably attributed to sampling variability. Put another way, one desires to test the hypothesis that the true regression coefficient differs from zero. Of course, zero might not be the only parameter value of interest. Past analyses or other studies may have suggested that a regression coefficient should have a particular value and one might wish to confirm or refute this value by analyzing a new data set. All of these situations suggest that the investigator should perform a test of hypothesis on the true value of the regression coefficient.

Test procedures for (two-sided) tests of hypothesis on individual model parameters are outlined in Table 6.4. These tests all utilize the coefficient estimate from the original model and the probability distributions that were defined in the previous subsections. Suppose now that one wishes to test the hypothesis that the jth regression coefficient equals some specified constant β_0 (e.g., zero) against the alternative hypothesis that it is not equal to β_0:

$$H_0: \beta_j = \beta_0 \quad \text{vs.} \quad H_a: \beta_j \neq \beta_0 \, .$$

From Table 6.4, the precise test procedure depends on whether the error variance is known. If so, the following statistic is a standard normal random variable:

$$Z = \frac{\hat{\beta}_j - \beta_0}{\sigma C_{jj}^{1/2}} \, .$$

Once Z has been calculated, its value can be compared with critical values from normal tables as is indicated in the last column of Table 6.4.

If σ^2 is unknown, a t statistic can be formed using the least squares estimate of the error standard deviation, $\hat{\sigma} = (\text{MSE})^{1/2}$:

$$t = \frac{\hat{\beta}_j - \beta_0}{\hat{\sigma} C_{jj}^{1/2}} \, .$$

Table 6.4. Tests of Hypothesis for Individual Model Parameters in Multiple-Variable Regression Analyses; Significance Level = γ

Hypotheses	Test Statistic*	Reject H_0 if:
$H_0: \alpha = \alpha_0$ $\quad$ $H_a: \alpha \neq \alpha_0$ (σ^2 known)	$Z = \dfrac{\hat{\alpha} - \alpha_0}{\sigma C_{00}^{1/2}}$	$Z < -Z_{1-\gamma/2}$ $\quad$ or $\quad$ $Z_{1-\gamma/2} < Z$
$H_0: \beta_j = \beta_0$ $\quad$ $H_a: \beta_j \neq \beta_0$ (σ^2 known)	$Z = \dfrac{\hat{\beta}_j - \beta_0}{\sigma C_{jj}^{1/2}}$	$Z < -Z_{1-\gamma/2}$ $\quad$ or $\quad$ $Z_{1-\gamma/2} < Z$
$H_0: \sigma^2 = \sigma_0^2$ $\quad$ $H_a: \sigma^2 \neq \sigma_0^2$	$\chi^2 = (n-p-1)\hat{\sigma}^2/\sigma_0^2$ $= \text{SSE}/\sigma_0^2$	$\chi^2 < \chi_{\gamma/2}^2$ $\quad$ or $\quad$ $\chi_{1-\gamma/2}^2 < \chi^2$
$H_0: \alpha = \alpha_0$ $\quad$ $H_a: \alpha \neq \alpha_0$ (σ^2 unknown)	$t = \dfrac{\hat{\alpha} - \alpha_0}{\hat{\sigma} C_{00}^{1/2}}$	$t < -t_{1-\gamma/2}(n-p-1)$ $\quad$ or $\quad$ $t_{1-\gamma/2}(n-p-1) < t$
$H_0: \beta_j = \beta_0$ $\quad$ $H_a: \beta_j \neq \beta_0$ (σ^2 unknown)	$t = \dfrac{\hat{\beta}_j - \beta_0}{\hat{\sigma} C_{jj}^{1/2}}$	$t < -t_{1-\gamma/2}(n-p-1)$ $\quad$ or $\quad$ $t_{1-\gamma/2}(n-p-1) < t$

*In all these tests, $\underline{\hat{B}} = \begin{pmatrix} \hat{\alpha} \\ \underline{\hat{\beta}} \end{pmatrix} = (X^{*'}X^{*})^{-1}X^{*'}\underline{Y}$ $\quad$ and $\quad$ $C = (X^{*'}X^{*})^{-1}$.

The computed t value can be compared with critical values from tables of Student's t distribution for $(n-p-1)$ degrees of freedom.

Table 6.5 shows the simplification of formulas for single-variable regression models. It also outlines the modifications needed to perform one-sided tests of hypothesis. The only alteration in either Table 6.4 or 6.5 is that one compares the calculated value of the test statistic with *either* (but not both of) the upper or lower critical point of the appropriate probability distribution. (For an example of a one-sided test, see the introduction to Appendix C.)

None of the test statistics for the centered and standardized predictor variables are displayed in either Table 6.4 or 6.5. Test statistics utilizing centered, normal deviate, or unit length transformations of the predictor variables can readily be obtained from these tables by replacing the appropriate quantities in any of the test statistics by the corresponding quantities from Table 6.3. For example, in the unit length standardization a test of the hypothesis

$$H_0: \beta_j^* = \beta_0 \qquad \text{vs.} \qquad H_a: \beta_j^* \neq \beta_0$$

can be performed by comparing the value of

$$t = \frac{\hat{\beta}_j^* - \beta_0}{\hat{\sigma} q_{jj}^{1/2}}$$

to a critical point from a t table having $(n-p-1)$ degrees of freedom. In this formula, q_{jj} is the jth diagonal element of $Q = (W'W)^{-1}$.

In order to illustrate these tests of hypotheses, let us begin a closer examination of the police applicant data, Data Set A.6. Initial two-variable plots of each predictor variable with the response do not indicate any obvious nonlinear trends so no transformations of the response or predictor variables are performed. On the contrary, many of the plots exhibit almost no trends at all, suggesting that the data are highly variable and that a good fit may be difficult to obtain.

Standardized coefficient estimates using the unit length standardization are exhibited in Table 6.6 along with other summary information. The standardized coefficient estimates appear to indicate that FAT is the most influential determinant of the reaction time, followed by HEIGHT, PELVIC, RECVR, CHEST, and SPEED; however, we have not analyzed possible interrelationships among the predictor variables as yet so these conclusions can only be regarded as tentative.

An interesting facet of multiple-variable parameter estimation that is illustrated with these estimates is that, unlike single-variable estimation, the

Table 6.5. Tests of Hypotheses for Model Parameters in Single-Variable Regression Analyses; Significance Level = γ

Hypotheses	Test Statistic	Reject H_0 if:
$H_0: \alpha = \alpha_0$ $H_a: \alpha \neq \alpha_0$ (σ^2 known)	$Z = \dfrac{\hat{\alpha} - \alpha_0}{\sigma[n^{-1} + (\bar{X}/d_x)^2]^{1/2}}$	$Z < -Z_{1-\gamma/2}$ or $Z_{1-\gamma/2} < Z$
$H_0: \beta = \beta_0$ $H_a: \beta \neq \beta_0$ (σ^2 known)	$Z = \dfrac{\hat{\beta} - \beta_0}{\sigma/d_x}$	$Z < -Z_{1-\gamma/2}$ or $Z_{1-\gamma/2} < Z$
$H_0: \sigma^2 = \sigma_0^2$ $H_a: \sigma^2 \neq \sigma_0^2$	$\chi^2 = (n-2)\hat{\sigma}^2/\sigma_0^2$ $= SSE/\sigma_0^2$	$\chi^2 < \chi^2_{\gamma/2}$ or $\chi^2_{1-\gamma/2} < \chi^2$
$H_0: \alpha = \alpha_0$ $H_a: \alpha \neq \alpha_0$ (σ^2 unknown)	$t = \dfrac{\hat{\alpha} - \alpha_0}{\hat{\sigma}[n^{-1} + (\bar{X}/d_x)^2]^{1/2}}$	$t < -t_{1-\gamma/2}(n-2)$ or $t_{1-\gamma/2}(n-2) < t$
$H_0: \beta = \beta_0$ $H_a: \beta \neq \beta_0$ (σ^2 unknown)	$t = \dfrac{\hat{\beta} - \beta_0}{\hat{\sigma}/d_x}$	$t < -t_{1-\gamma/2}(n-2)$ or $t_{1-\gamma/2}(n-2) < t$

One-Sided Tests: Let θ be any of the above model parameters and W be the corresponding test statistic.

Hypothesis		Reject H_0 if:
$H_0: \theta \leq \theta_0$	$H_a: \theta > \theta_0$	$W > W_{1-\gamma}$
$H_0: \theta \geq \theta_0$	$H_a: \theta < \theta_0$	$W < W_{1-\gamma}$

Table 6.6. Summary Information on the Regression of Reaction Time on 14 Predictor Variables in Data Set A.6

Variable	Corr(X_j, Y)	Coefficient Estimates	Diagonal of $(W'W)^{-1}$	t Statistics	95% Confidence Interval
Intercept		.3162			
HEIGHT	.222	.1894	3.35	2.34	(.025, .354)
WEIGHT	.056	-.0852	17.06	-.47	(-.456, .286)
SHLDR	-.093	.0261	3.92	.30	(-.152, .204)
PELVIC	-.056	-.1664	3.31	-2.07	(-.330,-.003)
CHEST	-.032	-.1469	8.92	-1.11	(-.415, .121)
THIGH	.132	-.0933	6.30	-.84	(-.319, .132)
PULSE	.163	.0278	1.93	.45	(-.097, .153)
DIAST	.147	.0831	1.44	1.56	(-.025, .191)
CHNUP	-.159	.0461	3.48	.56	(-.122, .214)
BREATH	.160	.0497	2.07	.78	(-.080, .179)
RECVR	-.076	-.1504	2.30	-2.24	(-.287,-.014)
SPEED	-.149	-.1255	2.41	-1.83	(-.265, .014)
ENDUR	-.053	-.0140	1.40	-.27	(-.120, .092)
FAT	.165	.2998	14.63	1.77	(-.044, .644)

estimated coefficients do not necessarily have the same signs as the correlation coefficients between response and predictor variables. Coefficient estimates for WEIGHT, SHLDR, THIGH, and CHNUP differ in sign from their correlation coefficients with REACT. This is not necessarily surprising since the estimator for each coefficient is actually a function of the correlations between the response and each of the predictor variables.* Still, the changes in sign could be due to interrelationships among the predictor variables; i.e., multicollinearities could be distorting the signs on the coefficient estimates. We will return to this point in Chapter 9.

The individual t statistics for testing whether one of the standardized parameters is zero are shown in the fifth column of Table 6.6. Observe that testing whether $\beta_j^* = 0$ is the same as testing whether $\beta_j = 0$ since $d_j > 0$ implies that

$$\beta_j^* = 0 \quad \leftrightarrow \quad \beta_j d_j = 0 \quad \leftrightarrow \quad \beta_j = 0.$$

To construct the t statistics one needs $\hat{\beta}_j$, the jth diagonal element of $(W'W)^{-1}$, and MSE. Obtaining the mean squared error from the ANOVA

---

*The beta weights, eqn. (5.1.19), when using the unit length standardization, are a function of the vector of correlations between the response and predictor variables, $W'Y^0$. Since the original coefficient estimates and each of the various centered and standardized estimates can be written as transformations of the beta weights, each is a function of all the correlations between the response variable and the p predictor variables.

table, (see Table 6.7, Section 6.3.2), one finds the t statistic for the regression coefficient for HEIGHT to be

$$t = \frac{\hat{\beta}_1^*}{(MSE \cdot q_{11})^{1/2}} = \frac{.1894}{[(.00196)(3.35)]^{1/2}} = 2.34 \, .$$

This t statistic is significant (d.f. = 35, $.01 < p < .05$) and the hypothesis that $\beta_1 = 0$ is rejected. Other individual hypotheses can be judged to be significant or not based on the t statistics in Table 6.6, although one should not use these results to simultaneously test two or more hypotheses since the test statistics are all correlated.

The ability to perform statistical tests of hypotheses does not add credibility to an otherwise careless analysis of regression data. For example, the problems cited in Chapter 3 regarding the fitting of a single-variable model to the female suicide rate data are not overcome by testing statistical hypotheses. The estimated regression coefficient for fitting the female suicide rates over the age groups 25-29 to 50-54 is $\hat{\beta} = 0.23$ and the t statistic used to test H_0: $\beta = 0$ vs. H_a: $\beta \neq 0$ is $t = 30.10$. One could conclude from this test that $\beta \neq 0$ and from an examination of residuals, etc. feel confident that a correct, comprehensive analysis of the model confirmed its validity. This conclusion is defensible if one restricts it to the age groups analyzed; namely 25-29 to 50-54.

Carelessness or a lack of time could lead one to proceed directly to tests of hypothesis without examination of two-variable plots, residuals, R^2, and estimates of residual variability, σ. In doing so with the entire data set on female suicide rate one would find $\hat{\beta} = 0.10$ and $t = 3.63$, still a highly significant result. Such a scant analysis would let pass unnoticed the pronounced curvature seen in Figure 3.7. It should be apparent that regression data analysis must consist of much more than routine tests of hypotheses, but that hypothesis testing does play an important role in drawing inferences on regression models.

6.3.2   Analysis of Variance Tests

Under appropriate hypotheses which we will describe shortly, each of the sums of squares in the partitioning of TSS, when divided by σ^2, follow chi-square probability distributions. The degrees of freedom associated with each chi-square statistic are the same as those listed in the second column of Table 5.4; i.e.,

$$SSM/\sigma^2 \sim \chi^2(1) \, , \qquad SSR/\sigma^2 \sim \chi^2(p) \, ,$$

and

$$SSE/\sigma^2 \sim \chi^2(n-p-1).$$

It is also true that SSM is distributed independently of SSR, SSM and SSE are independent, and SSR and SSE are independent. Consequently, under appropriate hypotheses, the ratio of any two of the mean squares corresponding to these sums of squares follows an F probability distribution.

Two hypotheses that can be tested directly from the ANOVA table are

$$H_0: \alpha^* = 0 \qquad \text{vs.} \qquad H_a: \alpha^* \neq 0$$

and

$$H_0: \underline{\beta} = \underline{0} \qquad \text{vs.} \qquad H_a: \underline{\beta} \neq \underline{0},$$

with the test statistics being the two F statistics (i.e., MSM/MSE and MSR/MSE, respectively) that were shown in Table 5.4. Note that the latter hypothesis asserts that the regression coefficients on the predictor variables are all zero and, therefore, that the predictor variables are of no value in predicting the response. Unlike the individual t tests discussed in the last subsection, MSR/MSE allows one to simultaneously test that all $\beta_j = 0$ against the alternative that at least one β_j is not zero.

The hypothesis $H_0: \alpha^* = 0$ is actually

$$H_0: \alpha + \beta_1 \overline{X}_1 + \beta_2 \overline{X}_2 + \ldots + \beta_p \overline{X}_p = 0.$$

It is difficult to interpret this hypothesis because there are many reasons why H_0 could be true: (i) all the regression coefficients could be zero, (ii) the last $p - 1$ β_j could be zero and $\alpha = -\beta_1 \overline{X}_1$, etc. In addition, the hypothesis is affected by the predictor variables through the X_j. For these reasons, $H_0: \alpha^* = 0$ is frequently left untested in a regression analysis, although $H_0: \alpha = 0$ is often of interest and can be tested with the statistics shown in Table 6.4.

If one desires to test $H_0: \underline{\beta} = \underline{\beta}_0$, where $\underline{\beta}_0$ is some nonzero vector of specified constants, $F = \text{MSR/MSE}$ cannot be used. Instead, one can form the F statistic

$$F = \text{MSR}(\underline{\beta}_0)/\text{MSE} \qquad\qquad (6.3.1)$$

where

$$SSR(\underline{\beta}_0) = (\hat{\underline{\beta}} - \underline{\beta}_0)' M'M(\hat{\underline{\beta}} - \underline{\beta}_0)$$

and

$$MSR(\underline{\beta}_0) = SSR(\underline{\beta}_0)/p.$$

This F statistic, like the one in an ANOVA table, has degrees of freedom $\nu_1 = p$ and $\nu_2 = (n-p-1)$. Observe that if $\beta_0 = \underline{0}$, the test statistic (6.3.1) reduces to the F statistic in the ANOVA table; viz, MSR/MSE.

Let us now return to the police-applicant data for an examination of the overall significance of the regression coefficients. The ANOVA table associated with this data set is shown in Table 6.7. Note that the F statistic for testing whether $\alpha^* = 0$ turns out to be extremely large. This is common but it again sheds no light as to the reason for its large value.

Table 6.7. Analysis of Variance Table for the Reaction Times of White Male Police Applicants, Data Set A.6

ANOVA					
Source	d.f.	S.S.	M.S.	F	R^2
Mean	1	4.99912	4.99912	2549.282	.397
Regression	14	.04527	.00323	1.649	
Error	35	.06863	.00196		
Total	50	5.11303			

The F statistic for testing whether $\beta = \underline{0}$ is 1.65. The significance probability for this test is about $p = 0.10$. Should the hypothesis be rejected; i.e., should one conclude that at least one of the β_j is not zero? We prefer to remain conservative and retain predictor variables if there is a reasonable chance that they have predictive value. Accordingly, we choose as a cutoff significance probability for this test and for any individual t tests of value of 0.25. If the significance probability is less than 0.25 we will conclude that $\beta \neq \underline{0}$ or $\beta_j \neq 0$, depending on which type of test we are making; otherwise, we will not reject the hypothesis of zero regression coefficients.

One disheartening feature of this data set is that the coefficient of determination is so low. Accounting for only 40% of the adjusted response variability limits the conclusions that can be drawn from this regression analysis. One must seriously question whether all the relevant predictor variables have been included in the initial model. If the main goal of the investigation is model specification or parameter estimation, any conclusions drawn may be in error. On the other hand, if one merely desires to be able to predict reaction times based on these anthropometric and fitness variables, the estimated standard deviation of 0.044 may indicate that prediction will be accurate enough. Interval estimation (see Section 6.4) provides one gauge of whether the prediction equation is sufficiently accurate for the purposes intended. Finally, if one's intent is to assess the relative influence of *only* the predictor variables in Data Set A.6 and no interest is expressed in whether other variables may influence the reaction times, analysis of this data can elicit correct conclusions.

6.3.3 Repeated Predictor Variable Values (Lack of Fit Test)

Correct model specification is always of great concern to data analysts. The visual assessment of two-variable plots discussed in Chapter 2 are valuable for detecting misspecified models. A very important analytic technique (historically referred to as a "lack of fit" test) for assessing model inadequacies can be performed if two or more responses are recorded for each of several predictor variable values. We illustrate the procedure for single-variable regression models but the analysis is generalizable to multiple regression models for which *rows* of X are repeated more than once.

Repeated predictor variable values occur in Table 6.8 for the grade 13 averages from high school 1 in Data Set A.2. The repeated values arise from the original observations in Data Set A.2 by rounding the grade 13 averages to the nearest whole number—a common practice. Two responses are now available for predictor values 61, 66, and 75 and three responses appear for values 69 and 72. These 12 responses are the basis of the specification test we now derive.

Table 6.8. Rounded Grade 13 Averages
for High School 1

Grade 13 Average	First Year Average	Grade 13 Average	First Year Average
61	49.6	72	64.8
61	54.6	72	60.8
62	47.4	74	67.8
66	51.8	75	69.4
66	51.2	75	68.6
69	60.2	76	65.8
69	64.2	77	70.0
69	57.4	78	73.4
72	50.0	82	73.0

Source: Ruben and Stroud (1977). (See Appendix, Data Set A.2.)

Let $Y_1, Y_2, ..., Y_{n_1}$, denote responses that are recorded for the same value of the predictor variable and let X_1 denote the common value of the predictor variable for all n_1 responses. For the data in Table 6.8, $Y_1 = 49.6$ and $Y_2 = 54.6$ for the two responses corresponding to $X_1 = 61$. All the responses for a

particular value of the predictor variable are presumed to be generated by
the model

$$Y_i = \alpha + \beta X_{i1} + \varepsilon \qquad i = 1, 2, \ldots, n_1 \ . \qquad (6.3.2)$$

Under Assumptions 1-4 of Table 6.2, one can equivalently write that $Y_i \sim$
NID(μ, σ^2), $i = 1, 2, \ldots, n_1$, where the population mean μ is $\alpha + \beta X_1$. An
unbiased estimator of σ^2 can be calculated from these n_1 responses:

$$\hat{\sigma}_1^2 = \sum_{i=1}^{n_1} \frac{(Y_i - \overline{Y}_1)^2}{n_1 - 1} \ , \qquad (6.3.3)$$

where $\overline{Y}_1$ is the average of the n_1 responses. The estimator, $\hat{\sigma}_1^2$, is the familiar sample variance of the responses.

Now suppose the model is incorrectly specified in that it is not simply a
linear function of X. Perhaps the true relationship is quadratic or logarithmic or, in general,

$$Y_i = f(X_1) + \varepsilon_i \qquad i = 1, 2, \ldots, n_1 \qquad (6.3.4)$$

where $f(X)$ is an unknown function of the predictor variable X. So long as Y
is only influenced by X, *regardless of the true functional relationship
between Y and X*, $\hat{\sigma}_1^2$ is still an unbiased estimator of σ^2 . This unbiasedness
is due to the fact that $Y_i \sim$ NID(μ, σ^2) as before, only now $\mu = f(X_1)$. Unbiasedness is no longer a property of $\hat{\sigma}^2 = $ MSE if eqn. (6.3.2) is not the correct
model specification; i.e., if Assumption 1 is violated, MSE is a biased estimator of σ^2.

If a data set consists of k values of the predictor variable for which there
are at least two responses, there are k unbiased estimators of σ^2 similar to
eqn. (6.3.3). Denote these estimators by $\hat{\sigma}_1^2, \hat{\sigma}_2^2, \ldots, \hat{\sigma}_k^2$. Since $E[\hat{\sigma}_j^2] = \sigma^2$,

$$E[(n_j - 1)\hat{\sigma}_j^2] = (n_j - 1)\sigma^2$$

and

$$E\left[\sum_{j=1}^{k} (n_j - 1)\hat{\sigma}_j^2 \right] = \sigma^2 \sum_{j=1}^{k} (n_j - 1). \qquad (6.3.5)$$

From eqn. (6.3.5), a combined estimator of σ^2, denoted by $\tilde{\sigma}^2$, using all
the $\hat{\sigma}_j^2$ is

$$\tilde{\sigma}^2 = \frac{\sum\limits_{j=1}^{k} (n_j - 1)\hat{\sigma}_j^2}{\sum\limits_{j=1}^{k} (n_j - 1)}. \tag{6.3.6}$$

The computation of $\tilde{\sigma}^2$ for the repeated predictor variable values listed in Table 6.8 is summarized in Table 6.9. Individual $\hat{\sigma}_j^2$ estimates of σ^2 are highly variable due to the extremely small number of degrees of freedom for each (1 or 2 d.f.). The combined estimate is based on 7 degrees of freedom and is much more accurate than any of the individual estimates. The combined estimate $\tilde{\sigma}^2 = 21.94$ compares favorably with the least squares estimate, $\hat{\sigma}^2 = 19.23$, from an ANOVA table.

Table 6.9. Combined Estimate of σ^2 from Repeated Values in Table 6.8

j	X_j	Y_i	$n_j - 1$	$\hat{\sigma}_j^2$
1	61	49.6, 54.6	1	12.50
2	66	51.8, 51.2	1	.18
3	69	60.2, 64.2, 57.4	2	11.68
4	72	50.0, 64.8, 60.8	2	58.62
5	75	69.4, 68.6	1	.32
			7	$\tilde{\sigma}^2 = 21.94$

This comparison of $\tilde{\sigma}^2$ with $\hat{\sigma}^2$ is, however, misleading. Since SSM, SSR, and SSE constitute an exact additive partitioning of the response variability, $\tilde{\sigma}^2$ cannot be independent of all three statistics. Actually, $\sum (n_j - 1)\hat{\sigma}_j^2$ is a portion of SSE. This can be appreciated by recalling that SSM measures whether $E[\overline{Y}] = \alpha + \beta \overline{X}$ is zero and SSR measures whether β is zero. Both MSE and $\hat{\sigma}_j^2$ are estimators of the error variance: MSE is unbiased if $E[Y_i] = \alpha + \beta X_i$; $\hat{\sigma}_j^2$ is unbiased if $E[Y_i] = f(X_i)$.

A measure of the lack of fit, or misspecification of $f(X)$, can be obtained by partitioning SSE as

$$SSE = SSE_M + SSE_P$$

where $SSE_P = \sum (n_j - 1) \hat{\sigma}_j^2$ is referred to as the "pure" sum of square due to error since it is unaffected by the true functional form of $f(X)$. The misspecification sum of squares, SSE_M, is obtained as the difference between SSE and SSE_P. Since SSE is affected by misspecification, the effect is entirely contained in SSE_M. Also, if $f(X) = \alpha + \beta X$, MSE_M and MSE_P both estimate σ^2; if $f(X) \neq \alpha + \beta X$, MSE_M will tend to overestimate σ^2 while

MSE_P remains unbiased. A test of hypothesis that the model is correctly specified is provided by MSE_M/MSE_P, which is an F statistic under this hypothesis.

The ANOVA table can be adapted to indicate the partitioning of SSE into misspecification and pure-error components. Each line of the partitioning is indented to stress that the components are a part of the overall residual error and that under the hypothesis of no misspecification both mean squares estimate the same parameter as the overall mean square due to error, they both estimate σ^2. The partitioned ANOVA table for the above data is:

ANOVA					
Source	d.f.	S.S.	M.S.	F	R^2
Mean	1	67,222.22	67,222.22	3,495.01	.758
High School Score	1	964.28	964.28	50.13	
Error	16	307.74	19.23		
Misspecification	9	154.15	17.13	.78	
Pure	7	153.59	21.94		
Total	18	68,494.24			

The two error mean squares, MSE_M and MSE_P, show a seemingly surprising result. One might expect that MSE_M would always be larger than MSE_P. If the model is misspecified this is generally true. If the model is correctly specified both mean squares estimate σ^2 and either could be larger than the other. The small number of degrees of freedom for each (9 and 7) indicate that the mean squares could be quite variable. The two mean squares are not inconsistent with a correct specification for X; in particular, the F statistic of 0.78 is not significant at any reasonable significance level.

Had the specification test proven nonsignificant, the subsequent tests for $\alpha^* = 0$ and $\beta = 0$ would have been made with MSE_P in the denominator of the F statistic rather than MSE, provided that there are a sufficient number of degrees of freedom for MSE_P. This is, of course, because the lack of fit test may not be sensitive enough to detect some types of misspecification but MSE_P is still a valid estimator of σ^2 in such cases.

The lack of evidence of misspecification in the above test and the few degrees of freedom associated with MSE_P leads us to recommend using MSE rather than MSE_P in the remaining F statistics in this ANOVA table. By doing so, we recognize that MSE could be a biased estimator of σ^2, but there is no strong evidence to suggest that the model is misspecified. The larger number of degrees of freedom (16 vs. 7) compensate for the variability of MSE_P and constitute our main reason for preferring MSE.

Two final comments should be made regarding the specification test proposed in this section. First, if the model is misspecified due to the exclusion of important predictor variables this test may not detect such model errors. Misspecification of this type biases both MSE_M and MSE_P since neither eqn. (6.3.2) nor eqn. (6.3.4) are the correct models. The second point we wish to make is that the experimenter should, if possible, plan the investigation so that repeated predictor variable values occur in the data base. If this is not possible, nearby points* can be used to perform an approximate specification test or perhaps a slight rounding of predictor variable values (like Table 6.8) can accomplish the same function. One must be somewhat cautious, however, not to round too much nor to force repeated values in the data set by grouping discrepant predictor values lest the misspecification test be invalidated.

6.4 INTERVAL ESTIMATION

Throughout much of the discussion in this book we have alluded to concerns for the accuracy of least squares estimators. Confidence intervals for model parameters incorporate the standard deviations of the estimators to provide a measure of their accuracy. In this way they provide more information about the estimates than does just the numerical value of a test statistic. Both confidence intervals and test statistics are derived from the same probability distributions, so both provide essentially the same information about the model parameters. The explicit reference of confidence intervals to the accuracy of the estimators through the bounds, however, frequently allow the confidence intervals to be more informative to the data analyst.

6.4.1   Confidence Intervals and Regions

A confidence interval for a parameter θ is an interval

$$a < \theta < b,$$

where a and b are numbers calculated partially from sample data, within which we feel reasonably certain the unknown parameter θ lies. A confidence interval is derived from a probability statement that involves the unknown parameter θ and, hopefully, no other unknown parameters. An ample will clarify the concept.

---

*Daniel and Wood (1971, Section 7.5) expand on this notion and use the residuals from the fitted model along with points having similar predictor variable values to obtain a "near-neighbor" estimate of σ.

If one wishes to find a confidence interval for β_j and be, say, 95% confident that the numerical bounds on the resulting interval will include the true value of β_j, one begins with a probability statement such as:

$$Pr(- Z_{.975} \leqslant Z \leqslant Z_{.975}) = 0.95 \qquad (6.4.1)$$

where $Z_{.975}$ is the percentage point of a standard normal distribution corresponding to a probability of 0.975. The standard normal random variable Z associated with $\hat{\beta}_j$ is

$$Z = \frac{\hat{\beta}_j - \beta_j}{\sigma_\beta} \qquad \text{where} \quad \sigma_\beta^2 = \sigma^2 C_{jj}.$$

In order to employ eqn. (6.4.1), σ_β^2 reveals that the error variance, σ^2, must be known; if not, one would begin with a probability statement based on a t statistic, e.g., eqn. (6.2.12). Now by isolating β_j between the inequalities in eqn. (6.4.1), the probability statement becomes

$$Pr(\hat{\beta}_j - Z_{.975}\sigma_\beta \leqslant \beta_j \leqslant \hat{\beta}_j + Z_{.975}\sigma_\beta) = 0.95. \qquad (6.4.2)$$

The probability statement (6.4.1) involves the random variable $\hat{\beta}_j$ through Z and indicates that the probability that the random variable Z takes on values between $-Z_{.975}$ and $+Z_{.975}$ is 0.95. The interpretation of eqn. (6.4.2) is slightly altered. Here it is the bounds $\hat{\beta}_j - Z_{.975}\sigma_\beta$ and $\hat{\beta}_j + Z_{.975}\sigma_\beta$ that are random; hence, eqn. (6.4.2) asserts that these random bounds will include β_j with a probability of 0.95. Once numerical values are assigned to $\hat{\beta}_j$, $Z_{.975}$, and σ_β, the resulting interval is referred to as a confidence interval and we conclude that we are 95% confident that the bounds do include the true value of β_j.

The center of the confidence interval (6.4.2) is $\hat{\beta}_j$. The length of the interval is $(\hat{\beta}_j + Z_{.975}\sigma_\beta) - (\hat{\beta}_j - Z_{.975}\sigma_\beta) = 2Z_{.975}\sigma_\beta$. The length depends on the standard deviation of $\hat{\beta}_j$, σ_β. If σ_β is large, the confidence interval will be wide, implying that there is a wide range possible of values β_j in the interval. This in turn indicates that $\hat{\beta}_j$, since it is only one point in a wide confidence interval, should not be considered a precise estimate of β_j. On the other hand, if the interval is narrow, all the points in the interval will be close to the center, $\hat{\beta}_j$, and the least squares estimate can be considered precise. Confidence intervals are very powerful methods for assessing the accuracy of least squares estimates.

Confidence intervals for model parameters using raw predictor variables are shown in Table 6.10. These intervals are valid inference procedures only if Assumptions 1-4 (or 1, 3', and 4') are correct for the assumed model. By

**Table 6.10. Confidence Intervals for Individual Model Parameters in
Multiple-Variable Regression Analyses; Confidence
Coefficient = $100(1-\gamma)\%$**

Parameter	Confidence Interval
α (σ^2 known)	$\hat{\alpha} - Z_{1-\gamma/2}\sigma C_{00}^{1/2} \leqslant \alpha \leqslant \hat{\alpha} + Z_{1-\gamma/2}\sigma C_{00}^{1/2}$
β_j (σ^2 known)	$\hat{\beta}_j - Z_{1-\gamma/2}\sigma C_{jj}^{1/2} \leqslant \beta_j \leqslant \hat{\beta}_j + Z_{1-\gamma/2}\sigma C_{jj}^{1/2}$
σ^2	$SSE/\chi_{1-\gamma/2}^2(n-p-1) \leqslant \sigma^2 \leqslant SSE/\chi_{\gamma/2}^2(n-p-1)$
α (σ^2 unknown)	$\hat{\alpha} - t_{1-\gamma/2}(n-p-1) \cdot \hat{\sigma} C_{00}^{1/2} \leqslant \alpha \leqslant \hat{\alpha} + t_{1-\gamma/2}(n-p-1) \cdot \hat{\sigma} C_{00}^{1/2}$
β_j (σ^2 unknown)	$\hat{\beta}_j - t_{1-\gamma/2}(n-p-1) \cdot \hat{\sigma} C_{jj}^{1/2} \leqslant \beta_j \leqslant \hat{\beta}_j + t_{1-\gamma/2}(n-p-1) \cdot \hat{\sigma} C_{jj}^{1/2}$

inserting standardized coefficient estimators and their corresponding standard deviations, these same formulas can be used for centered, normal deviate, and unit length transformations of the predictor variables.

Simultaneous confidence intervals for the regression coefficients cannot be computed using the individual intervals in Table 6.10. The confidence coefficient is correct only if a confidence interval for a single parameter is desired since each interval is constructed from the probability distribution of one of the parameter estimates and not from the joint probability distribution of all the estimators. Any simultaneous confidence interval procedures must make use of the joint probability distribution of the estimators in order for the probabilities (and hence confidence coefficients) to be correct.

Fortunately, there is a statistic that will enable simultaneous confidence regions for two or more model parameters to be calculated. Similar to eqn. (6.3.1), define

$$SSR(\underline{\beta}) = (\hat{\underline{\beta}} - \underline{\beta})'M'M(\hat{\underline{\beta}} - \underline{\beta}).$$

If $\underline{\beta}$ is the true regression coefficient vector,

$$SSR(\underline{\beta})/\sigma^2 \sim \chi^2(p) .$$

It then follows that

$$Pr[SSR(\underline{\beta})/\sigma^2 \leqslant \chi_{1-\gamma}^2(p)] = 1 - \gamma$$

so that a $100(1-\gamma)\%$ confidence region for $\underline{\beta}$ includes all values of $\underline{\beta}$ for which

$$(\hat{\underline{\beta}} - \underline{\beta})'M'M \, (\hat{\underline{\beta}} - \underline{\beta}) \leqslant \chi^2_{1-\gamma}(p) \cdot \sigma^2 \,. \tag{6.4.3}$$

If σ^2 is unknown, $MSR(\beta)/MSE \sim F(p, n-p-1)$ and a $100(1-\gamma)\%$ confidence region for $\underline{\beta}$ contains all p-dimensional vectors of regression coefficients, $\underline{\beta}$, which satisfy

$$(\hat{\underline{\beta}} - \underline{\beta})'M'M(\hat{\underline{\beta}} - \underline{\beta}) \leqslant p\text{MSE} \cdot F_{1-\gamma}(p, n-p-1) \,. \tag{6.4.4}$$

In general the regions formed by all values of β satisfying eqns. (6.4.3) or (6.4.4) are ellipsoids, generalizations of a two-dimensional ellipse. If p separate confidence intervals similar to those in Table 6.10 are erroneously used to construct a simultaneous confidence region for β, many points that are not contained in the regions (6.4.3) or (6.4.4) will be included in the separate intervals. While the correct regions are more difficult to use, one's certitude about the correctness of the confidence coefficient may make the extra effort worthwhile. Particularly if one is only concerned with whether several specific choices of β are contained in the confidence regions (such as estimates of β from other research studies), the computations may not be too laborious. If one has no idea of the values of β that are included in the region, use of the individual intervals in Table 6.10 can provide a rough bound to the region. One should not, however, consider these bounds as the true limits to a $100(1-\gamma)\%$ confidence region for $\underline{\beta}$.

Confidence intervals also provide a means of performing a test of hypothesis. For example, suppose one wishes to test H_0: $\beta_j = \beta_0$ vs. H_a: $\beta_j \neq \beta_0$, where β_0 is some specified constant, often zero. A confidence interval can be found for β_j using either of the two expressions in Table 6.10, depending on whether σ^2 is known. Once a confidence interval for β_j is calculated from a data base, any value of β_j included in the interval can be considered concordant with the data base since the interval is constructed so that the bounds will include β_j with a probability of $1 - \gamma$. If γ is selected small enough, any value of β_j not included in the confidence interval is judged to be discordant with the data base. Consequently, if β_0 is not included in the confidence interval the null hypothesis is rejected and the alternate is accepted. This generalizes to situations wherein the null hypothesis includes a range of parameter values: if any of the values of the null hypothesis are excluded from the confidence interval the null hypothesis is rejected and the alternate hypothesis is accepted.

Separate 95% confidence intervals for individual regression coefficients for the police-applicant data are given in the last column of Table 6.6 (see

Section 6.3.1). Eleven of the fourteen intervals contain zero, indicating that if one of these eleven coefficients were tested at the 5% significance level the hypothesis $\beta_j = 0$ would not be rejected. If γ was raised to 0.25 some of these intervals would no longer include zero since the intervals would become smaller; nevertheless, it appears that several of the predictor variables are not contributing substantially to the prediction of the response. It is also important to note that many of these individual confidence intervals appear to be wide relative to the numerical values of the corresponding coefficient estimates. This suggests that the standard errors of the estimators are large, at least partially because the error variance estimate is large. Recall that $R^2 = 0.40$, suggesting that the responses might be poorly fit; i.e., that important predictor variables are not accounted for and, hence, their variability is still included in SSE (and thus in $\hat{\sigma}^2$).

Before leaving this example an additional comment on misspecification can be made. The total adjusted sum of squares

$$\text{TSS(adj)} = \Sigma Y_i^2 - n\bar{Y}^2$$

does not involve the predictor variables, only the response variable. Adding predictor variables to or deleting them from the prediction equation does not alter the value of TSS(adj). Now suppose one or more important predictor variables were added to the model so that SSR and $R^2 = \text{SSR}/\text{TSS(adj)}$ increased dramatically. Since the total sum of squares is fixed and TSS(adj) = SSR + SSE, a dramatic increase in SSR is accompanied by an equivalent decrease in SSE. If changes in the degrees of freedom for error ($n-p-1$-the number of new predictor variables) do not overcompensate for the drop in SSE, $\hat{\sigma}^2$ = MSE should be reduced as should the width of the confidence intervals in Table 6.6. Thus, with the addition of relevant predictor variables to the prediction equation, individual tests in Table 6.10 could very well be altered so that some of the intervals that include zero would not do so. These comments again point out the effect that misspecification can have on model inferences: conclusions about the importance of individual predictor variables depend on how well the prediction equation is fit.

6.4.2   Response Intervals

Just as confidence intervals are important measures of the accuracy of parameter estimates, they are also valuable in assessing how precisely one can predict. Two types of intervals related to the response variable can be derived similarly to the confidence intervals for α and β_j: confidence intervals on expected responses and prediction intervals for future responses.

Suppose now one wishes to estimate the expected value of the response variable for a specified set of values of the predictor variables. Let u_1, u_2, ..., u_p be the predictor variable values for which the expected response is desired. Then since $\hat{\underline{B}}$ is an unbiased estimator of $\underline{B}$ (i.e., $\hat{\alpha}$, $\hat{\beta}_1$, ..., $\hat{\beta}_p$ are unbiased estimators of the corresponding parameters),

$$\hat{Y} = \hat{\alpha} + \hat{\beta}_1 u_1 + \hat{\beta}_2 u_2 + ... + \hat{\beta}_p u_p$$

is an unbiased estimator of the expected response

$$E[Y] = \alpha + \beta_1 u_1 + \beta_2 u_2 + ... + \beta_p u_p .$$

In order to obtain an interval estimator of $E[Y]$, let $\underline{u}^{*\prime} = (1, u_1, u_2, ..., u_p)$. We can then write

$$\hat{Y} = \underline{u}^{*\prime} \hat{\underline{B}} \quad , \qquad E[Y] = \underline{u}^{*\prime} \underline{B}$$

and from the results given in Appendix 6.C, it follows that

$$E[\hat{Y}] = \underline{u}^{*\prime} E[\hat{\underline{B}}] = \underline{u}^{*\prime} \underline{B} = E[Y]$$

$$\text{Var}[\hat{Y}] = \underline{u}^{*\prime} \text{Var}[\hat{\underline{B}}] \underline{u}^* = \sigma^2 \underline{u}^{*\prime} (X^{*\prime} X^*)^{-1} \underline{u}^* .$$

Predicted responses are normally distributed under Assumptions 1-4 so that

$$\hat{Y} \sim N(E[Y], \text{Var}[\hat{Y}])$$

and a $100(1-\gamma)\%$ confidence interval for $E[Y]$ when σ^2 is known is

$$\hat{Y} - Z_{1-\gamma/2}(\text{Var}[\hat{Y}])^{1/2} \leqslant E[Y] \leqslant \hat{Y} + Z_{1-\gamma/2}(\text{Var}[\hat{Y}])^{1/2} . \qquad (6.4.5)$$

If σ^2 is unknown, $Z_{1-\gamma/2}$ and σ would be replaced in eqn. (6.4.5) by $t_{1-\gamma/2}(n-p-1)$ and $(\text{MSE})^{1/2}$, respectively.

To obtain a $100(1-\gamma)\%$ prediction interval for a future observation, Y, on the response note that

$$Y \sim N(E[Y], \sigma^2)$$

$$Y \sim N(E[Y], \sigma^2 \underline{u}^{*\prime} (X^{*\prime} X^*)^{-1} \underline{u}^*)$$

and that both Y and $\hat{Y}$ are independently distributed. The distribution of $Y - \hat{Y}$ is thus normally distributed with mean zero and variance $\sigma^2(1 + \underline{u}^{*\prime}(X^{*\prime}X^*)^{-1}\underline{u}^*)$. Following the above development, a $100(1-\gamma)\%$ prediction interval for a future response is

$$\hat{Y} - Z_{1-\gamma/2}(\text{Var}[Y - \hat{Y}])^{1/2} \leqslant Y \leqslant \hat{Y} + Z_{1-\gamma/2}(\text{Var}[Y - \hat{Y}])^{1/2} . \qquad (6.4.6)$$

If σ^2 is unknown the same modifications that were mentioned for eqn. (6.4.5) would be made in eqn. (6.4.6).

Confidence intervals for an expected response and prediction intervals for a future response can also be constructed using centered, normal deviate, or unit length transformations of the predictor variables. As was detailed in Section 5.1.4, all these predictor variable modifications yield the same predicted response, only the appearances of eqns. (6.4.5) and (6.4.6) are changed. For example, with centered predictor variables

$$\hat{Y} = \overline{Y} + \underline{u}'_m \hat{\beta},$$

where $\underline{u}'_m = (u_1 - \overline{X}_1, u_2 - \overline{X}_2, ..., u_p - \overline{X}_p)$. Since

$$E[\hat{Y}] = (\alpha + \sum_{j=1}^{p} \beta_j \overline{X}_j) + \underline{u}'_m \underline{\beta}$$

$$= \alpha + \sum_{j=1}^{p} u_j \beta_j$$

$$= \underline{u}^{*'} \underline{\beta}$$

and

$$\text{Var}[\hat{Y}] = \sigma^2/n + \sigma^2 \underline{u}'_m (M'M)^{-1} \underline{u}_m,$$

eqn. (6.4.5) becomes

$$\hat{Y} - Z_{1-\gamma/2}\sigma[n^{-1} + \underline{u}'_m(M'M)^{-1}\underline{u}_m]^{1/2} \leqslant E[Y] \leqslant \hat{Y}$$

$$+ Z_{1-\gamma/2}\sigma[n^{-1} + \underline{u}'_m(M'M)^{-1}\underline{u}_m]^{1/2}.$$

Despite their different appearances, this interval and eqn. (6.4.5) yield identical numerical bounds for $E[Y]$.

For single-variable prediction equations, confidence intervals for $E[Y] = \alpha + \beta X$ become

$$(\hat{\alpha} + \hat{\beta}u) - Z_{1-\gamma/2}\sigma[n^{-1} + (u - \overline{X})^2/d_x^2]^{1/2} \leqslant \alpha + \beta u \leqslant$$

$$(\hat{\alpha} + \hat{\beta}u) + Z_{1-\gamma/2}\sigma[n^{-1} + (u - \overline{X})^2/d_x^2]^{1/2} \tag{6.4.7}$$

if σ^2 is known and

$$(\hat{\alpha} + \hat{\beta}u) - t_{1-\gamma/2}(n-2) \cdot \hat{\sigma}[n^{-1} + (u - \bar{X})^2/d_x^2]^{1/2} \leqslant \alpha + \beta u \leqslant$$
$$(\hat{\alpha} + \hat{\beta}u) + t_{1-\gamma/2}(n-2) \cdot \hat{\sigma}[n^{-1} + (u - \bar{X})^2/d_x^2]^{1/2}$$

(6.4.8)

if σ^2 is unknown.

A 95% confidence interval for expected tuition and fee costs for the data in Table 3.1 at year u is calculated as follows:

$$\hat{Y} = 393.80 + 9.74u$$

$$t_{.975}(3)\hat{\sigma} = (3.182)(9.86) = 31.37$$

and so

$$(393.80 + 9.74u) - 31.37[0.20 + (u - 5.40)^2/53.20]^{1/2} \leqslant \alpha + \beta u$$

$$\leqslant (393.80 + 9.74u) + 31.37[0.20 + (u - 5.40)^2/53.20]^{1/2}.$$

Table 3.1 only includes observed responses for years 1, 3, 5, 8, 10. Using this confidence interval, a 95% confidence interval can be constructed for any of the missing years:

Year	$\hat{Y}$	Confidence Interval for $E[Y]$
2	413.28	$393.02 \leqslant \alpha + 2\beta \leqslant 433.54$
4	432.76	$417.49 \leqslant \alpha + 4\beta \leqslant 448.03$
6	452.24	$437.98 \leqslant \alpha + 6\beta \leqslant 466.50$
7	461.98	$446.35 \leqslant \alpha + 7\beta \leqslant 477.61$
9	481.46	$460.57 \leqslant \alpha + 9\beta \leqslant 502.35$

The behavior of the confidence interval (6.4.8) as a function of u is illustrated in Figure 6.5 for the tuition data. The value of u that results in the shortest confidence interval is $u = \bar{X}$. From eqn. (6.4.8), when $u = \bar{X}$, $(u - \bar{X})^2/d_x^2 = 0$. All other values of u cause $(u - \bar{X}^2)/d_x^2$ to be a positive value and lengthens the confidence interval. Values of u equidistant above and below $u = \bar{X}$ produce the same values of $(u - \bar{X})^2/d_x^2$ and, hence, the same width for the respective confidence intervals. Figure 6.5 also shows the symmetry in the confidence interval widths for points on either side of $\bar{X}$. The closer u is to $\bar{X}$, the wider the confidence intervals. Estimation of $\alpha + \beta u$ near the extremes of the observed predictor variable values, much worse

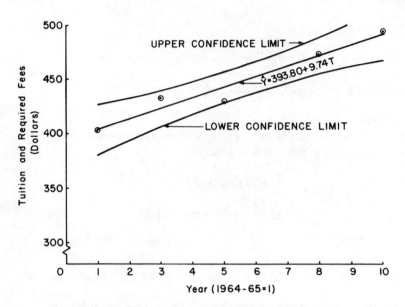

Figure 6.5 Confidence limits (95%) on $E[Y]$ for educational expenditure data.

extrapolation beyond the extreme values, can be highly inadequate if the curvature in the upper and lower limits is pronounced.

APPENDIX

6.A Expectation of Least Squares Estimators

The expectation of a random variable W is mathematically defined as

(a) $E[W] = \sum u_j Pr[W = u_j]$ if W is discrete

where the summation is taken over all possible values, u_j, that can be observed by the variable W, and $Pr[W = u_j]$ is the probability that the random variable W will take on the value of u_j;

or

(b) $E[W] = \int u f(u) du$ if W is continuous

where the integral is taken over all possible values (indicated by the dummy variable of integration, u) of the continuous random variable W, and $f(u)$ is the probability density function of W at the value u.

In particular, if W is a normal random variable with population mean μ and standard deviation σ,

$$E[W] = \mu = \int_{-\infty}^{\infty} u \cdot \frac{1}{\sqrt{2\pi}\,\sigma} e^{-\frac{(u-\mu)^2}{2\sigma^2}} \, du$$

and if $g(w)$ is some function of W,

$$E[g(w)] = \mu = \int_{-\infty}^{\infty} g(u) \cdot \frac{1}{\sqrt{2\pi}\,\sigma} e^{-\frac{(u-\mu)^2}{2\sigma^2}} \, du \, .$$

Several properties of expectations are needed to derive the results of this chapter. If a_0, a_1, ..., a_n are constants and W_1, W_2, ..., W_n are random variables with population means μ_1, μ_2, ..., μ_n, then

(i) $E[a_i] = a_i$ $i = 0, 1, 2, ..., n$

(ii) $E[a_i W_i] = a_i E[W_i] = a_i \mu_i$ $i = 1, 2, ..., n$

(iii) $E[a_0 + a_1 W_1 + ... + a_n W_n] = a_0 + a_1 E[W_1] + ... + a_n E[W_n]$

$$= a_0 + a_1 \mu_1 + ... + a_a \mu_n.$$

As mentioned in Section 6.2.2, the expected value (population mean) of a random vector is a vector containing the expected values of each of the random variables in the vector. Notationally, if W_1, W_2, ..., W_n are random variables having individual population means μ_1, μ_2, ..., μ_n, respectively, then the expectation of $\underline{W}$ is defined to be $E[\underline{W}] = \underline{\mu}$.

Now let A be any matrix of constants,

$$E[A\underline{W}] = E \begin{bmatrix} a_{11} W_1 + a_{12} W_2 + ... + a_{1n} W_n \\ a_{21} W_1 + a_{22} W_2 + ... + a_{2n} W_n \\ \vdots \\ a_{n1} W_1 + a_{n2} W_2 + ... + a_{nn} W_n \end{bmatrix}$$

The expectation of the ith element of the vector $A\underline{W}$ is, from property (iii)

$$a_{i1} \mu_1 + a_{i2} \mu_2 + ... + a_{in} \mu_n.$$

Consequently, $E[A\underline{W}] = AE[\underline{W}] = A\underline{\mu}$ as in eqn. (6.2.1).

In addition to this property of expectations for random vectors, two other properties are needed to derive the results in this appendix. They can be derived similar to the above derivation. If $\underline{b}$ is an $r \times 1$ vector of constants, A_1 ..., A_n are $r \times p$ matrices of constants, and $\underline{W_1}, \underline{W_2}$, ..., $\underline{W_n}$ are p-dimensional random vectors with means μ_1, μ_2, ..., μ_n, respectively, the three properties can be summarized as

(iv) $E[A_j] = A_j$ $j = 1, ..., n$

(v) $E[A_j\underline{W_j}] = A_jE[\underline{W_j}] = A_j\mu_j$ $j = 1, 2, ..., n$

(vi) $E[\underline{b} + A_1\underline{W_1} + ... + A_n\underline{W_n}] = \underline{b} + A_1E[\underline{W_1}] + ... + A_nE[\underline{W_n}]$

$$= \underline{b} + A_1\mu_1 + ... + A_n\mu_n.$$

Under Assumptions 1-3 of Table 6.2 (or just Assumptions 1, 2, and $E[\underline{\varepsilon}] = \underline{0}$), the above properties can be used to show that

$$E[\underline{Y}] = E[X^*\underline{B} + \underline{\varepsilon}]$$

$$= X^*\underline{B} + E[\underline{\varepsilon}]$$

$$= X^*\underline{B}$$

since $E[\underline{\varepsilon}] = \underline{0}$. Also,

$$E[\underline{\hat{B}}] = E[(X^{*'}X^*)^{-1}X^{*'}\underline{Y}]$$

$$= (X^{*'}X^*)^{-1}X^*E[\underline{Y}]$$

$$= (X^{*'}X^*)^{-1}X^{*'}X^*\underline{B} = \underline{B}$$

i.e., the least squares estimators are unbiased.

Using these same properties one can show that the least squares estimators for the centered, normal deviate, and unit length prediction equations are unbiased for their respective parameters. For example, the transformed model for the unit length standardization is

$$\underline{Y} = \alpha^*\underline{1} + W\underline{\beta^*} + \underline{\varepsilon}.$$

Consequently,

$$E[\underline{\hat{\beta}^*}] = (W'W)^{-1}W'E[\underline{Y}]$$

$$= (W'W)^{-1}W'(\alpha^*\underline{1} + W\underline{\beta^*})$$

$$= \underline{\beta^*}$$

since $W'\underline{1} = \underline{0}$. Similarly,

$$E[\hat{\alpha}*] = E[\overline{Y}]$$

$$= n^{-1}\underline{1}'E[\underline{Y}]$$

$$= n^{-1}\underline{1}'(\alpha*\underline{1} + W\underline{\beta}*)$$

$$= \alpha* .$$

One can also verify that transforming the standardized estimators back to estimators of the original model parameters produces unbiased estimators of the elements of $\underline{B}$.

6.B   Variances and Covariances of Least Squares Estimators

The variance of a random variable W is mathematically defined as $E[(W - \mu)^2]$. The variance is evaluated similarly to $E[W]$:

(a) $\text{Var}[W] = E[(W - \mu)^2] = \Sigma (u_j - \mu)^2 Pr[W = u_j]$

   if W is discrete

or

(b) $Var[W] = E[(W - \mu)^2] = \int (u - \mu)^2 f(u) du$

   if W is continuous,

where in both cases $\mu = E[W]$. Some properties of variances are

(i) $\text{Var}[a_i] = 0$ $i = 0, 1, 2, ..., n$

(ii) $\text{Var}[a_i W_i] = a_i^2 \text{Var}[W_i] = a_i^2 \sigma_i^2$ $i = 1, 2, ..., n$.

If all the W_i are pairwise uncorrelated (see below) and $\sigma_i^2 = \text{Var}[W_i]$, then

(iii) $\text{Var}[a_0 + a_1 W_1 + ... + a_n W_n] = a_1^2 \text{Var}[W_1] + ... + a_n^2 \text{Var}[W_n]$

$$= a_1^2 \sigma_1^2 + ... + a_n^2 \sigma_n^2 .$$

Under Assumptions 1-3, and invoking these three properties

$$\text{Var}[Y_i] = \text{Var}[\varepsilon_i] = \sigma^2$$

and

$$\text{Var}[\overline{Y}] = (n)^{-2} \Sigma \, \text{Var}[Y_i] = \frac{\sigma^2}{n} .$$

Before deriving the variances of $\hat{\alpha}$ and $\hat{\beta}$, we must define one more type of expectation. The covariance between two random variables V and W, with respective population means v and μ, is

(c) $\text{Cov}[V, W] = \underset{\text{all } t_i, \, u_i}{\Sigma \; \Sigma} \; (t_i - v)(u_j - \mu)Pr[V = t_i \text{ and } W = u_j]$

   if V and W are discrete

or

(d) $\text{Cov}[V, W] = \underset{\text{all } t, \, u}{\int \int} \; (t - v)(u - \mu)f(t, u) \, dt \, du$

   if V and W are continuous.

Here $Pr[V = t_i \text{ and } W = u_j]$ is the joint probability that $V = t_i$ and $W = u_j$. Similarly, $f(t, u)$ is the joint probability density of the random variables V and W. If V and W are uncorrelated, $\text{Cov}[V, W] = 0$ and conversely. Another way of writing the covariance between V and W is $E[(V - v)(W - \mu)]$.

   Property (iii) for variances must be altered when the random variables are correlated. It then becomes

(iv) $\text{Var}[a_0 + a_1 W_1 + a_2 W_2 + ... + a_n W_n]$

   $= a_1^2 \, \text{Var}[W_1] + a_2^2 \, \text{Var}[W_2] + ... + a_n^2 \, \text{Var}[W_n]$

   $+ 2a_1 a_2 \, \text{Cov}[W_1, W_2] + 2a_1 a_3 \, \text{Cov}[W_1, W_3]$

   $+ ... + 2a_{n-1} a_n \, \text{Cov}[W_{n-1}, W_n] .$

Another important property needed below is

(v) $\text{Cov}[a_1 W_1 + ... + a_n W_n, b_1 W_1 + b_2 W_2 + ... + b_n W_n]$

   $= a_1 b_1 \text{Var}[W_1] + ... + a_n b_n \text{Var}[W_n] + a_1 b_2 \text{Cov}[W_1, W_2]$

   $+ ... + a_n b_{n-1} \text{Cov}[W_n, W_{n-1}] .$

From properties (iv) and (v), one can show that if A is an $r \times n$ $(r \leqslant n)$ matrix of constants and $\underline{b}$ is an $r \times 1$ vector of constants,

$$Var[AW + \underline{b}] = A\Sigma_W A' , \qquad (6.B.1)$$

where Σ_W is the covariance matrix of the random vector $\underline{W}$. To prove this result is straightforward but somewhat tedious. We will outline the proof without filling in all the steps. Let $\underline{t} = A\underline{W} + \underline{b}$ so that $t_j = \underline{a}_j'\underline{W} + b_j$, where $\underline{a}_j'$ is the jth row of A. The diagonal elements of $\Sigma_t = Var[\underline{t}]$ contain the individual variances of the t_j:

$$Var[t_j] = Var[a_{1j}Y_1 + a_{2j}Y_2 + \ldots + a_{nj}Y_n + b_j]$$

$$= a_{1j}^2\sigma_{11} + a_{2j}^2\sigma_{22} + \ldots + a_{nj}^2\sigma_{nn}$$

$$+ 2a_{1j}a_{2j}\sigma_{12} + 2a_{1j}a_{3j}\sigma_{13} + \ldots + 2a_{n-1,j}a_{nj}\sigma_{n-1,n}$$

from property (iv). By direct multiplication one can verify that the above expression is identical to

$$Var[t_j] = \underline{a}_j'\Sigma_W \underline{a}_j .$$

Similar algebraic derivations utilizing property (v) show that the off-diagonal elements of Σ_t are the covariances between pairs of elements of $\underline{t}$ and that

$$Cov[t_j, t_k] = \underline{a}_j'\Sigma_W \underline{a}_k .$$

Combining these last two results yields

$$Var[A\underline{W} + \underline{b}] = \begin{bmatrix} \underline{a}_1'\Sigma_W\underline{a}_1 & \underline{a}_1'\Sigma_W\underline{a}_2 & \ldots & \underline{a}_1'\Sigma_W\underline{a}_r \\ \underline{a}_2'\Sigma_W\underline{a}_1 & \underline{a}_2'\Sigma_W\underline{a}_2 & \ldots & \underline{a}_2'\Sigma_W\underline{a}_r \\ \vdots & \vdots & & \vdots \\ \underline{a}_r'\Sigma_W\underline{a}_1 & \underline{a}_r'\Sigma_W\underline{a}_2 & \ldots & \underline{a}_r'\Sigma_W\underline{a}_r \end{bmatrix} = A\Sigma_W A' .$$

Under Assumptions 1-3, $Var[\underline{Y}] = \sigma^2 I$. Since $\underline{\hat{B}} = (X^{*'}X^*)^{-1}X^{*'}\underline{Y}$, invoking this last result one finds that

$$Var[\underline{\hat{B}}] = (X^{*'}X^*)^{-1}X^{*'}(\sigma^2 I)X^*(X^{*'}X^*)^{-1}$$

$$= (X^{*'}X^*)^{-1}\sigma^2 .$$

If one employs standardized or centered predictor variables, the covariance matrices can be similarly derived. For example, with centered predictor variables

$$\text{Var}[\hat{\alpha}^*] = \text{Var}[\overline{Y}] = \frac{\sigma^2}{n}.$$

$$\text{Var}[\hat{\beta}] = (M'M)^{-1}M'(\sigma^2 I)M(M'M)^{-1} = (M'M)^{-1}\sigma^2,$$

and

$$\text{Cov}(\hat{\alpha}^*, \hat{\beta}) = n^{-1}\underline{1}'(\sigma^2 I)M(M'M)^{-1} = \underline{0}'$$

since $\underline{1}'M = \underline{0}$.

6.C Distributions of Least Squares Estimators

When inferences are desired on least squares parameter estimators, the joint probability distribution of all the estimators must be utilized. The importance of the joint probability distribution over the distributions of individual estimators stems from the lack of independence of the least squares estimators; i.e., in general the least squares estimators are correlated, implying that probabilities associated with two or more estimators cannot be calculated simultaneously without taking into account their correlations with one another. Under Assumptions 1-4 of Table 6.2, the least squares estimators follow a multivariate normal probability distribution.

The multivariate normal distribution is a generalization of the univariate normal distribution. Its theoretical properties are covered comprehensively in many texts, including Anderson (1958), Graybill (1976), Searle (1971), and Tatsuoka (1971). Fortunately, we do not have to be directly concerned with the multivariate normal distribution per se as much as with some of the properties of the random variables involved. Many of the t, chi-square, and F statistics typically used in regression analyses are functions of random variables that follow a multivariate normal distribution.

Notationally, if the elements of a random vector, $\underline{Y}$, follow a multivariate normal probability distribution with population mean $\underline{\mu}$ and covariance matrix Σ, we write

$$\underline{Y} \sim \text{MVN}(\underline{\mu}, \Sigma).$$

Now under Assumptions 1-4 (or 1, 3', and 4'), $\varepsilon_i \sim \text{NID}(0, \sigma^2)$. Collectively, $\underline{\varepsilon} \sim \text{MVN}(\underline{0}, \sigma^2 I)$. The addition of fixed constants to $\underline{\varepsilon}$ does not change its probability distribution, only its mean vector. Hence,

$$\underline{Y} = \alpha \underline{1} + X\beta + \underline{\varepsilon} \sim MVN(\mu_Y, \Sigma_Y),$$

where $\mu_Y = E[\underline{Y}] = \alpha \underline{1} + X\beta$ and $Var[\underline{Y}] = Var[\underline{\varepsilon}] = \sigma^2 I$. The off-diagonal elements of Σ_Y are zero, implying that the Y_i are independent normal random variables since $Corr(Y_i, Y_j) = \sigma_{ij}/(\sigma_{ii} \cdot \sigma_{jj})^{1/2} = 0$.

An important property of a multivariate normal random variable is that if $\underline{Y} \sim MVN(\mu_Y, \Sigma_Y)$, then if A is a $r \times n$ matrix of constants with $r \leqslant n$

$$A\underline{Y} + \underline{b} \sim MVN(\mu_Y + \underline{b}, A\Sigma_Y A'). \tag{6.C.1}$$

The mean vector and covariance matrix of $A\underline{Y} + \underline{b}$ were derived in Appendix 6.B; the additional property to be noted here is that $A\underline{Y} + \underline{b}$ has a multivariate normal probability distribution since $\underline{Y}$ has one.

From eqn. (6.C.1) we can immediately deduce the distribution of the least squares estimators:

$$\underline{\hat{B}} = (X^{*'}X^{*})^{-1}X^{*'}\underline{Y} \sim MVN(\mu_B, \Sigma_B), \tag{6.C.2}$$

where $\mu_B = \underline{B}$ and $\Sigma_B = (X^{*'}X^{*})^{-1}\sigma^2$ as in Table 6.3. It also follows that, individually,

$$\hat{\alpha} \sim N(\alpha, C_{00}\sigma^2) \quad \text{and} \quad \hat{\beta}_j \sim N(\beta_j, C_{jj}\sigma^2)$$

since each of these individual estimators can be written as in eqn. (6.C.1) with $\underline{\hat{B}}$ replacing $\underline{Y}$. For example, if

$$A = (1, 0, 0, ..., 0) \quad \text{and} \quad b = 0,$$

then $\hat{\alpha} = A\underline{\hat{B}} + b$ and eqn. (6.C.1) can be used to verify the distribution of $\hat{\alpha}$. The distribution of $\hat{\beta}_j$ follows by letting A be a row vector with 1 in the $(j+1)$th position and zeros elsewhere and letting $b = 0$.

All the remaining distributional properties cited in Sections 6.2.4 and 6.3.2 follow from additional properties of the multivariate normal probability distribution, properties which are beyond the scope of this book to cover. Interested readers are referred to Graybill (1976) or Searle (1971) for complete details. Two properties of particular importance we mention without proof.

First, the error sum of squares [and hence $\hat{\sigma}^2 = SSE/(n-p-1)$] is distributed independently of the regression coefficient estimators as

$$\frac{SSE}{\sigma^2} \sim \chi^2(n-p-1);$$

i.e., a chi-square random variable having $(n-p-1)$ degrees of freedom. Second, since $(n-p-1)\hat{\sigma}^2/\sigma^2$ is distributed as a chi-square random variable and is independent of all the regression coefficient estimators,

$$\frac{\hat{\alpha} - \alpha}{\hat{\sigma}C_{00}^{1/2}} \sim t(n-p-1) \quad \text{and} \quad \frac{\hat{\beta}_j - \beta_j}{\hat{\sigma}C_{jj}^{1/2}} \sim t(n-p-1).$$

Thus each of these t statistics can be utilized to obtain probabilities for the regression coefficient estimators when σ^2 is unknown, as is usually the case. Similar probability distributional properties hold for centered and standardized predictor variables. For example, using the unit length standardization with $Q = (W'W)^{-1}$,

$$\frac{\hat{\alpha} - \alpha}{\hat{\sigma}/n^{1/2}} \sim t(n-p-1) \quad \text{and} \quad \frac{\hat{\beta}_j^* - \beta_j^*}{\hat{\sigma}q_{jj}^{1/2}} \sim t(n-p-1).$$

EXERCISES

1. For a single-variable regression model, $Y_i = \alpha + \beta X_i + \varepsilon_i$, derive the correlation between the least squares estimators of α and β. Show that if $X_i = C$, a constant, for every observation, $\text{Corr}(\hat{\alpha}, \hat{\beta}) = 1$ or -1.

2. Show that although centering produces uncorrelated estimators of α and β, the transformation from $\hat{\alpha}^*$ and $\hat{\beta}$ to $\hat{\alpha}$ produces correlated estimators of α and β. [Hint:$\text{Cov}(\hat{\alpha}, \hat{\beta}) = E(\hat{\alpha} - \alpha)(\hat{\beta} - \beta)$; now express $\hat{\alpha}(\alpha)$ in terms of $\hat{\alpha}^*$ and $\hat{\beta}(\alpha^*$ and $\beta)$.]

3. Consider a prediction equation for NOX (Data Set A.5) using only HUMID. Numerically calculate the variance (apart from σ^2) of the least squares estimator for the coefficient of HUMID (a) when only a constant term and HUMID are used in the prediction equation, (b) when a constant term, HUMID, and $(\text{HUMID})^2$ are included in the prediction equation. What do these answers suggest about variances of highly correlated predictor variables?

4. Show that for single-variable models t statistics used to test H_0: $\beta = 0$ vs. H_a: $\beta \neq 0$ are numerically identical regardless of whether one uses raw, centered, normal deviate, or unit length predictor variables. (Note: this property is also true for multiple-variable models.) Numerically evaluate these t statistics using all four types of predictor variables for the salary data (Table 2.1).

5. Analyze a three-variable prediction equation for NOX using HUMID, TEMP, and BPRES. After testing an appropriate hypothesis, would you conclude that BPRES is an important predictor variable in this three-variable model?

6. Construct an ANOVA table for the emissions data in Data Set A.5 using only the three predictor variables listed. Test the hypotheses H_0: $\alpha* = 0$ vs. H_a: $\alpha* \neq 0$ and H_0: $\underline{\beta} = \underline{0}$ vs. H_a: $\underline{\beta} \neq \underline{0}$. Examine the effect of adding quadratic terms to the model on the ANOVA table and these tests.

7. In the experiment conducted to predict NOX, TEMP was controlled to be near 68°, 77°, and 86°. Round all temperature values in Data Set A.5 to one of these three values. For each of the three sets of rounded temperatures, select the following observation as "repeated" measurements:

   TEMP = 68   Nos. 20, 21

   TEMP = 77   Nos. 3, 10, 27; Nos. 1, 8, 15; Nos. 17, 22

   TEMP = 86   Nos. 13, 14; Nos. 26, 29; Nos. 32, 33, 34; Nos. 35, 38, 39, 40.

   Use these $k = 8$ groups of repeated predictor-variable values to perform a lack of fit test on the three-variable prediction equation for NOX.

8. Construct confidence intervals for each of the regression coefficients in Data Set A.5 using a confidence coefficient of 75%. How do these confidence intervals compare with the corresponding tests of hypothesis?

9. Let $X_1 = $ HUMID, $X_2 = $ TEMP, and $X_3 = $ BPRES. Does $\underline{\beta}' = (0, 0, .3)$ fall within a 95% confidence region for $\underline{\beta}$? What about $\underline{\beta}' = (.1, .1, 0)$?

10. Test the hypothesis H_0: $\sigma^2 = .01$ vs. H_a: $\sigma^2 \neq .01$ for the police applicant data, Data Set A.6. (Note: an ANOVA table for this data appears in Table 6.7.) Find a 95% confidence interval for σ^2.

11. As a measure of the adequacy of the fits to the suicide data (Table 3.4), construct prediction intervals for the female suicide rate using (a) only the 25-29 to 50-54 age groups, and (b) all the age groups. Evaluate the limits on the prediction interval for ages 17.5, 32.5, and 62.5. How do the intervals differ for the two prediction equations?

CHAPTER 7

RESIDUAL ANALYSIS

Examining residuals is one of the most important tasks in any regression analysis. A residual analysis involves the careful inspection of the differences between the observed and predicted values of the response variable after a prediction equation is fit to the data. In doing so, one hopes to spot any anomalies in the data which might cause poor prediction or poor parameter estimation.

While the most familiar techniques for examining residuals involve graphical methods, such as the plotting of the residuals against their corresponding fitted values, there are also several numerical approaches, including the calculation of statistics to detect certain types of data defects. Because of the ease with which residual plots can be analyzed and interpreted, the emphasis in this chapter will be primarily on graphical techniques.*

Residual plots are utilized chiefly for observing patterns in the data. Hence, one may plot residuals against fitted values, against values of the individual predictor variables, and even against values of new variables. The plots, in turn, help the analyst to detect trends and extreme measurements, such as outliers; to identify potential problems, such as changes in the error variance; to assess model assumptions, such as the shape of the

---

*For more detailed discussion on graphical and analytic residual analyses see Anscombe (1960, 1973), Anscombe and Tukey (1963), Behnken and Draper (1972), Cook (1977), Draper and Smith (1966), Glejser (1969), Goldfeld and Quandt (1965), Hoaglin and Welsch (1978), Larson and McCleary (1972), and Wood (1973).

error distribution; and to determine the importance of predictor variables. The interpretation of residual plots is, however, somewhat subjective and requires careful scrutiny by the investigator. A peculiar pattern in the data may lead one to choose a number of alternative actions including keeping the current model, using a different prediction equation, or eliminating or modifying the data and redoing the analysis.

An appreciation of the type of information provided by residuals can be obtained by reexamining the data presented in Table 2.1 on the average annual salaries of teachers in public elementary and secondary schools in the U. S. during the period 1964-65 to 1974-75. At least three different prediction equations could be suggested for this data after studying Figure 2.1. These include a linear fit, a quadratic fit, and an exponential fit that initially rises then flattens out as t increases. The resulting equations are [observe that the exponential fit is intrinsically linear and can be fit with a single variable regression of $(13,500 - Y)/Y$ on t]:

LINEAR: $\hat{Y} = 10,927 + 174t$

QUADRATIC: $\hat{Y} = 9,998 + 603t - 36t^2$

EXPONENTIAL: $\hat{Y} = 13,500/[1 + e^{(1.399 + 0.126t)}]$.

The corresponding residuals from each of these fits are tabulated in Table 7.1. A quick glance at this table reveals that, ignoring all other considerations, the quadratic prediction equation appears to provide the best fit to the salary data since its residuals are the smallest for the majority of the

Table 7.1. Comparison of Residuals for Prediction Equations of Annual Salaries

Year	Average Salary	Residuals From		
(t)	(Y)	Linear Fit	Quadratic Fit	Exponential Fit
1	10,606	-495	41	-481
2	11,249	-26	189	-78
3	11,186	-263	-297	-361
4	12,035	412	201	287
5	11,911	114	-202	-20
6	12,123	152	-197	25
7	12,480	335	25	232
8	12,713	394	195	329
9	12,835	342	326	328
10	12,496	-171	68	-121
11	12,070	-771	-205	-645

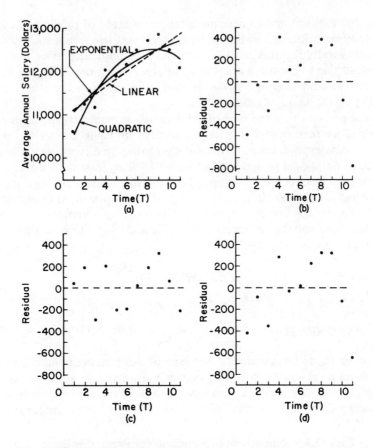

Figure 7.1 Residuals from prediction equations for average annual salaries: (a) Prediction equations, (b) Linear fit, (c) Quadratic fit, (d) Exponential fit.

observations. Further, the linear equation offers the worst fit since its residuals are the largest. Unfortunately, it is difficult to visualize any other relationships or trends in the residuals by this simple perusal of the data.

Another informative technique consists of plotting the residuals against the predictor variable, time. This has been done in Figure 7.1 which illustrates the resultant patterns. In Figure 7.1(a) the original data is plotted against time and the three prediction equations are superimposed over the plotted points. All three fits appear useful and could be utilized as representations of the data. However, Figure 7.1(b), which plots the linear fit residuals against time clearly shows the need for some transformation of the predictor variable, possibly by including a quadratic term in time or using a logarithmic or exponential transformation. Note how the points tend to take the shape of a parabola and how large the residuals are in magnitude.

Figure 7.1(c) shows the residuals of the quadratic fit versus time. Note the stability of the resultant points: the residuals appear to be randomly centered around the value zero. The range of the residuals in Figure 7.1(c) is half of that for the linear fit and no detectable patterns appear on the graph. In short, the addition of the quadratic term has greatly improved the fit.

The final plot, Figure 7.1(d), graphs the exponential residuals versus time. The observable pattern is similar to that displayed in the linear fit plot only now the residuals are somewhat smaller in magnitude. The fit has been improved though not as markedly as that provided by the quadratic fit.

If one were mainly concerned with obtaining a good fitting predictor equation for *these eleven points*, the quadratic fit is best according to the residual plots. However, as noted in Chapter 2, it may be unrealistic to imagine that teacher salaries will decrease with time, even if salaries are measured in 1974-75 dollars and the value of the dollar has declined in recent years. The better fit from a prediction viewpoint would appear to be the exponential equation since it does lead to reduced residuals relative to the linear fit while still allowing for increases in salaries. Regardless of which fit is ultimately selected, the value of examining residual plots is evident.

There are many different methods of analyzing residuals and the above example is a very simplified approach to such usage. The present chapter will present a more detailed examination of residuals in multiple variable prediction equations. Several different types of residuals will be defined and discussed and particular attention will be given to how residuals can aid the data analyst in checking the error assumptions, determining model misspecification, and detecting extreme observations and outliers.

7.1 TYPES OF RESIDUALS

There are many different types of residuals, although each is a function of the difference between the observed and predicted responses,

$$r_i = Y_i - \hat{Y}_i \quad , \quad i = 1, 2, ..., n . \tag{7.1.1}$$

There are, for example, raw residuals, deleted residuals, standardized residuals, and studentized residuals. These transformed residuals have different properties and each can be useful in detecting different patterns in the data.

This section contains definitions and discussions of the more commonly used residual types. While not exhaustive, the types considered should provide the data analyst with enough varieties to effectively evaluate the regression model and its corresponding assumptions as well as to detect data anomalies.

7.1.1 Distinction Between Residuals and Errors

An error is generally thought of as a mistake. In a regression analysis it is more closely related to inexactness or imprecision. If an object moves at a constant velocity (v) it is well known that the distance (d) it travels is linearly related to the time (t) it travels by

$$d = vt.$$

Empirically, however, this relationship is not exact. That is, if we move an object at a supposed constant velocity, the distance traveled will not exactly equal the velocity multiplied by time. Reasons for this are many: the velocity cannot be maintained exactly at some constant value, time and distance cannot be measured precisely, in the absence of a vacuum wind resistance and friction impede movement, etc. So while $d = vt$ is the correct theoretical functional relationship, it is not the correct observational one. To account for these observational errors we specify the empirical model

$$d = vt + \varepsilon,$$

where ε denotes the magnitude of the error or imprecision in observing the exact relationship between d and t.

The errors in a regression model are generally stated as being random. They need not be entirely so. If the timer malfunctions and hesitates for, say, one second after it is started, the correct observational relationship between distance and the time recorded is

$$d = v(t + 1) + \varepsilon$$
$$= vt + \varepsilon^*$$

where $\varepsilon^* = v + \varepsilon$. If v is large relative to ε for all measurements, the error in the assumed model, ε^*, will have a large constant component and a small random one. Similarly, if two operators are used to time the experiment and the first operator hesitates before engaging the timer, the correct relationship between distance and time is

$$d = vt + \varepsilon^{**},$$

where $\varepsilon^{**} = v + vX + \varepsilon$ and

$$X = \begin{cases} t_1 & \text{if operator 1 times the experiment} \\ 0 & \text{if operator 2 times the experiment.} \end{cases}$$

Here t_1 is the amount of hesitation induced by operator 1 timing the experiment.

In this discussion we have assumed that the coefficient of t, namely v, is known and that the correct theoretical relationship between d and v is known. If the latter assumption is incorrect we have seen that the error term of the model will not contain just a random component. In either case, the exact magnitude of the error can be found by subtraction:

$$\varepsilon = d - vt.$$

Now suppose that the timer does not malfunction and the first operator does not hesitate but that v is unknown. If data are available for several measurements of d and t, the least squares estimate of v, say $\hat{v}$, can be found. The difference between an observed distance, d, and its predicted value, $\hat{v}t$, is a residual, r:

$$r = d - \hat{v}t.$$

A residual is not identical to an error in the original model unless (for this example) $v = \hat{v}$, since

$$r = (vt + \varepsilon) - \hat{v}t$$
$$= (v - \hat{v})t + \varepsilon.$$

So residuals contain both random and nonrandom components even if the true theoretical model is correctly specified. If the model is incorrectly specified as when both the first operator and timer hesitate, then

$$r = (v - \hat{v})t + v + vX + \varepsilon.$$

Thus residuals can be explored to assess model specifications, model assumptions, and the accuracy of prediction.

7.1.2 Raw and Scaled Residuals

Three common types of residuals are *raw* or unscaled residuals, *standardized* residuals, and *studentized* residuals. Raw residuals are simply those represented by eqn. (7.1.1), or, in vector notation, by $\underline{r}' = (r_1, r_2, ..., r_n)$ where

$$\underline{r} = \underline{Y} - \hat{\underline{Y}} \qquad\qquad (7.1.2)$$

$$= \underline{Y} - X^*\hat{\underline{B}}$$

$$= (I - H)\underline{Y}, \qquad\qquad (7.1.3)$$

where $H = X^*(X^{*'}X^*)^{-1}X^{*'}$. Inserting $X^*\underline{B} + \underline{\varepsilon}$ for $\underline{Y}$ in eqn. (7.1.3) and noting that $(I - H)X^* = \Phi$, an expression for $\underline{r}$ in terms of the errors $\underline{\varepsilon}$ is

$$\underline{r} = (I - H)\underline{\varepsilon} . \qquad (7.1.4)$$

Equations (7.1.3) and (7.1.4) demonstrate that raw residuals can be expressed as linear transformations of the response, $\underline{Y}$, or the unknown errors, $\underline{\varepsilon}$. Thus, they can be very useful in detecting model misspecification, incorrect error assumptions, and extreme data values.

The ability to detect problem data with raw residuals can often be enhanced by using a scaling factor. This becomes evident when one examines the distributional properties of the raw residuals. Assuming that the error terms in the proposed regression model are uncorrelated, have zero mean, and have constant variance (Assumption 3), it follows from equation (7.1.4) that

$$E[\underline{r}] = \underline{0} \quad \text{and} \quad \text{Var}(\underline{r}) = (I - H)\sigma^2. \qquad (7.1.5)$$

Hence, the raw residuals also have zero means but, unlike the error terms, these residuals have unequal variances, since the diagonal elements of $(I - H)\sigma^2$ are generally not all equal, and they are correlated since $(I - H)\sigma^2$ is generally not a diagonal matrix.

The variance of an individual residual is [from eqn. (7.1.5)]

$$\text{Var}[r_i] = (1 - h_{ii})\sigma^2 , \qquad i = 1, 2, ..., n$$

where h_{ii} is the ith diagonal element of the matrix H. Similarly, the covariance between r_i and r_j $(i \neq j)$ is $\text{Cov}(r_i, r_j) = -h_{ij}\sigma^2$. One can show that since H is symmetric $(H = H')$ and idempotent $(H = H^2)$,

i) $0 \leqslant h_{ii} \leqslant 1$

ii) $-1 \leqslant h_{ij} \leqslant 1 \qquad i \neq j .$ \qquad (7.1.6)

Thus, the residual variances can range from zero to σ^2, and the covariances from $-\sigma^2$ to $+\sigma^2$.

The relationships in eqn. (7.1.6) are useful in assessing the influence of the jth observed response on the ith predicted response. Using the definition of H in eqn. (7.1.3),

$$\hat{\underline{Y}} = X^*\hat{\underline{B}}$$

$$= H\underline{Y} . \qquad (7.1.7)$$

Thus,

$$\hat{Y}_i = \sum_{j=1}^{n} h_{ij} Y_j . \qquad (7.1.8)$$

Large h_{ij} values indicate observed responses that most influence the ith predicted response. In particular, if a single h_{ij} value is large relative to all the others in eqn. (7.1.8), the jth *observed* response dominates the ith *predicted* response. The key role of h_{ii} in determining the variance of the ith residual and in measuring the influence of the ith observed response on its predicted values has resulted in it being referred to as a "leverage" value.

Table 7.2 lists the predicted responses and residuals for the linear fit to the salary data of Table 2.1 (and Table 7.1). Due to the differences in the leverage values, the variances of the residuals range from $0.682\sigma^2$ to $0.909\sigma^2$, over a thirty percent difference in magnitude. These differences in variances make comparison of the magnitude of the residuals difficult.

Table 7.2. Prediction and Residual Summary for Linear Fit to Annual Salary Data

Year	Annual Average Salary	Predicted Average Salary	Residual (r_i)	Leverage Value (h_{ii})	Residual Variance $(Var[r_i])$
1	10,606	11,101	-495	.318	$.682\sigma^2$
2	11,249	11,275	-26	.236	$.764\sigma^2$
3	11,186	11,449	-263	.173	$.827\sigma^2$
4	12,035	11,623	412	.127	$.873\sigma^2$
5	11,911	11,797	114	.100	$.900\sigma^2$
6	12,123	11,971	152	.091	$.909\sigma^2$
7	12,480	12,145	335	.100	$.900\sigma^2$
8	12,713	12,319	394	.127	$.873\sigma^2$
9	12,835	12,493	342	.173	$.827\sigma^2$
10	12,496	12,667	-171	.236	$.764\sigma^2$
11	12,070	12,841	-771	.318	$.682\sigma^2$

Scaling residuals serves much the same purpose as did standardizing predictor variables in Chapters 4 and 5. Standardization of predictor variables has been advocated as a means for eliminating differences in the variability of individual predictor variables. Estimated regression coefficients are thereby directly comparable. So too, the scaling of residuals enables the magnitudes of the residuals to be directly compared. For some residual analyses scaling is not necessary; for others, such as outlier detection, scaling is important.

Standardized residuals attempt to mimic standard normal deviates. If ε_i is a normal random variable with mean zero and variance σ^2 then ε_i/σ is a standard normal random variable. Hence, standardized residuals are defined to be

$$s_i = \frac{r_i}{\hat{\sigma}}, \qquad (7.1.9)$$

where

$$\hat{\sigma}^2 = \text{MSE} = \frac{r_i^2}{n-p-1}.$$

However, the standardized residuals, s_i, are not standard normal deviates since the denominator of eqn. (7.1.9) is not the standard deviation of r_i.

Studentized residuals are raw residuals that are scaled by dividing each by its estimated standard deviation. Since

$$r_i \sim N(0, \text{Var}[r_i]) \qquad \text{and} \qquad \text{Var}[r_i] = (1 - h_{ii})\sigma^2, \qquad (7.1.10)$$

studentized residuals are defined to be

$$t_i = \frac{r_i}{\hat{\sigma}\sqrt{1 - h_{ii}}}. \qquad (7.1.11)$$

It should be noted that the t_i are also not standard normal deviates since σ^2 still must be estimated; rather, they are t statistics (individually) with $n-p-1$ degrees of freedom. Because the residuals are divided by their standard errors, studentized residuals should behave more like standard normal deviates than should either the raw or standardized ones. In many data sets, particularly those with a large number of observations, the h_{ii} are close to zero so that t_i is approximately equal to s_i and both can be treated as approximate standard normal deviates.

A comparison of the third, fourth and fifth columns of Table 7.3 illustrates the differences observable in the three types of residuals just described. From an examination of the raw residuals it is apparent that year 11 ($r_{11} = -771$) and possibly year 1 ($r_1 = -495$) are being fit much more poorly than the other nine years by the linear prediction equation. The differences in the leverage values indicate that the standardized and studentized residuals in Table 7.3 might be moderately dissimilar. The studentized

Table 7.3. Raw and Scaled Residuals for Average Annual Salary Data

Year	Leverage Values (h_{ii})	Raw (r_i)	Standardized (s_i)	Studentized (t_i)	Deleted $[r_{(-i)}]$	Studentized Deleted $[t_{(-i)}]$
			Residual Type			
1	.318	-495	-1.201	-1.454	-726	-1.568
2	.236	-26	-.063	-.072	-34	-.068
3	.173	-263	-.638	-.702	-318	-.680
4	.127	412	1.000	1.070	472	1.080
5	.100	114	.277	.292	127	.276
6	.091	152	.369	.387	167	.368
7	.100	335	.813	.857	372	.843
8	.127	394	.956	1.023	451	1.026
9	.173	342	.830	.912	414	.903
10	.236	-171	-.415	-.475	-224	-.453
11	.318	-771	-1.871	-2.265	-1130	-3.257

residuals should therefore be preferred to the standardized residuals when assessing whether all eleven points are being adequately fit by the prediction equation. Examination of the t_i reveals that the last year is much more poorly fit than any other year including the first one. We have thus confirmed the visual impression given by Figure 7.1(b) and also determined that the magnitude of the residual for year 11 is not solely attributable to its having a large variance. We will return to this data set in the next subsection to provide even more dramatic evidence that the last data point is not consonant with the remaining ten.

7.1.3 Deleted Residuals

In addition to examining residual plots and individual residual values, another method of detecting outliers in a regression analysis is to determine how the predictor equation would change if the maverick point(s) were removed. With this motivation a different type of residual, termed the deleted residual, has recently been proposed for the purpose of detecting outliers.

The ith deleted residual is defined as the residual obtained from predicting the ith response, Y_i, when the prediction equation is derived with the ith observation deleted. Write the ith deleted residual, $r_{(-i)}$, as

$$r_{(-i)} = Y_i - \underline{u}_i^{*'} \hat{\underline{B}}_{(-i)} \qquad (7.1.12)$$

where $\underline{u}_i^{*'}$ is the ith row of X^* and $\hat{\underline{B}}_{(-i)}$ is the estimated regression coefficient vector obtained from eqn. (5.1.7) or eqn. (5.1.9) for the $n - 1$ observations with the ith case removed from the calculations. It can be shown

(Appendix 7.A) that $r_{(-i)}$ need not be obtained from a separate regression analysis using only $n - 1$ data points, since it can be calculated as

$$r_{(-i)} = \frac{r_i}{1 - h_{ii}}, \tag{7.1.13}$$

where r_i is the ith raw residual from the regression using all n observations and h_{ii} is the ith diagonal element of the H matrix defined in Section 7.1.2 (the ith leverage value). Hence, the deleted residual $r_{(-i)}$, is simply a scaling of the raw residual, r_i.

Corresponding to the raw deleted residual, $r_{(-i)}$, is a studentized deleted residual, $t_{(-i)}$, defined as $r_{(-i)}$ divided by its estimated standard deviation. From eqn. (7.1.13),

$$\text{Var}[r_{(-i)}] = \text{Var}\left[\frac{r_i}{1 - h_{ii}}\right]$$

$$= \frac{\sigma^2(1 - h_{ii})}{(1 - h_{ii})^2}$$

$$= \frac{\sigma^2}{1 - h_{ii}} \tag{7.1.14}$$

so that the estimated variance of $r_{(-i)}$ is

$$\hat{\text{Var}}[r_{(-i)}] = \frac{\hat{\sigma}^2_{(-i)}}{1 - h_{ii}} \tag{7.1.15}$$

where $\hat{\sigma}^2_{(-i)}$ is the residual mean square error from the fit obtained by excluding the ith observation. Thus, the ith studentized deleted residual is defined by

$$t_{(-i)} = \frac{r_{(-i)}}{(\hat{\text{Var}}[r_{(-i)}])^{1/2}}$$

$$= \frac{r_i}{\hat{\sigma}_{(-i)}(1 - h_{ii})^{1/2}}. \tag{7.1.16}$$

Hoaglin and Welsch (1978) showed that $\hat{\sigma}_{(-i)}$ is a function of elements obtained from the full prediction equation utilizing all n observations:

$$\hat{\sigma}^2_{(-i)} = \frac{SSE - (1 - h_{ii})^{-1}r_i^2}{n-p-2}. \qquad (7.1.17)$$

where $\hat{\sigma}^2$ = MSE. Substituting eqn. (7.1.17) into eqn. (7.1.16) yields

$$t_{(-i)} = r_i \left[\frac{(1 - h_{ii})SSE - r_i^2}{n-p-2} \right]^{-1/2}. \qquad (7.1.18)$$

Hence, the ith studentized deleted residual can be determined without actually removing any observations or recalculating the prediction equation. It can be obtained by a suitable transformation [eqn. (7.1.18)] of the raw residuals.

Returning now to Table 7.3 the poor fit of a linear prediction equation to the annual salary data becomes dramatically apparent by comparing the studentized deleted residuals for the eleven years. The exact distribution of individual studentized deleted residuals is that of a Student t statistic with $n-p-2$ degrees of freedom. Since there are only eight degrees of freedom for the $t_{(-i)}$ in Table 7.3, the Student t distribution provides more accurate tail probabilities than does the standard normal approximation. From Appendix C we see that $t_{(-11)}$ is in roughly the lower 0.5 percent (p = .005) tail of the t distribution with eight degrees of freedom.

The Student t distribution clearly identifies the last average annual salary as an extremely unusual observation. Whether this drop in average annual salary is due to miscalculation, a new trend in the salary data, or an unusual set of circumstances in this particular year cannot be answered from the information in Table 2.1. Nevertheless, the value of residual analysis in detecting the great disparity of this point from the other ten is evident.

7.2 VERIFICATION OF ERROR ASSUMPTIONS

Checking for the validity of the error assumptions includes the determination of whether the errors are uncorrelated, have the same constant variance, and are normally distributed. Since the raw and scaled residuals are multiples of the error terms, they should serve as good indicators of violations of these assumptions.

The following subsections contain some of the more common statistical tests and plotting techniques that can be used with residuals to verify properties of the error terms in a regression model. All discussions, unless otherwise noted, will be in terms of the raw residuals since these are the simplest to determine and interpret. However, any of the various residual types could as well be utilized since each is sensitive to incorrect error assumptions.

The use of plotting techniques, as was indicated in Chapter 2, is particularly helpful in detecting violations of error assumptions. Residuals, whether plotted by size, against a time order, or against the predicted average response, vary in systematic patterns when certain error assumptions are violated. Recognition of these patterns and learning how to combine them with results of statistical tests to evaluate the error terms are two major goals of this section.

7.2.1   Checks for Random Errors

One procedure for determining whether it is reasonable to assume that the errors are randomly distributed is to examine the arrangement of the signs (+ or -) of the corresponding residuals. Termed the "runs test," it is used when the time order in which the observations were collected is known. The test is based on observing sequences of plus and minus signs and deciding if the resulting pattern is abnormal under the assumption that the observations are a random sample.

An example of a set of observations whose time sequence is known is the homicide data given in Appendix A.3. The year each homicide rate was taken has been recorded so it is straightforward to determine the time order of the data. Suppose we now examine the residuals given in Table 7.4 that result from the regression of homicide rate (HOM) on the numbers of handgun licenses issued (LIC). The prediction equation utilized has been previously discussed in Chapter 3 and is illustrated in Figure 3.4.

The first step in the runs test is to determine the sequence of residual signs as well as the number of runs, r. A "run" is defined as a group of neighboring residuals having the same sign. From Table 7.4 the signs of the thirteen residuals are (in sequence)

$$(- - - - - - - - -)(+ + + +)$$

where the parentheses delineate two runs. Thus, in this example, there are $n = 13$ residuals with $n_1 = 9$ plus signs and $n_2 = 4$ minus signs, and there are $r = 2$ runs. (The use of r for the number of runs should not be confused with the use of r_i for a residual.)

The next step in the runs test is to decide whether this particular arrangement of residual signs is unusual enough to cause one to conclude that the residuals are not in a random order. This necessitates calculating the probability of obtaining a sign sequence as extreme or more extreme than the

Table 7.4. Residuals from Regression of Detroit Homicide Rate on Number of Handgun Licenses

Year	Homicide Rate*	Licenses Issued*	Residuals (r_i)	First Differences $(r_i - r_{i-1})$
1961	8.60	178.15	-3.01	
1962	8.90	156.41	-1.89	1.12
1963	8.52	198.02	-3.84	-1.95
1964	8.89	222.10	-4.37	-0.53
1965	13.07	301.92	-3.19	1.18
1966	14.57	391.22	-5.05	-1.86
1967	21.36	665.56	-8.58	-3.53
1968	28.03	1131.21	-19.41	-10.83
1969	31.49	837.60	-4.91	14.50
1970	37.39	794.90	2.59	7.50
1971	46.26	817.74	10.60	8.01
1972	47.24	583.17	20.40	9.80
1973	52.33	709.59	20.74	0.34

*Number of homicides and licenses per 100,000 population.

observed sequence given the belief that the signs are randomly observed. If the calculated probability is small it should lead us to conclude that the ordering is not random. Tables C.5a and C.5b of Appendix C are available for determining critical values at the 0.05 significance level (two-tailed significance level of 0.10) when the number of plus or minus signs are both less than or equal to twenty. In other situations (i.e., $n_1 > 20$ and $n_2 > 20$) a normal approximation can be made and the hypothesis of randomness rejected if

$$Z = \frac{r - \mu + 1/2}{\sigma} < -Z_{1-\gamma/2} \quad \text{or} \quad Z = \frac{r - \mu - 1/2}{\sigma} > Z_{1-\gamma/2} \quad (7.2.1)$$

where

$$\mu = \frac{2n_1 n_2}{n_1 + n_2} + 1, \qquad \sigma^2 = \frac{2n_1 n_2 (2n_1 n_2 - n_1 - n_2)}{(n_1 + n_2)^2 (n_1 + n_2 - 1)},$$

and r is the number of runs. The variable Z in eqn. (7.2.1) can be considered a standard normal deviate and the needed probability determined from a table of standard normal probabilities (e.g., Table C.1).

For the homicide data n_1 and n_2 are small so that the critical values can be determined exactly. From Table C.5a the probability of observing two or fewer runs is less than 0.05, indicating that an unusually low number of runs has occurred if the residuals are assumed randomly ordered. Therefore, with such a small probability associated with the observed number of runs, we conclude that the residual signs are not randomly ordered. Instead, there is some systematic pattern present which results in neighboring responses (homicide rates) being quite similar; i.e., there is evidence of a positive correlation among consecutive residuals.

Another simple technique for detecting correlations between time-consecutive residuals is to plot the residuals in their time sequence. This approach is illustrated in Figure 7.2(a) with the homicide data from Table 7.4. Note the obvious lack of random ordering among the residuals and the pattern of positive correlation between time-consecutive residuals; i.e., the first six residuals are all negative and have about the same magnitude, the next two decrease rapidly, and the last five increase in a steady fashion. The graph clearly supports the conclusion of the runs test in addition to providing a pictorial view of the time sequence of residuals.

7.2.2 Test for Serial Correlation

The graphical analysis of time series data was shown to be very useful in the last subsection. The homicide example contains correlated errors and the graph in Figure 7.2(a) reveals that the residuals of the same sign tended to cluster together. This was also verified through the runs test. In addition to these two techniques, time-correlated errors can also be detected using the Durbin-Watson test statistic*.

Suppose the error terms, ε_i, in a regression model are not necessarily independent of one another but may be time-related in the following manner:

$$\varepsilon_i = \varrho\varepsilon_{i-1} + \delta_i, \qquad -1 < \varrho < 1, \tag{7.2.2}$$

where δ_i are normally, independently distributed random variables with zero mean and constant variance, σ^2. Equation (7.2.2) models the ith error, ε_i, as a fraction, ϱ, of the previous error, ε_{i-1}, plus an independent error component, δ_i. If the error structure is more complex, eqn. (7.2.2) is still often useful as a simple approximation to the true relationship among the errors. The Durbin-Watson test statistic is defined by

---

*For more details on the Durbin-Watson test see the texts by Chatterjee and Price (1978), Seber (1977), or Theil (1971), or the papers by Durbin and Watson (1950, 1951, 1971).

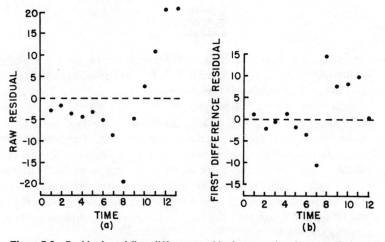

Figure 7.2 Residuals and first-difference residuals versus time from prediction of homicide rate from number of handgun licenses: (a) Raw residual versus time, (b) First-difference residual versus time.

$$d = \frac{\sum_{i=2}^{n}(r_i - r_{i-1})^2}{\sum_{i=1}^{n} r_i^2}, \qquad (7.2.3)$$

and is used for testing the null hypothesis $H_o: \varrho = 0$ versus the alternative $H_a: \varrho > 0$. If $\varrho = 0$, then $\varepsilon_i = \delta_i$ and the errors are uncorrelated; if $\varrho > 0$, the errors are positively time-correlated. To test $H_o: \varrho = 0$ versus $H_a: \varrho < 0$ the same procedure outlined below can be used but the statistic is $4.0 - d$.

Durbin and Watson (1950, 1951, 1971) formulated bounds (d_L, d_U) for the test statistic d. The decision rule was postulated as follows:

i)   if $d < d_L$ reject H_o,

ii)   if $d > d_U$ do not reject H_o

iii)   if $d_L < d < d_U$ draw no conclusion.

The bounds, d_L and d_U, are tabulated in Table C.6 of Appendix C for models with $p = 1$ to $p = 5$ predictor variables (other techniques for executing the Durbin-Watson test can be found in the above articles).

The homicide data and the corresponding first-difference residuals, $r_i - r_{i-1}$, from Table 7.4 will now be utilized to illustrate the Durbin-Watson test. Substituting these values into eqn. (7.2.3) yields

$$d = \frac{566.76}{1522.00} = 0.37$$

where $n = 13$ and $p = 1$. Although Table C.6 begins with a sample size of $n = 15$, it is evident that $d = 0.37$ is so small that $d < d_L$; hence, at the 5% significance level the null hypothesis is rejected and we conclude that there is positive correlation among neighboring error terms. The tendency for these residuals to be positively correlated is illustrated further in Figure 7.2(b) which plots the first-difference residuals against time. The presence of a positive correlation is evident here just as it was in Figure 7.2(a) using the raw residuals. Specifically, there are clearly two groups of first-difference residuals in Figure 7.2(b) rather than a random scatter of points. Hence, both when examined graphically and when tested statistically with the runs test or the Durbin-Watson test, the homicide residuals continue to demonstrate a strong correlation.

One difficulty with applying the Durbin-Watson test is deciding what to do when the statistic d falls in the range between d_L and d_U. This can be particularly troublesome because the interval (d_L, d_U) is often large. Many solutions have been suggested to resolve this dilemma such as reestimating the prediction equation, adding the interval (d_L, d_U) to the rejection region, or approximating the distribution of d with various probability distributions (see Durbin-Watson, 1971).

A second and perhaps more critical difficulty with the Durbin-Watson test is its dependence on the validity of the assumption that the model is correctly specified. Suppose the true regression model is

$$Y_i = \alpha + \beta_1 X_{i1} + \beta_2 X_{i2} + \varepsilon_i$$

but one erroneously assumes that the correct model is

$$Y_i = \alpha + \beta_1 X_{i1} + \varepsilon_i^* .$$

The error term ε_i^* now contains the original error term ε_i and the misspecified portion of the model, $\beta_2 X_{i2}$. If X_{i2} either increases or decreases with Y_i, the Durbin-Watson test could reject the hypothesis that $\varrho = 0$ because of the model misspecification.

7.2.3 Detecting Heteroscedasticity

There are several different statistical tests available for deciding whether the error terms in a regression model have unequal variances. One relatively simple approach is based on examining plots of the residuals versus the predicted responses.

Using the definition of residuals and predicted values it can be shown algebraically that the correlation coefficient between r_i and $\hat{Y}_i$ is always zero for prediction equations with intercepts. Hence, any graph of r_i versus $\hat{Y}_i$ should reflect a random scatter of points about a line with zero slope. Systematic patterns, however, would indicate an inadequate model or a changing variance and it is this property that makes plots of residuals against predicted responses so useful.

In detecting heteroscedasticity (error terms with unequal variances) it is also helpful to plot squared residuals versus $\hat{Y}_i$. This is because r_i^2 reflects the contribution of a given response to the error sum of squares, which is an estimate of σ^2. [Recall $\hat{\sigma}^2 = \Sigma r_i^2/(n-p-1)$.] If the squared residuals are varying in a systematic way, such as increasing or decreasing with $\hat{Y}$, a plot of r_i^2 versus $\hat{Y}_i$ should detect such trends. Many computer packages provide for plots of both r_i and r_i^2 against the predicted responses. The advantage of the r_i versus $\hat{Y}_i$ plot is that it can detect model inadequacies and outliers in addition to unequal error variances; the advantage to plotting r_i^2 versus $\hat{Y}_i$ is that such a plot can accentuate some types of trends existing between r_i and $\hat{Y}_i$.

One type of plot that should be avoided is that of r_i versus the observed response, Y_i. These variables are generally correlated and a plot of r_i versus Y_i usually indicates a linear trend even if the fit is excellent and all model assumptions are valid.

Detection of model inadequacies through residual plots requires a reasonable number of observations in order to determine any but the most obvious deficiencies. For this reason we turn to the police-applicant data, Data Set A.6, to illustrate the use of residual versus predicted response plots. Figure 7.3 is a graph of the residuals versus the predicted reaction times for the 50 male police applicants. A careful scrutiny of this figure reveals a slight tendency for the variation in the residuals to increase with the predicted reaction times. This tendency, if true, would be an example of a nonconstant error-variance which would require a transformation of the response variable. If one covers up the single largest positive and negative residuals, however, the trend disappears and there does not appear to be any reason to suspect nonconstancy of the error variances. Once again we must be careful not to let one or two plotted points unduly influence conclusions that are drawn from graphical techniques.

Figure 7.4, a plot of the squared residuals versus predicted reaction time, accentuates the above tendencies. If the variation of the residuals did indeed increase with the reaction time the trends should be clearer in this graph. Yet what is most evident from Figure 7.4 is that two of the residuals look suspiciously large in magnitude. Again covering these two points no systematic trends of the squared residuals with the predicted reaction times appear in the plot. The reader may wish to decide whether these two points are

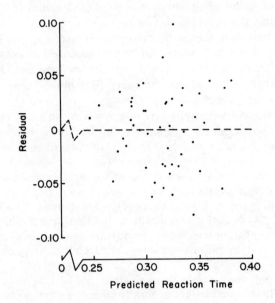

Figure 7.3   Residuals versus predicted reaction time for
fourteen-variable prediction equation.

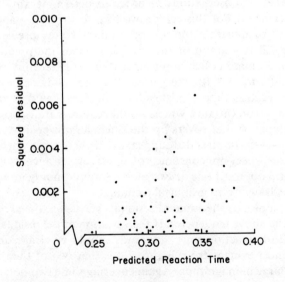

Figure 7.4   Squared residuals versus predicted reaction time
for fourteen- variable prediction equation.

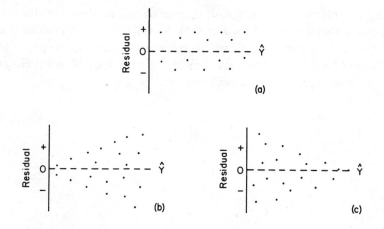

Figure 7.5 Residual patterns in detecting heteroscedasticity of error variances: (a) Equal error variances, (b) Increasing error variances, (c) Decreasing error variances.

outliers by computing and examining the various residual types discussed in Section 7.1.

Figure 7.5 is a schematic of residual versus $\hat{Y}$ plots for regression models with error variances that are a) constant, b) increasing with the magnitude of the response, and c) decreasing with the magnitude of the response. The trends in Figures 7.5(b) and 7.5(c) are signals which should alert the data analyst to inequalities among the error variances. Transformations of Y which are often recommended so the errors of the transformed responses have a constant variance include log Y, $\sqrt{Y}$, and $1/Y$.

7.2.4 Normal Probability Plots

Checking that the error terms are normally distributed must be an important concern of the data analyst if tests of hypotheses or confidence intervals are to be utilized as part of the regression analysis. While there are many different techniques for assessing the normality of the error terms, one of the most popular graphical approaches is the normal probability plot.

Let $r_{(1)} < r_{(2)} < ... < r_{(n)}$ be the n ordered raw residuals (i.e., the residuals ordered from largest negative to largest positive). In a normal probability plot one plots $r_{(i)}$ versus $100 \cdot (i - 1/2)/n$ using special graph paper called normal probability paper (see Figure 7.18). If the errors are normally distributed, these points should lie roughly on a straight line; otherwise, in the case of non-normality or in the presence of oversized residuals, the data will not follow a general linear trend. Some random fluctuations will occur in this type of plot since the residuals themselves are random variables. The variation is also a result of the fact that the residuals are correlated. Such wobbling reduces as the sample size increases and it has been recommended (Daniel and Wood, 1971) that the sample size be at least 20 and preferably larger than 50 in order to effectively evaluate the plot.

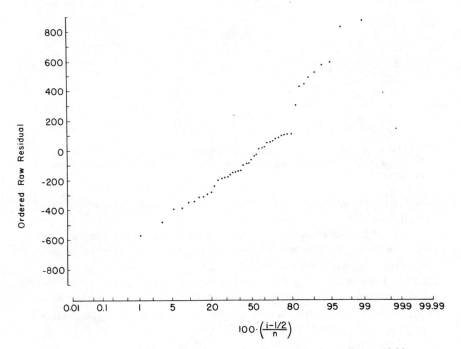

Figure 7.6   Normal probability plot for GNP residuals with six predictor variables.

An example of a normal probability plot is given in Figure 7.6 for the GNP residuals using the six predictor variables of Data Set A.1. The vertical, scale denotes possible values of the 47 ordered residuals while the horizontal scale contains values of $100 \cdot (i - 1/2)/n$, $i = 1, 2, ..., 47$. For example, the largest negative residual, $r_{(1)} = -575$, is plotted on the vertical scale versus $100 \cdot (1 - 0.5)/47 = 1.06$ on the horizontal scale; and similarly for the remaining 46 ordered residuals. The resultant plot fluctuates systematically away from any straight line one attempts to fit through the points. The variation from a linear trend is much more than should be expected from random fluctuations and leads one to conclude that the GNP errors are not normally distributed.

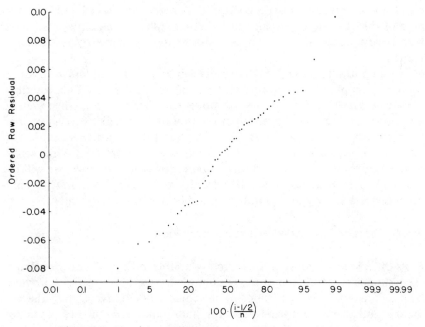

$$100 \cdot \left(\frac{i - 1/2}{n}\right)$$

Figure 7.7   Normal probability plot for reaction time residuals with
fourteen predictor variables.

In contrast Figure 7.7 is a normal probability plot for the residuals of the
reaction time data of Data Set A.6. This residual plot much more closely
resembles that of a straight line than does Figure 7.6. There is a slight
tendency to fluctuate away from a line, particularly for a few points at
either end, but a linear trend is characteristic of most of the data points.
Minor variations in the extremes and systematic fluctuations of three or
four points are common in normal probability plots even when the data is
truly normally distributed. Gross systematic departure, however, such as
those appearing in Figure 7.6 are evidence of violations of the assumption
of normality for the error terms.

7.3   MODEL MISSPECIFICATION

Correct model specification in a regression analysis involves two important
aspects. All relevant variables must be contained in the data base and the
proper functional form of each predictor must be defined in the prediction
equation. Both of these requirements have been discussed and examined
repeatedly in each of the prior chapters of this text. Apart from the lack of
fit outlined in Section 6.3.3, however, little mention has been made of the

problems of detecting misspecification in a regression model. It is the purpose of this section to present some useful graphical techniques associated with residual analysis that aid the data analyst in correctly specifying a model.

Two important types of residual plots will be examined: plots of residuals against each predictor variable and partial residual plots. Each of these graphs provides a different view of model specification and, when the two are combined, they yield a useful evaluation of the predictor variables and their correct functional expressions. The question of whether to use raw residuals or scaled residuals in these plots is always present. While scaled residuals are preferable due to their standardization properties, we will use the simple raw residuals since they are the most common output of computer regression programs and, hence, will be seen most frequently.

7.3.1 Plots of Residuals Versus Predictor Variables

Plots of residuals against each predictor variable indicate the distribution of the residuals as a function of the predictor variables. The patterns provided by the plots are helpful in choosing the correct functional form of the predictors and determining the need for additional terms in the prediction equation.

From the definition of the residuals, r_i, it can be shown that the correlation between r_i and each predictor variable, X_j, is zero. Hence, there should be no discernible trend in the plot of r_i versus any predictor, only a random scatter of points about the line $r_i = 0$. Such a horizontal band of residuals would be an indication that the specification of X_j is satisfactory.

The primary patterns resulting from a plot of the residuals against each predictor variable are similar to those of plots against the predicted response. Their interpretations, however, are not necessarily the same. If the plot is wedge-shaped like Figure 7.5(b) or 7.5(c), there is an indication that the error variances are not homogeneous and some type of transformation is needed, perhaps $Y/\sqrt{X_j}$ or $Y\sqrt{X_j}$. Curvilinear trends in the plots of residuals against X_j indicate the need for extra variables in the model or some type of transformation on the observations. For example, Figure 7.8 contains a plot of the residuals from the prediction of first year college grade averages. The prediction equation utilized is eqn. (2.2.1), which is discussed in Section 2.2.1. The need for a quadratic term in the grade 13 averages is more evident in Figure 7.8 than it was in Figure 2.6 (although it is suggested by Figure 2.6) and the curvilinear trend corroborates the suggestions made after viewing plots of the smoothed data, Figure 2.7.

Residual plots are also useful in deciding if an interaction term, such as X_iX_j, should be included in a regression model. Under the assumptions of

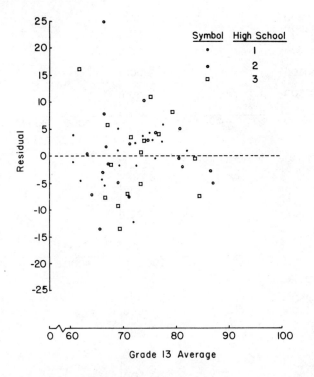

Figure 7.8   Residuals from prediction of first year of college
grade averages by grade 13 averages.

the model, the effects on the response of varying any predictor, X_i, should
be uninfluenced by changes in another predictor, X_j. Hence, a plot of the
residuals from a prediction equation that excludes X_iX_j against the interaction term, X_iX_j, should be a horizontal band of residuals. If there is some
other pattern in the plot, then the residuals and interaction term are correlated and some function of X_iX_j should be added to the prediction equation.

   An example of the use of plots of residuals against predictor variables to
detect model misspecification can be seen from an analysis of the GNP data.
Figure 7.9 contains scatter plots of GNP against each predictor variable. The
plots in Figures 7.9(a) and 7.9(b), and, to a lesser extent, in Figures 7.9(c) and
7.9(d), closely resemble the curve in Figure 2.8(d), $Y = 1/X$, or an inverted
Figure 2.8(f), $Y = -\ln(X)$. The graph of GNP against LIT is a mirror-image of
these curves (residual plots versus the predictor variables also show systematic trends). Since the patterns in these graphs are essentially the same, it would
appear that a transformation on the response variable, such as $\ln(GNP)$
or $(GNP)^{-1}$, is needed rather than some respecification of the six predictor

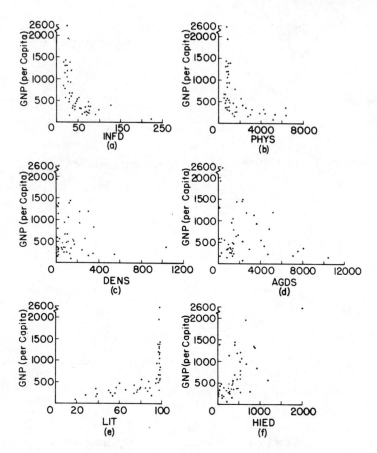

Figure 7.9 Scatter plots of GNP versus the following predictor variables: (a) INFD, (b) PHYS, (c) DENS, (d) AGDS, (e) LIT, (f) HIED.

variables. Since ln(Y) does not curve as dramatically as $1/Y$ (see Figure 2.8), we will first attempt to apply this response transformation.

The resultant prediction equation of ln(GNP) versus the six predictor variables is

$$\ln(\hat{GNP}) = 5.32 - .00846(INFD) + .00003(PHYS) + .00051(DENS)$$

$$- .00009(AGDS) + .01450(LIT) + .00059(HIED). \qquad (7.3.1)$$

In order to assess whether misspecification is still a problem, consider the plots in Figure 7.10 of the residuals from the above fit versus the predictor variables. No systematic patterns are evident in any of the graphs except a random scatter of points about the zero line. Hence, there appear to be no

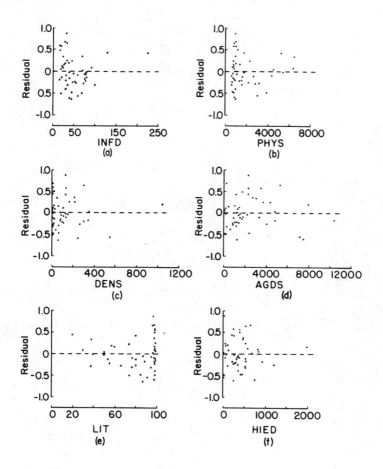

Figure 7.10   Residuals of ln(GNP) versus the following predictor variables:
(a) INFD, (b) PHYS, (c) DENS, (d) AGDS, (e) LIT, (f) HIED.

anomalies in the predictor variables; the chosen transformation has removed the nonlinearity present in the original data set.

A second possible transformation of GNP that was mentioned above is to fit $(GNP)^{-1}$ to the predictor variables in Data Set A.1. Using this new response variable, the prediction equation becomes

$$(GNP)^{-1} = .004103 + .000038(INFD) - .0000001(PHYS)$$

$$- .0000021(DENS) + .0000003(AGDS)$$

$$- .000042(LIT) - .0000004(HIED). \qquad (7.3.2)$$

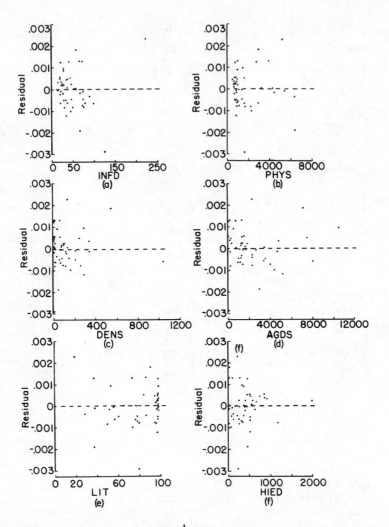

Figure 7.11   Residuals of $(GNP)^{-1}$ versus the following predictor variables:
(a) INFD, (b) PHYS, (c) DENS, (d) AGDS, (e) LIT, (f) HIED.

A plot of the residuals versus each of the six predictor variables is given in
Figure 7.11 and indicates no systematic departures from a horizontal scatter
of points. Both the inverse transformation and the logarithmic one* appear

---

*It is common in the economic and statistical literature to use ln(GNP), so this transformation
might be preferable to $(GNP)^{-1}$ due to the recurring use of ln(GNP) on similar data sets.

to substantially improve the fit while removing the nonlinearity existing between GNP and the predictor variables.

As a check on the error assumptions of the fit provided by eqn. (7.3.2), the resultant residuals were plotted against the predicted responses in Figure 7.12. The data is scattered uniformly about the zero line and no obvious systematic patterns are evident. Further, the normal probability plot in Figure 7.13 of the same residuals appears to have a linear trend (compare this plot with Figure 7.6). The indication from these two plots then, is that the inverse transformation does not lead to any apparent violations of the error assumptions but instead provides an adequate fit to the GNP data.

7.3.2 Partial Residual Plots

Useful supplements to plots of residuals against each predictor variable are partial residual plots. A partial residual is defined as

$$r_i^* = Y_i - (\hat{Y}_i - \hat{\beta}_j X_{ij}) \; = r_i + \hat{\beta}_j X_{ij} \,. \tag{7.3.3}$$

Recall that

$$\hat{Y}_i = \hat{\beta}_0 + \hat{\beta}_1 X_{i1} + \hat{\beta}_2 X_{i2} + \dots + \hat{\beta}_p X_{ip}, \qquad i = 1, 2, \dots, n \,.$$

Thus $\hat{Y}_i - \hat{\beta}_j X_{ij}$ is a predictor of the ith response using all predictor variables *except* X_j; consequently, $Y_i - (\hat{Y}_i - \hat{\beta}_j X_{ij})$ is called a *partial* residual. The plot of r_i^* against the predictor variable, X_j, allows one to examine the relationship between Y_i and X_j after the effects of the other predictors have been removed. Viewed in this context, partial residual plots offer much the same information as Y versus X_j plots, only a great amount of the variability of Y that is due to the other predictor variables is first removed.

Partial residual plots are especially useful in specifying a regression model. While the usual residual plots indicate *deviations* of each predictor variable from linearity, the partial residual plots can be used to access the *extent* and the *direction* of the linearity. These plots can therefore be used to determine the importance of each predictor in the presence of all the others and assess the extent of nonlinearity in a given predictor. This is in addition to providing information on correct transformations and extreme data points.

One useful property of partial residuals is that the regression of r_i^* versus X_j, through the origin, yields a slope equal to $\hat{\beta}_j$, the coefficient of X_j from the fit of the full model. The prediction equation of r_i^* on X_j is

$$\hat{r}_i^* = \hat{\beta}_j X_{ij} \,. \tag{7.3.4}$$

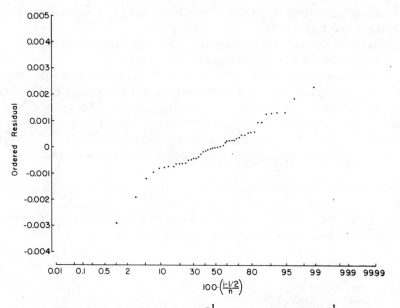

Figure 7.12   Residuals of $(GNP)^{-1}$ against predicted $(GNP)^{-1}$.

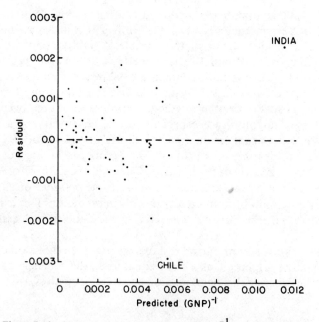

Figure 7.13   Normal probability plot for $(GNP)^{-1}$ ordered residuals.

This additional feature of a slope of $\hat{\beta}_j$, compared to the zero slope resulting from the residual plot against X_j, is the property that allows partial residual plots to provide information on the direction and magnitude of the linearity as well as the nonlinearity of X_j.

As an example of partial residual plots consider the homicide data in Data Set A.3. In Section 5.1.2 a prediction equation from the regression of homicide rate (Y) on full-time police (X_1), handgun licenses (X_2), and handgun registrations (X_3) was given by eqn. (5.1.11):

$$\hat{Y} = -67.95217 + 0.28033X_1 + 0.01176X_2 + 0.00255X_3. \quad (7.3.5)$$

The resultant residuals and partial residuals from eqn. (7.3.5) are contained in Table 7.5. Figure 7.14 contains a plot of the raw residuals versus each of three predictor variables, while Figure 7.15 illustrates the corresponding partial residual plots. Note immediately how the two figures differ.

In the raw residual plots of Figure 7.14 the data are scattered about the line $r_i = 0$ so that only *deviations* from linearity are apparent. Figure 7.14(a) suggests an exponential transformation of full-time police might be desirable, while the addition of quadratic terms due to handgun licenses and handgun registrations is suggested in Figures 7.14(b) and 7.14(c). The strengths of the associations of the three predictors with homicide rate, however, are not indicated in any of the figures.

Consider now the partial residual plots. In Figure 7.15(a) FTP and the partial residuals appear to be strongly correlated since the plot indicates a

Table 7.5. Residuals and Partial Residuals from Prediction of Homicide Rate

Year	Homicide Rate Residual	Partial Residual Full-Time Police	Handgun License	Handgun Registration
1961	.92	73.91	3.02	1.47
1962	-1.08	74.55	.76	-.62
1963	-2.65	73.61	-.32	-2.12
1964	-2.88	73.64	-.27	-2.29
1965	.32	76.71	3.87	1.08
1966	3.72	76.98	8.32	4.66
1967	4.53	79.91	12.36	6.10
1968	-2.93	80.04	10.37	-.30
1969	-2.09	87.58	7.76	-.08
1970	-1.54	94.17	7.81	.28
1971	2.71	102.68	12.33	4.62
1972	.11	105.71	6.97	2.73
1973	.85	110.23	9.20	2.55

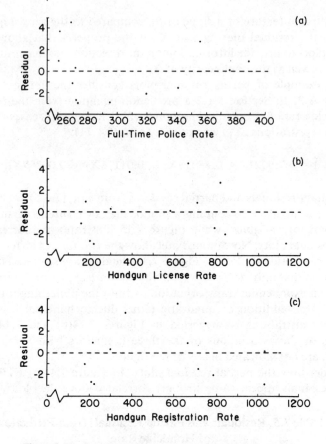

Figure 7.14   Residuals of homicide rate against the following
predictor variables: (a) Full-time police rate, (b) Handgun license
rate, (c) Handgun registration rate.

tight fit of a straight line with slope 0.280. Recall from eqn. (7.3.4) that the indicated slope of the best fitting line through these points is simply the coefficient of FTP in eqn. (7.3.5). This corresponds closely to the results in Table 5.1 where it was found that the coefficient of determination between HOM and FTP is 0.929. No outliers nor deviations from linearity are apparent in this plot so one can conclude that FTP is properly specified in the original regression model.

The fit between HOM and LIC in Figure 7.15(b) is much weaker. Although the best linear fit to the partial residuals is a line through the origin with slope 0.012, the relationship is only moderately strong due to the scatter of the data. On closer examination it appears that a logarithmic transformation of LIC might be needed rather than an additional quadratic term

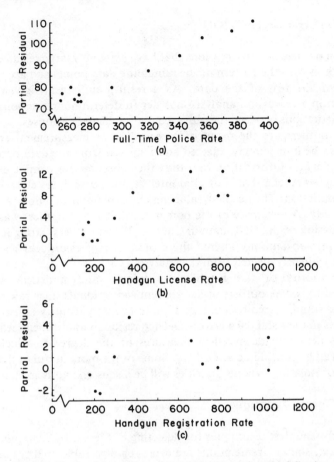

Figure 7.15 Partial residuals of homicide rate against the following predictor variables: (a) Full-time police rate, (b) Handgun license rate, (c) Handgun registration rate.

as was indicated in the residual plot in Figure 7.14(b). Hence, while the usual residual plots did detect the possible need for a curvilinear term in LIC, the partial residual plots allow the analyst to view the linearity and nonlinearity of LIC simultaneously so that the correct transformation can be chosen more precisely.

The final partial residual plot in Figure 7.15(c) illustrates the relationship between HOM and GR. The regression line through the thirteen points has a slope of 0.003, and the small magnitude of the slope is a direct consequence of the large scatter in the points and the apparent need for a transformation of GR. In the residual plot in Figure 7.14(c) a quadratic term was indicated; the partial residual plot appears to confirm this need.

7.4   OUTLIER DETECTION

Potential outliers are observations that have extremely large residuals. They do not fit in with the pattern of the remaining data points and are not at all typical of the rest of the data. As a result, outliers are given careful attention in a regression analysis in order to determine the reasons for the large fluctuations between the observed and predicted responses.

Some outliers are the result of a recording or measurement error and hence can be immediately rejected and removed from the data base. If the outlier provides information that the other observations cannot due to an unusual phenomena that is of vital interest, then it requires closer scrutiny and examination. The peculiar value may be explaining an important aspect of the study. An example of the potential influence of outliers was seen in the discussion of the GNP data in Chapter 2: two outliers (Hong Kong and Singapore) substantially altered the estimated regression coefficients and were subsequently deleted from the analysis.

The majority of the graphical techniques and statistical measures employed to isolate outliers utilize studentized residuals since raw residuals as well as standardized residuals can fail to properly identify outliers due to variations in their standard errors. Deleted residuals and studentized deleted residuals also provide excellent measures of the degree to which observations can be considered as outliers. Some of the more useful of these techniques for isolating data peculiarities will be discussed in this section.

7.4.1   Plotting Techniques

One of the simplest techniques for detecting outliers is by examining a plot of the data, since extreme points are often readily visible on a graph. While all the different types of plots discussed in this chapter can be used to isolate outliers, the more common ones include scatter plots of the original response variable against each predictor variable, relative frequency histograms of the studentized residuals, and plots of residuals against either the fitted response or each predictor variable. In addition, a relatively new graphical technique that plots residuals against deleted residuals can be extremely helpful in detecting outliers.

Scatter plots were shown in Chapter 2 to be complimentary to residual plots in that extreme observations for the response or predictor variables are often easily recognized. The scatter plot in Figure 2.5 between PHYS and AGDS was used to isolate Hong Kong and Singapore as outliers. Another example was given in Figure 2.4 which illustrated a plot of GNP versus INFD. Three data points corresponding to Canada, India, and the United States were labelled as potential extreme values but were later not considered to be outliers. A third example was plotted in Figure 2.3. In all these cases the outlying observations were determined not from residuals but from the location of the data points in the scatter plots.

Graphs of residuals can sometimes be much more informative than simple scatter plots in that they reflect the relationship between the observed and predicted responses. One such graph is the frequency histogram of studentized residuals. Since these residuals are approximate standard normal variates (under the assumptions of the regression model), observations with residuals that are larger than three in absolute value (corresponding to three standard errors) can be considered for rejection if the sample size is moderately large. Residuals that are between two and three in absolute magnitude may not necessarily be outliers but should warrant close scrutiny by the data analyst.

An example of a histogram of residuals is given in Figure 7.16 for the studentized residuals from the prediction of $(GNP)^{-1}$ using the six predictor variables in Data Set A.1. There are two large residuals corresponding to India ($t = 3.49$) and Chile ($t = -3.31$) that indicate potential outliers and these observations need to be examined more closely before considering rejection. Two moderately large residuals associated with Barbados ($t = 2.20$) and Malaya ($t = -2.29$) may also be troublesome and these data points also should be scrutinized.

Other useful residual plots for detecting outliers are plots of residuals against the predicted response and against each predictor variable. Figure

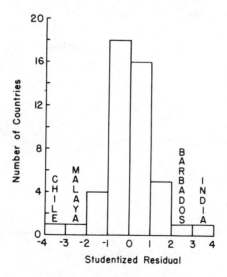

Figure 7.16   Relative frequency histogram of studentized residuals of $(GNP)^{-1}$.

7.12 is an example of a residual plot against the predicted $(GNP)^{-1}$ data discussed above. Two observations, Chile and India, could be considered as outliers since they have relatively large residuals. Once again, however, we must be careful not to judge outliers only on the basis of a single plot since most plots have one or two points that will appear, at first glance, to be somewhat discrepant from the remaining points. In the $(GNP)^{-1}$ residual plots against the six predictors illustrated in Figure 7.11, Chile and India are again isolated as having large residuals relative to the remaining points. This adds corroboration to the earlier indication that these two countries are outliers.

A relatively new graphical technique for detecting outliers is to plot raw residuals against the deleted residuals. The ith deleted residual, $r_{(-i)}$, as defined in Section 7.1, is the residual obtained from the prediction equation with the ith observation deleted. In other words, $r_{(-i)}$ reflects the change in the fit when the ith case is removed. If there is no change, the plot of r_i versus $r_{(-i)}$ should be approximately a straight line through the origin with a slope of one. If any points are relatively far from this line the observations are potentially outliers and need to be examined more closely.

An example of such a plot is illustrated in Figure 7.17 for the prediction of $(GNP)^{-1}$. Six countries are labelled on the graph to indicate points that

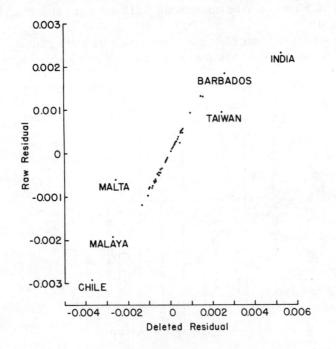

Figure 7.17   Raw residuals versus deleted residuals from prediction of $(GNP)^{-1}$.

deviate from the 45° line through the origin. Of these, the observations corresponding to India, Malta, and Taiwan are far enough away to require more careful scrutiny. The remaining three points which include Barbados, Chile, and Malaya, do not appear to be of sufficient magnitude to severely distort the regression coefficients. In fact, these appear to be extensions of the data patterns and not outliers.

As is obvious from this example, deciding to discard or accept the outlying observation is subjective. Also, model misspecification, incorrect error assumptions, or the presence of several outliers can distort the magnitude of the residuals. In all three situations the model coefficients will be poorly estimated and the residuals might not truly reflect the problems with the data. For this reason we still recommend that two variable plots be constructed so that potential difficulties can be better identified by examining both types of plots.

7.4.2 Statistical Measures

Reviewing the various conclusions arrived at from the plotting techniques applied to the $(GNP)^{-1}$ data, it is evident that the results are not uniform. While six countries—Barbados, Chile, India, Malaya, Malta, and Taiwan—appear to be outliers, it is uncertain which ones, if any, should be eliminated from the regression analysis. There is a need for a sensitive outlier measure which assesses both the degree to which an observation is an outlier as well as the relative sensitivity of the estimated regression coefficients to such outliers.

One suggested measure reflects the distance between the estimated regression coefficient vector, $\hat{\underline{B}}$, from the full model with all observations, and the corresponding estimated coefficient vector, $\hat{\underline{B}}_{(-i)}$, from the model with the ith observation removed. This statistic indicates the extent to which the prediction equation would change if the ith observation was deleted. It is based on the concept of confidence regions as expressed in eqn. (6.4.4). A $100(1 - \gamma)\%$ confidence region for $\underline{B}$ is

$$\frac{(\hat{\underline{B}} - \underline{B})' X^{*\prime} X^* (\hat{\underline{B}} - \underline{B})}{(p + 1)\text{MSE}} \leqslant F_{1-\gamma}(p + 1, n - p - 1) \qquad (7.4.1)$$

where $F_{1-\gamma}(p + 1, n - p - 1)$ is the $1 - \gamma$ probability point of the F distribution with $p + 1$ and $n - p - 1$ degrees of freedom; i.e., all coefficient vectors $\underline{B}$ which satisfy equation (7.4.1) are reasonably consistent with the least squares coefficient estimates.

The following distance measure is similar to a confidence region for $\hat{\underline{B}}_{(-i)}$,

$$D_i = \frac{[\hat{\underline{B}} - \hat{\underline{B}}_{(-i)}]' X^{*\prime} X^* [\hat{\underline{B}} - \hat{\underline{B}}_{(-i)}]}{(p + 1)\text{MSE}} \qquad (7.4.2)$$

By calculating D_i, finding a tabled F value, $F_{1-\gamma}(p + 1, n - p - 1)$, corresponding to the numerical value of D_i and observing the corresponding value, $1 - \gamma$, one can determine the size* of the confidence region containing $\hat{\underline{B}}_{(-i)}$. Removing the ith data point should keep $\hat{\underline{B}}_{(-i)}$ close to $\hat{\underline{B}}$, unless the ith observation is an outlier. Cook (1977) suggests retaining an observation if, upon its removal, $\hat{\underline{B}}_{(-i)}$ stays within a 10% confidence region; i.e., close to $\hat{\underline{B}}$. If $\hat{\underline{B}}_{(-i)}$ is in a region larger than 10%, this indicates that deleting the observation has changed $\hat{\underline{B}}_{(-i)}$ relative to $\hat{\underline{B}}$; the ith point, therefore, is an outlier and should be considered for rejection.

The distance measure, D_i, can be expressed in several equivalent forms; for example,

$$D_i = \frac{h_{ii} r_{(-i)}^2}{(p + 1)\text{MSE}} . \qquad (7.4.3)$$

Equation (7.4.3) demonstrates that D_i is a function of the deleted residuals, $r_{(-i)}$, and the diagonal elements, h_{ii}, of the H matrix defined in Section 7.1. A further simplification for D_i is

$$D_i = \left(\frac{t_i^2}{p + 1}\right)\left(\frac{h_{ii}}{1 - h_{ii}}\right) \qquad (7.4.4)$$

where t_i is the ith studentized residual and a measure of the degree to which an observation is considered an outlier. The ratio $h_{ii}/(1 - h_{ii})$ indicates the influence that the associated response, Y_i, has in the determination of $\hat{\underline{B}}$, with a large value implying the observation is critical to the calculation. Combined in eqn. (7.4.4) these two measures indicate the overall impact of the ith observation in a prediction equation.

An observation can have a large influence on the prediction equation if it has either a large studentized residual or a large $h_{ii}/(1 - h_{ii})$ ratio, or if it has a moderately large t_i value and ratio. One precaution to observe occurs when h_{ii} is one. In such cases, the measure, D_i, is defined to be zero. Also, additional information can be gained by separately examining t_i, $h_{ii}/(1 - h_{ii})$, and D_i. An example will help to illustrate these points.

Consider the prediction equation for $(\text{GNP})^{-1}$ and the six countries previously considered to be potential outliers. Table 7.6 lists the values of h_{ii}, t_i, $h_{ii}/(1 - h_{ii})$, D_i, and $t_{(-i)}$ for each of these six observations. The response

---

*The value of γ is a "descriptive" significance level and does not have the familiar "p-value" interpretation. This is because the distance measure D_i is not an F statistic but is compared to an F table because of the similarity of eqn. (7.4.2) to eqn. (7.4.1). Thus D_i is not used as a test statistic for outliers but only as an indicator of the closeness of $\hat{\underline{B}}_{(-i)}$ to $\hat{\underline{B}}$.

Table 7.6. Distance Measures for Prediction of $(GNP)^{-1}$

Country	h_{ii}	t_i	$h_{ii}/(1 - h_{ii})$	D_i	$t_{(-i)}$
Barbados	.297	2.228	.423	.300	2.351
Chile	.214	-3.347	.270	.432*	-3.895
India	.564	3.534	1.295	2.310***	4.207
Malaya	.285	-2.315	.398	.305	-2.456
Malta	.756	-1.275	3.093	.718**	-1.285
Taiwan	.614	1.549	1.593	.546*	1.578

$*F_{.10}(7,40) = .360$
$**F_{.25}(7,40) = .603$
$***F_{.95}(7,40) = 2.25$

associated with India has the greatest impact on the determination of the regression coefficients according to the magnitude of the D_i. Removal of India will move the estimated coefficient vector $\hat{B}_{(-40)}$, to the edge of a 95% confidence region around $\hat{B}$, since $D_{40} = 2.253$ exceeds $F_{.95}(7,40) = 2.25$.* The second largest D_i value corresponds to Malta and is due to its large $h_{ii}/(1 - h_{ii})$ ratio. Its removal will place the deleted estimate of $\underline{B}$ at approximately the edge of a 25% confidence region around $\hat{B}$. The problem with India is its extremely large value for INFD coupled with its low value for GNP, while the problem with Malta is its unusually large DENS value. Two other countries, Chile and Taiwan, have distance measures that exceed the critical value for a 10% confidence region, but the distance measures for Barbados and Malaya are not sufficiently large to classify these observations as outliers.

The studentized deleted residuals in Table 7.5 clearly point to the removal of both Chile and India. Barbados and Malaya also have large studentized residuals but they are not as extreme as India or Chile when the $t_{(-i)}$ values are compared with the Student t table in Appendix C. Malta and Taiwan would not be judged to be outliers on the basis of their studentized deleted residuals.

Combining the results of Figures 7.12 and 7.17 with the statistics in Table 7.6 it appears that India and Chili should be deleted from the data set. We hesitate in deleting any of the other countries because (i) the outlier indicators in this and the preceding subsection do not consistently identify any of the other countries and (ii) the more countries that are deleted the more limited are the conclusions from the analysis. Thus, from the original 49

---

*The tabled values of $F(7,40)$ are sufficiently close approximations to those of $F(7,39)$ for our purposes.

countries in Data Set A.1, the discussions in this chapter and in Chapter 2 lead to the deletion of Chile, Hong Kong, India, and Singapore.

There are many other techniques available for isolating outliers (see the references in the introduction to this chapter). The leverage values, studentized residuals and studentized deleted residuals are most appealing. Combined with the distance measure, D_i, they offer excellent diagnostic ability. Graphical techniques serve as valuable compliments to these numerical rules.

APPENDIX

7.A Derivation of Deleted Residual

The ith deleted residual is defined for $i = 1, 2, \ldots, n$ as

$$
\begin{aligned}
r_{(-i)} &= Y_i - \underline{u}_i^{*\prime} \underline{B}_{(-i)} \\
&= Y_i - \underline{u}_i^{*\prime} [X_{(-i)}^{*\prime} X_{(-i)}^*]^{-1} X_{(-i)}^{*\prime} \underline{Y}_{(-i)}
\end{aligned}
\tag{7.A.1}
$$

where $X_{(-i)}^*$ is obtained by removing the ith row, $\underline{u}_i^{*\prime}$, from X^* and $\underline{Y}_{(-i)}$ is found by discarding the ith observation, Y_i, from $\underline{Y}$. It can be show (Hoaglin and Welsch, 1978), that

$$
[X_{(-i)}^{*\prime} X_{(-i)}^*]^{-1} = (X^{*\prime} X^*)^{-1} + \frac{(X^{*\prime} X^*)^{-1} \underline{u}_i^* \underline{u}_i^{*\prime} (X^{*\prime} X^*)^{-1}}{1 - h_{ii}}
$$

where $h_{ii} = \underline{u}_i^{*\prime} (X^{*\prime} X^*)^{-1} \underline{u}_i^*$. Hence,

$$
\begin{aligned}
r_{(-i)} &= Y_i - \underline{u}_i^{*\prime} \left[(X^{*\prime} X^*)^{-1} + \frac{(X^{*\prime} X^*)^{-1} \underline{u}_i^* \underline{u}_i^{*\prime} (X^{*\prime} X^*)^{-1}}{1 - h_{ii}} \right] X_{(-i)}^{*\prime} \underline{Y}_{(-i)} \\
&= Y_i - \underline{u}_i^{*\prime} (X^{*\prime} X^*)^{-1} X_{(-i)}^{*\prime} \underline{Y}_{(-i)} \\
&\qquad - \frac{\underline{u}_i^{*\prime} (X^{*\prime} X^*)^{-1} \underline{u}_i^* \underline{u}_i^{*\prime} (X^{*\prime} X^*)^{-1} X_{(-i)}^{*\prime} \underline{Y}_{(-i)}}{1 - h_{ii}} .
\end{aligned}
$$

Multiplying and dividing the first two terms of this equation by $(1 - h_{ii})$ yields

$$
r_{(-i)} = \frac{(1 - h_{ii}) Y_i - (1 - h_{ii}) \underline{u}_i^{*\prime} (X^{*\prime} X^*)^{-1} X_{(-i)}^{*\prime} \underline{Y}_{(-i)} - h_{ii} \underline{u}_i^{*\prime} (X^{*\prime} X^*)^{-1} X_{(-i)}^{*\prime} \underline{Y}_{(-i)}}{1 - h_{ii}}
$$

$$
= \frac{(1 - h_{ii}) Y_i - \underline{u}_i^* (X^{*\prime} X^*)^{-1} X_{(-i)}^{*\prime} \underline{Y}_{(-i)}}{1 - h_{ii}} .
$$

Now, $X^{*'}\underline{Y} = X^{*'}_{(-i)}\underline{Y}_{(-i)} + \underline{u}_i Y_i$ so that

$$r_{(-i)} = \frac{(1 - h_{ii})Y_i - \underline{u}_i^{*'}(X^{*'}X^*)^{-1}(X^{*'}\underline{Y} - \underline{u}_i^* Y_i)}{1 - h_{ii}}$$

$$= \frac{Y_i - h_{ii}Y_i - \underline{u}_i^{*'}(X^{*'}X^*)^{-1}X^{*'}\underline{Y} + \underline{u}_i^{*'}(X^{*'}X^*)^{-1}\underline{u}_i^* Y_i}{1 - h_{ii}}$$

$$= \frac{Y_i - h_{ii}Y_i - \underline{u}_i^{*'}\hat{\underline{B}} + h_{ii}Y_i}{1 - h_{ii}}$$

$$= \frac{r_i}{1 - h_{ii}} \tag{7.A.2}$$

where r_i is the ith raw residual.

EXERCISES

1. Derive the exponential fit to the salary data (Table 2.1) by regressing $\ln[(13,500 - Y)/Y]$ on time and making the inverse transformation once the regression coefficients are calculated. Recalculate the fit using an asymptote of 12,500 instead of 13,500. Does this latter fit improve or worsen the prediction of the observed responses over the fits displayed in Table 7.1 and Figure 7.1?

2. Calculate the leverage values, residuals and studentized residuals for the female suicide rates (Table 3.4) (a) only over the age groups 25-29 to 50-54 and (b) for the entire data set. By comparing the magnitudes and sign patterns in the two sets of residuals, is the poor fit to the entire data set apparent?

3. Compute leverage values, studentized residuals, deleted residuals, and studentized deleted residuals for the prediction of GNP from HIED in Table 2.5 using (a) the initial ten countries and Puerto Rico, and (b) the original ten countries and the United States. Interpret these statistics for each of the analyses and compare the interpretations with the visual impressions left by Figures 2.3(c) and (d).

4. Analyze the validity of the error assumptions for the salary data of Table 2.1. (Note: a sheet of normal probability paper is included in Figure 7.18.)

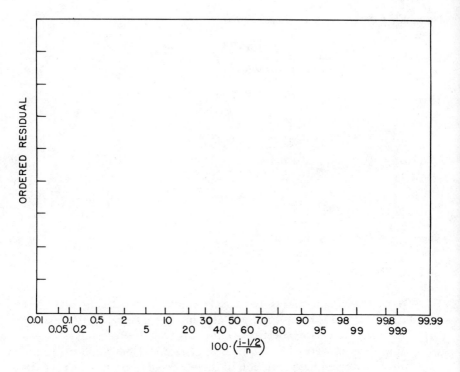

Figure 7.18. Normal probability plot paper.

5. In Figures 7.3 and 7.4, two residuals were identified as potential out-
    liers. Using any of the techniques of this chapter which you feel are
    relevant, determine whether Data Set A.6 contains any outliers which
    might substantially harm the fit of the response variable.

6. Plot the raw residuals of the linear fit to the salary data (Table 7.1)
    versus the original responses. Observe the strong but erroneous sug-
    gestion that a linear term in time is needed (the residuals have already
    been adjusted for a linear time term). Compare the trends in this
    graph with those of Figure 7.1(b). Which plot more clearly indicates
    the possible misspecification of the linear fit?

7. Make scatter plots of NOX versus each of the predictor variables in
    Data Set A.5. Next make scatter plots of NOX versus the square of
    each of the predictor variables. What strong similarities occur in the
    scatter plots? Why? Do the same similarities occur in
    residual plots vs. each predictor variable and residual plots vs. the
    square of each predictor variable?

8.   In conjunction with Exercise 7, graph partial residuals versus each of the predictor variables in Data Set A.5. What do the results of these two problems suggest regarding the fit of the emissions data beyond the need for linear terms (recall the discussion in the introduction to Chapter 6)?

9.   Plot raw residuals versus deleted residuals for the two eleven point data sets described in Exercise 3. Calculate distance measures D_i for the deletion of Puerto Rico in part (a) and the U. S. in part (b). Are the results of these analyses consistent with the discussion in Section 2.1.3 and the conclusions of Exercise 3?

10.   Perform the runs test and Durbin-Watson test on the homicide residuals when both FTP and LIC are used as predictor variables (Note: the residuals for the least squares fit are given in column 5 of Table 5.2). Are the conculsions that were drawn in the text from the fit of only LIC to HOM altered? Why or why not?

CHAPTER 8

VARIABLE SELECTION TECHNIQUES

Selection of a subset of predictor variables for use in a prediction equation when there are many potential variables available for use is a common problem encountered in regression analysis. Often a data analyst collects observations on a large number of predictor variables when it is unknown which specific ones are most influential on a response. Because of practical or economic reasons the large pool of variables must then be reduced to a smaller, more manageable set. One's goal in the variable selection process is to include only the more important predictor variables in the final prediction equation.

In order to illustrate the difficulties encountered in selecting subsets of predictor variables, consider the investigation reported in Data Set A.7. The purpose for the study from which the data are taken was to determine how low-income families use government-provided cash allowances to acquire adequate housing. Summary information is provided on 17 housing predictor variables [tenure characteristics ($X_1 - X_4$), dwelling unit features ($X_5 - X_{13}$), and neighborhood quality ($X_{14} - X_{17}$)] and one of the response variables, housing rent, investigated for 455 households located in and around Phoenix, Arizona. Although this is one of the largest data sets included in Appendix A, the original data base consists of several hundred variables and several thousand households. Inclusion of this data set has already necessitated some variable selection.

Table 8.1 contains the least squares regression coefficients for the complete seventeen variable model in column three. A preliminary examination

Table 8.1. Regression Coefficients for Three Prediction Equations, Housing Data

Variable Number	Variable Name	Full Model	Forward Selection	Backward Elimination
	CONSTANT	-123.879	-274.701	-121.063
X_1	RELLL	-24.970	-29.268	-26.795
X_2	RES15	-9.034		
X_3	RES510	-14.639		-10.635
X_4	RESG10	-15.038		
X_5	AREAPR	.348		.454
X_6	LTOTRMS	66.583	38.426	82.096
X_7	LTOTARE	10.411	43.577	
X_8	CENH	16.740	18.121	19.552
X_9	BADH	-8.828		
X_{10}	STOREF	9.965	14.674	13.095
X_{11}	APPL	10.835	11.843	
X_{12}	QUAL2	15.751	22.096	23.754
X_{13}	MULTI5	5.273		
X_{14}	CENFO1	5.720	4.732	5.159
X_{15}	CENFO2	3.052		
X_{16}	CENFO3	-3.275		
X_{17}	CNHF13	-2.320		
	R^2	.764	.731	.730
	$\hat{\sigma}^2$	21.981	23.216	23.275

of residuals as suggested in Chapter 7 revealed no obvious error assumption violations. The normal probability plot is approximately linear and the other residual plots show horizontal scatter.

No large pairwise correlations exist between the predictor variables except for LTOTRMS and LTOTARE ($r = 0.833$) and AREAPR and LTOTARE ($r = 0.729$). On closer scrutiny of the latent roots and vectors of the predictor variable correlation matrix, these relationships combine to form a three variable multicollinearity between AREAPR, LTOTRMS and LTOTARE. This is not surprising since all three variables are a measure of the area of the dwelling unit.

While the selected seventeen predictor variables appear to be useful in describing housing rent, there remains the question whether all are needed, or whether fewer variables could still offer an adequate fit. The costs and difficulty as well as the paperwork involved in maintaining information on housing characteristics for low-income families are sufficient reasons for desiring to reduce the size of the predictor variable set. The advantage of such a reduction would more than offset a small loss in the predictive ability of the regression equation.

Table 8.1 contains the regression coefficients for two eight variable equations derived from the original seventeen variable housing model. The variables associated with the values in column 4 were obtained through a forward selection (FS) procedure, i.e., variables were added to the equation in a stepping fashion, one-at-a-time, until eight were included. The variables with coefficients in column 5 were chosen through a backward elimination (BE) approach: all seventeen predictor variables were initially included in the equation and then one-by-one variables were eliminated until only eight remained. (Forward selection and backward elimination will be discussed more fully in Section 8.3.)

The coefficients of determination for both of these reduced equations are almost identical (0.731 versus 0.730); however, different sets of variables have been chosen. In the FS procedure LTOTARE and APPL are selected in addition to RELLL, LTOTRMS, CENH, STOREF, QUAL2, and CENFO1. In the BE approach these latter six predictor variables are included along with RES510 and AREAPR. The least squares coefficients for five of the six common variables are approximately equal but LTOTRMS has differing coefficients for each of the three prediction equations, possibly as a result of its multicollinearity with LTOTARE and AREAPR.

Despite the fact that these two methods, FS and BE, do not yield identical predictor variable subsets, both appear to provide an adequate fit to the housing data. After eliminating nine of the original seventeen predictor variables, the overall R^2 only decreases 0.03. Hence, using R^2 as a criterion, either set could be utilized for prediction; the choice would depend on both the purpose for which the prediction equation was constructed and the cost and difficulty of measuring the variables in each set.

In this chapter several of the more popular variable selection techniques will be discussed. Attention will be devoted not only to the choice of an appropriate technique but also to criteria needed for implementation of the technique (i.e., criteria for indicating when variable selection should terminate).

8.1 BASIC CONSIDERATIONS

Variable selection and variable specification are two closely-related considerations in fitting regression models. Variable selection comprises decisions to include or exclude predictor variables but these decisions are necessarily made on only the specifications determined by the researcher. If a predictor variable, say X_1, is related to a response variable through its logarithmic form, i.e.,

$$Y = \ln(X_1) + \text{other variables},$$

then failure to include the correct specification of X_1 can lead to its removal from the prediction equation. One would conclude that X_1 does not appreciably influence the response when, in fact, it does but through a different functional relationship than the one specified.

In this chapter we focus on variable selection techniques when a pool of predictor variables are available, some of which will collectively enable adequate prediction of the response variable. Included in the pool of potential predictor variables are any transformations of the raw predictor variables that are necessary. We thus assume the researcher has investigated various specifications of the predictor variables and retained any that are deemed worthy of further study. Our emphasis is on selecting an appropriate subset of these variables for use in a final prediction equation.

8.1.1 Problem Formulation

From the pool of potential predictor variables a theoretical regression model can be written in the usual form:

$$Y_i = \alpha + \beta_1 X_{i1} + \beta_2 X_{i2} + \ldots + \beta_p X_{ip} + \varepsilon_i \qquad i = 1, 2, \ldots, n \qquad (8.1.1)$$

or in matrix notation as

$$\underline{Y} = X^* \underline{B} + \underline{\varepsilon}. \qquad (8.1.2)$$

Both of these regression models are potentially misspecified in that there may be predictor variables included that have little or no ability to aid in the prediction of the response variable. These predictor variables, therefore, have regression coefficients that are zero and they should be removed from eqns. (8.1.1) and (8.1.2). The obvious question that follows is: which ones?

If there are, say, r predictor variables that should be removed from eqn. (8.1.2) and $k = (p + 1) - r$ variables that should be retained, one would like to partition eqn. (8.1.2) as

$$\underline{Y} = X_1^* \underline{B}_1 + X_2^* \underline{B}_2 + \underline{\varepsilon},$$

where X_1^* contains the n values of the k predictor variables that are to be retained and X_2^* contains the n values of the r predictor variables that are to be deleted from eqn. (8.1.2). As a general rule, the intercept term could be included in either X_1^* or X_2^* but it is usually retained in X_1^*. The coefficient vector $\underline{B}$ in eqn. (8.1.2) is partitioned similarly to X^* and if the variables in X_2^* do not aid in predicting the response, $\underline{B}_2 = \underline{0}$; i.e., the true regression model is

$$Y = X_1^* \underline{B}_1 + \underline{\varepsilon} . \tag{8.1.3}$$

Now let SSR and SSE be, respectively, the regression and error sums of squares from the least squares fit to model (8.1.2). Similarly, let SSR_1 and SSE_1 be the regression and error sums of squares from the least squares fit to the k variables in eqn. (8.1.3). When $\underline{B}_2 = \underline{0}$, models (8.1.2) and (8.1.3) are identical and both should produce about the same regression sum of squares [although one can show that $SSR_1 \leqslant SSR$ for any reduced model of the form given in eqn. (8.1.3)]. If $\underline{B}_2 \neq \underline{0}$, SSR should be considerably larger than SSR_1. The hypothesis H_0: $\underline{B}_2 = \underline{0}$ vs. H_a: $\underline{B}_2 \neq \underline{0}$ can be tested with the following statistic:

$$F = \frac{(SSR - SSR_1)/r}{MSE} \tag{8.1.4}$$

where $MSE = SSE/(n-p-1)$. The statistic in eqn. (8.1.4) follows an F distribution with degrees of freedom $v_1 = r$ and $v_2 = n-p-1$ when H_0 is true. Large values of F lead one to reject the hypothesis that $\underline{B}_2 = \underline{0}$.

As an illustration of this test procedure, let us collectively test whether the 9 predictor variables that are excluded in the FS column of Table 8.1 can be excluded from the prediction equation. Table 8.2 contains an ANOVA table in which SSR is calculated from the full prediction equation, SSR_1 is calculated from a second regression analysis using only the eight predictor variables in the FS column of Table 8.1, and $SSR-SSR_1$ is obtained by subtraction. The F statistic, eqn. (8.1.4), is 6.735 and is highly significant ($p <$ .001). Despite the fact that R^2 is about the same for the full model and the reduced one, the F statistic rejects the hypothesis $\underline{B}_2 = \underline{0}$ and one must conclude that at least one of the nine predictor variables should be retained.

This example dramatizes the need to be careful with selection criteria and aware of their strengths and weaknesses. As we shall see throughout this chapter, R^2 is a valuable measure of the adequacy of prediction but is not as sensitive to model misspecification as some of the other criteria to be discussed. The F statistic is more sensitive to model misspecification but it too must be used with care. In essence, $SSR-SSR_1$ measures the *additional* variability of the response that is accounted for by the variable in X_2^* above that which is accounted for by those in X_1^*. This is why we have labeled the regression sum of squares in the fourth row of Table 8.2 "X_2^* Adjusted for X_1^*." The F statistic (8.1.4) measures whether the predictor variables in X_2^* are extraneous when those in X_1^* are also included in the prediction equation. Still unanswered is the question of how one decides which variables to include in X_1 and which to place in X_2. We will delay addressing this question momentarily while additional selection criteria are introduced.

Table 8.2. Partitioned ANOVA Table for a Subset of the Housing Predictor Variables

Source	d.f.	S.S.	M.S.	F
Mean	1	7,674,339.080		
Regression	17	683,734.126		
Due to X_1^*	8	654,503.622		
X_2^* Adjusted for X_1^*	9	29,230.504	3,247.834	6.721
Error	437	211,160.041	483.204	
Total	455	8,569,233.247		

8.1.2   Additional Selection Criteria

The F statistic (8.1.4) is perhaps the most popular criterion for selecting variables in a regression model. It can be employed when X_2^* contains one or more predictor variables. Since the F statistic has a known probability distribution, cutoff values for F (or, equivalently, significance probabilities) can be calculated and exact statistical tests can be performed. It can also be adapted for inclusion in "automatic" variable selection procedures such as forward selection and backward elimination (see Section 8.3).

Other popular criteria for variable selection are functions of the error sum of squares, SSE_1, for reduced models of the form (8.1.3). The error mean square

$$MSE_1 = \frac{SSE_1}{n-k} \qquad (8.1.5)$$

and the coefficient of determination

$$R_1^2 = \frac{TSS(adj) - SSE_1}{TSS(adj)}$$

$$(8.1.6)$$

$$= 1.0 - \frac{SSE_1}{TSS(adj)}$$

of the reduced model (8.1.3) are often utilized as variable selection criteria. Subsets of predictor variables for which MSE_1 and/or R_1^2 do not substantially differ from MSE and R^2 for the full model (8.1.2) are viable alternatives to the full model since (i) fewer predictor variables are used, and (ii) measures of adequacy of prediction, MSE_1 and R_1, reveal that the predictive ability of the reduced models are equivalent to the full model.

Another criterion for assessing the adequacy of the fit of a reduced model is the C_k statistic*:

$$C_k = \frac{SSE_1}{MSE} - (n - 2k) .$$ (8.1.7)

This statistic is a function of the error sums of squares for both the reduced and full models and, like the F statistic (8.1.4), allows a direct comparison of the reduction in variability of the response that can be attributable to the two models.

An important feature of C_k is that it is an unbiased estimator of

$$\Gamma_k = \frac{E(SSE_1)}{\sigma^2} - (n - 2k) .$$ (8.1.8)

If $\underline{B}_2 = \underline{0}$, $E(SSE_1) = (n - k)\sigma^2$ and $\Gamma_k = k$. Thus, values of C_k that are near k (the number of predictor variables, including the constant term, retained in X_1) indicate that the subset of predictor variables has not induced bias into the reduced prediction equation. On the other hand, if $\underline{B}_2 \neq \underline{0}$, $E[SSE_1] > (n - k)\sigma^2$ and $\Gamma_k > k$. Values of C_k that are much greater than k , therefore, indicate that the subset of variables in the reduced prediction equation has induced bias through the erroneous elimination of important predictor variables. Hence it is desirable to select subsets of predictor variables for which $C_k \approx k$.

8.2   SUBSET SELECTION METHODS

Implementation of subset selection procedures requires not only the statistics described in the previous section but also some technique for partitioning X into X_1^* and X_2^*. One method of accomplishing such a partitioning is to examine all possible partitionings, the "all possible regressions" procedure. Another technique is to seek the "best" possible subset regression according to one of the aforementioned criteria. Both of these variable selection techniques are widely used and are detailed below. Also discussed

---

A complete discussion of the theoretical basis for C_k can be found in Gorman and Toman (1966), Daniel and Wood (1971), and Mallows (1973). Hocking (1976) compares C_k with R^2 and F statistics and concludes that C_k is more sensitive to changes in the adequacy of prediction.

are graphical displays of C_k statistics and alternative ways of finding adequate subsets of predictor variables.

Throughout this section it is assumed that a constant term is desired in all prediction equations. Although it is not necessary to do so, we include a constant term for convenience and uniformity of presentation.

8.2.1 All Possible Regressions

One of the most comprehensive yet cumbersome ways of selecting subset regressions is to compute all possible subset regressions. Summary statistics or selection criteria can be examined for each subset of the p predictor variables and subset regressions that possess acceptable values for the criteria are considered candidates for the final prediction equation. The major drawback to this approach is the number of regressions which must be calculated: 2^p for a pool of p predictor variables.

Although computational shortcuts can greatly reduce the computing effort involved if one only desires calculations of the selection criteria for an initial screening of the possible subsets, all possible regressions are generally not attempted for $p > 10$. For example, with the housing data introduced earlier in this chapter $2^{17} = 131,072$ values for any of the selection criteria would have to be calculated. Most of this information would later be discarded when the better subsets were identified.

Contrasting with the computational effort involved is the valuable information provided by examining all possible subsets. Most other variable selection procedures yield only one or perhaps a few candidates for the final prediction equation. By examining summary information on all possible subsets of predictor variables, a researcher can identify all the better subsets and can choose one or more according to the nature and purpose of the investigation—a definite advantage when some predictor variables are difficult or costly to acquire.

The housing data in Data Set A.7 can be used to illustrate the analysis of subset regressions using selection criteria from all possible regressions. For convenience of exposition only the following four predictor variables are examined: CNHF13(X_1), RES510(X_2), LTOTARE(X_3), and QUAL2(X_4). Table 8.3 contains a summary of the $2^4 = 16$ equations that were fit. Values of the four selection criteria introduced in Section 8.1.2 are displayed.

As measured by R_1^2, all the prediction equations that include both LTOTARE(X_3) and QUAL2(X_4) provide acceptable fits to the housing rent. In particular, the prediction equation with just these two variables accounts for nearly as much of the response variability as the four variable predictor ($R_1^2 = 0.627$ vs. $R^2 = 0.643$). No prediction equation that excludes either of these predictor variables retains a reasonably large R_1^2.

Table 8.3. All Possible Regressions for Four Variable Housing Data

Variables In Equation	R_1^2	C_k	MSE_1	F
Mean	.000	3440.3	1971.1	3893.3
X_1	.006	800.6	1964.4	2.5
X_2	.023	778.3	1929.3	10.8
X_3	.337	383.8	1310.2	230.0
X_4	.350	366.9	1283.7	244.1
X_1, X_2	.028	773.9	1923.7	6.6
X_1, X_3	.340	381.8	1306.8	116.4
X_1, X_4	.351	368.4	1285.8	122.0
X_2, X_3	.360	356.9	1267.7	127.0
X_2, X_4	.365	349.9	1256.7	130.0
X_3, X_4	.627	20.5	738.4	380.0
X_1, X_2, X_3	.363	355.4	1265.0	85.5
X_1, X_2, X_4	.366	351.5	1258.9	86.6
X_1, X_3, X_4	.627	22.4	740.0	252.8
X_2, X_3, X_4	.642	3.0	709.5	270.1
X_1, X_2, X_3, X_4	.643	5.0	711.0	202.2

Comparing C_k values, however, suggests that even the subsets (X_3, X_4) and (X_1, X_3, X_4) induce bias into the prediction equations since the C_k values are much greater than 3 and 4, respectively. The C_k values are more sensitive to bias in the prediction equation than is R_1^2 (see Section 8.2.3). A similar conclusion is drawn from the MSE_1 values and the F statistics. If one prediction equation is to be selected for further study or as a final fitted model, the summary values in Table 8.3 suggest that the three variable predictor with RES510, LTOTARE, and QUAL2 $(X_2, X_3,$ and $X_4)$ should be chosen.

Again, the advantage of an examination of all possible subsets as in Table 8.3 is that several of the better subsets can be identified and subjected to close scrutiny. The best alternative subsets, whether determined on the basis of selection criteria or other criteria such as cost or theoretical relationships, can then be chosen.

8.2.2 Best Subset Regression

The prohibitive cost and effort involved in computing all possible regressions, or even the selection criteria for all the regressions, has led to research for more efficient means of determining "best" subsets. Considerable attention (e.g., Hocking and Leslie, 1967; La Motte and Hocking, 1970; and Furnival and Wilson, 1974) has been directed at identifying the best (largest R_1^2 or, equivalently, smallest SSE_1) subsets of a given size; i.e., of all subsets having a fixed number of predictor variables, identifying the one that has the largest R_1^2.

While the computational procedures are somewhat technical, the above articles provide the necessary information to allow the best subset of a given size to be computed by evaluating only a fraction of the 2^p regressions. It is feasible to evaluate the best subsets of all sizes for as many as 30 or 40 predictor variables [see LaMotte (1972), and Furnival and Wilson (1974), for computational details]. Using these computing algorithms, we can now begin to study the 17 variable housing rent data in more detail.

Table 8.4. Best Subset Regressions for 17 Variable Housing Data

Predictor Variables Added to the Prediction Equation	p	R_1^2	Cumulative C_k
X_{12}	1	.350	752.46
X_7	2	.627	241.73
X_8	3	.675	154.57
X_{10}	4	.693	124.46
X_1,	5	.703	106.98
$-X_7, X_5, X_6$	6	.715	87.38
X_{14}	7	.725	70.72
X_{11}	8	.732	59.04
X_{16}	9	.737	51.32
$-X_{16}, X_2, X_3$	10	.744	41.48
X_9	11	.749	33.56
X_{16}	12	.753	27.88
X_4	13	.757	22.74
X_{15}	14	.759	20.50
X_{17}	15	.763	16.72
X_{13}	16	.764	16.15
X_7	17	.764	18.00

Table 8.4 summarizes the best subsets for 1 to 17 predictor variables. They were obtained by implementing LaMotte's (1972) SELECT program. From Table 8.4, the best 5 predictor variables for the housing rent data are X_1, X_7, X_8, X_{10}, and X_{12}. This 5 variable prediction equation has a coefficient of determination of 0.703 and a C_6 value of 106.98. Proceeding to the best subset of size 6, Table 8.4 reveals that X_7 must be removed from the best subset of size 5 and X_5 and X_6 need to be added, yielding X_1, X_5, X_6, X_8, X_{10}, and X_{12}. For this subset $R_1^2 = 0.715$ and $C_7 = 87.38$.

It is clear from Table 8.4 that there are five or six subsets for which R_1^2 is only slightly less than $R^2 = 0.764$. There are about as many subsets for which C_k is not drastically larger than k, although it appears that the closest equivalence between C_k and k occurs for $p = 15$ and 16 (i.e., subsets with X_7 and X_{13} removed and with only X_7 removed, respectively). In this case the researcher must weigh the relative consequences of reducing the number of predictor variables with the accuracy of prediction. If a high degree of accuracy is demanded or precise parameter estimates are desired, it appears from an examination of the C_k statistics that only X_7 and X_{13} can be safely removed from the prediction equation; however, R_1^2 is not greatly diminished if X_4, X_7, X_{13}, X_{15}, X_{16}, and X_{17} are eliminated.

Two defects of the best subset selection are frequently cited. First, the number of predictor variables for which the available computational procedures are efficient is limited, although many more can be analyzed than with all possible regressions. Second, the tradeoff with all possible regressions is that only the single best subset of a given size is identified. Other good subsets that are only trivially inferior to the best ones might not be identified, yet these other subsets may be preferable due to problem-related considerations: theoretical arguments might specify important variables that are included, say, in the second best subset but not the best one, economic constraints might make some subsets more desirable than others, etc. The great value of all possible regressions is that all competing subsets can be identified.

8.2.3 C_k Plots

C_k was introduced in Section 8.1.2 as a selection criterion that enables one to choose subsets of predictor variables that do not overly bias prediction of the observed response variables. Subsets for which C_k is small and approximately equal to k are said to be adequate reduced prediction equations. Just as the minimum SSE_1 value in the best possible regression procedure only identifies a single subset of a given size and ignores whether other subsets with only trivially larger SSE_1 values might be viable alternatives, so too the minimum C_k does not necessarily identify the only desirable subset of predictor variables. A reexamination of the formula for C_k, viz.

$$C_k = \frac{SSE_1}{MSE} - (n - 2k) , \qquad (8.2.1)$$

reveals that a subset of predictor variables of size k that has a trivially larger SSE_1 than the best subset of size k also should have only a slightly larger C_k than the minimum C_k. Consequently, where feasible, all possible C_k values should be plotted and not just the best one of a given size.

Another point that was noticed in previous examples is that C_k is more sensitive than R_1^2 or MSE_1 to alterations in the accuracy of prediction. The reason for the increased sensitivity can be partially explained by rewriting C_k as

$$C_k = (n - k)\left(\frac{\text{MSE}_1}{\text{MSE}} - 1\right) + k, \qquad (8.2.2)$$

where $\text{MSE}_1 = \text{SSE}_1/(n - k)$. As is apparent from eqn. (8.2.2), the sample size is a factor in the sensitivity of C_k. Small changes in MSE_1 (or R_1^2) as variables are added to or deleted from subset models can produce large changes in C_k because of the multiplier $(n - k)$. Naturally, large changes in C_k can also be due to the addition of bias and several authors (e.g., Hocking, 1972) conclude that C_k more accurately reflects the adequacy of subset models than does R_1^2 or MSE_1. Here again one must weigh the need for precision with the costs and uses of the final prediction equation. Particularly if $(n - k)$ is large, C_k could be overly sensitive to small changes in the fit.

Figure 8.1 is a C_k plot of the four predictor variable housing data. Since the constant term is included in the prediction equation the largest value

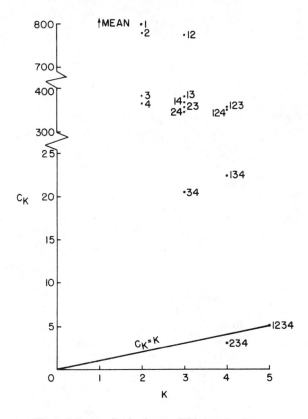

Figure 8.1 C_k plot for four variable housing data.

of k is $(p + 1) = 5$. Also, since $\text{MSE}_1 = \text{MSE}$ when all predictor variables are included, $C_5 = 5$. The line $C_k = k$ has been drawn in Figure 8.1 so that the C_k values can be visually compared with k.

It is immediately apparent from this graph, more so than from scanning the C_k values in Table 8.3, that the subset (X_2, X_3, X_4) is the only subset prediction equation for which $C_k \approx k$. Its closest competitors, the other two subsets containing both X_3 and X_4, have C_k values that are 5 to 7 times larger than k. All other subsets have C_k values over one hundred times larger than k.

A second illustration of C_k plots is given in Figure 8.2 where each of the best subset C_k values are graphed for the 17 variable housing data. The best subset of a given size produces the minimum C_k for that size. It is somewhat more difficult to select a precise cutoff for the best subset in Figure 8.2 than it was in Figure 8.1. It seems reasonable to require that no more than five and perhaps only two predictor variables should be deleted.

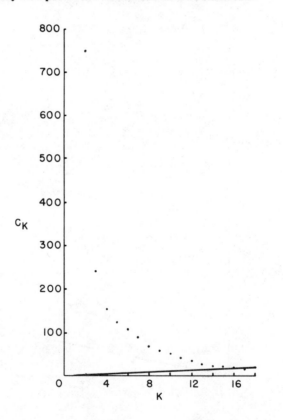

Figure 8.2   Plot of minimum C_k for 17 variable housing data.

Due to the impossibility of calculating all 2^{17} C_k statistics for the housing rent data, only the minimum C_k statistics are plotted in Figure 8.2. Other potentially useful subsets could be evaluated once one decides what a cutoff for the number of predictor variables should be. Clearly ten or fewer predictor variables are unacceptable for this data since C_k is so much larger than k for the best subsets. Since the best subsets of size 10 or smaller have such large C_k values, no subsets of size 10 or smaller need be considered. On the other hand, subsets of size $k = 14$ or larger could be investigated further to determine whether other subsets yield C_k values as small as the best one. The number of regressions that need to be computed for subsets of size k ($k = 14, ..., 17$) is greatly reduced from 2^{17}. Techniques presented in the next subsection can aid us in further reducing the number of subsets that need to be evaluated.

8.2.4 Additional Search Routines

Both the all possible regression and the best subset regression methods can be considered routines that search for reduced prediction equations that predict with close to the same accuracy as the full prediction equation. When the pool of predictor variables is large all possible regressions cannot be evaluated and best subset regression may not reveal important competing subsets. One compromise between these two search procedures was mentioned at the end of the last subsection.

Other search routines have been advocated that do not necessarily identify all the better subset prediction equations—although they generally do so—but which identify most of the better ones and greatly reduce the computational effort over the all possible regressions search. One such procedure is the "t-directed search" (Daniel and Wood, 1971).

One type of t-directed search begins by calculating the t statistics for elimination of each of the p predictor variables from the full model, statistics that are generally provided when one analyzes the full prediction equation with most standard computer programs. Predictor variables with large t statistics, say larger than 3.0 in magnitude, should be included in a basic subset. Selection criteria such as C_k are then calculated for all combinations of the basic subset and the remaining predictor variables.

Table 8.5 shows results of a t-directed search of the four variable housing data. First note that the predictor variables are ordered according to the magnitude of their t statistics. If we use as the basic set (X_3, X_4) due to their extremely large t statistics, the t-directed search requires that only the subsets (X_3, X_4), (X_1, X_3, X_4), (X_2, X_3, X_4) and (X_1, X_2, X_3, X_4) be examined. This means that only 4 of 16 possible regressions have had to be computed and these are indeed the four best subsets. If one uses $|t| > 3.0$ to

Table 8.5. t Statistics, Four Variable Housing Data

Predictor Variable	X_1	X_2	X_3	X_4
t Statistic	.19	-4.40	18.67	18.77

form the basic set, one immediately forms the best subset of size 3, (X_2, X_3, X_4).

A second type of t-directed search forms the basic subset by adding predictor variables one-by-one according to the magnitude of their t statistics, starting with the predictor variable having the largest (in magnitude) t statistic. The basic set is determined according to whether the smallest C_k is less than or equal to k. If so, the basic subset includes all the predictor variables corresponding to the smallest C_k; if not, one retreats two or three steps prior to the smallest C_k and forms the basic subset from those variables. Once the basic set is determined, C_k statistics for all possible combinations of the basic set with the remaining predictor variables are calculated.

If this alternate search is performed on the four variable housing data, variable X_4 is the first to enter the basic set since its t statistic is the largest one in Table 8.5. The C_k statistic, from Table 8.6, corresponding to a subset containing only the constant term and X_4 is equal to 366.93. Next X_3 is added to the basic set and C_3 is 20.47. When X_2 is included C_4 drops to 3.04. Again the best subset of size $k = 4$ ($p = 3$) has been identified with only four regression runs.

Table 8.6. t-Directed Search, Four Variable Housing Data

Predictor Variables	k	Cumulative C_k
X_4	2	366.93
X_3, X_4	3	20.47
X_2, X_3, X_4	4	3.04
X_1, X_2, X_3, X_4	5	5.00

The real value in t-directed searches can be seen from their application to the full seventeen variable housing rent analysis. Table 8.7 displays the t statistics for the full model. Eight of the predictor variables have t statistics that are larger than 3.0 in magnitude. All eight of these predictor variables are included in the best subset of size $p = 10$ ($k = 11$) in Table 8.4.

The t statistics in Table 8.7 were obtained from one analysis of the full model and have led us to the best subset of size 10. From here one could perform all possible regressions of these ten predictor variables with subsets

Table 8.7. t **Statistics for 17 Variable Housing Data**

Predictor Variable	X_1	X_2	X_3	X_4	X_5	X_6	X_7	X_8	X_9
t Statistic	-3.82	-3.89	-4.06	-2.71	1.30	2.33	.38	5.69	-2.71

Predictor Variable	X_{10}	X_{11}	X_{12}	X_{13}	X_{14}	X_{15}	X_{16}	X_{17}
t Statistic	3.38	3.20	6.10	1.57	4.53	2.72	-2.79	-2.28

of the remaining seven. All the better regressions with ten or more predictor variables should be identified, although this technique cannot guarantee that they will be. From this example, however, it is apparent that good subsets can be obtained with much less effort than all possible regressions and several alternate subsets to the best ones can be found.

If the second type of t-directed search is performed, Table 8.8 reveals that the smallest C_k occurs for $k = 17$. The sixteen predictor variables listed in the first sixteen rows of column two of Table 8.8 form the basic subset. There is only one additional subset to examine, the full prediction equation. As is to be expected, some of the better subsets of Table 8.4 are not identified in Table 8.8. The reason for their exclusion is that X_5 enters so late in Table 8.8. If, however, one forms the basic subset from, say, three steps earlier ($k = 14$) the best subsets of size $p = 15$ and 16 would be identified

Table 8.8. t **Statistics and Cumulative C_k Values for 17 Variable Housing Data**

k	Predictor Variable Added	t	Cumulative C_k
2	X_{12}	6.10	752.46
3	X_8	5.69	566.89
4	X_{14}	4.53	556.14
5	X_3	4.06	536.81
6	X_2	-3.89	533.74
7	X_1	-3.82	526.06
8	X_{10}	3.38	525.73
9	X_{11}	3.20	506.63
10	X_{16}	-2.79	478.73
11	X_{15}	2.72	459.19
12	X_9	-2.71	424.69
13	X_4	-2.71	406.72
14	X_6	2.33	88.10
15	X_{17}	-2.28	84.30
16	X_{13}	1.57	82.05
17	X_5	1.30	16.15
18	X_7	.38	18.00

when one analyzed all possible regressions of the basic set of variables with X_5, X_7, X_{13}, and X_{17}.

8.3   STEPWISE SELECTION METHODS

Stepwise regression procedures are selection techniques that sequentially add or delete single predictor variables to the prediction equation. Since a series of steps are involved before a final equation is obtained and since each step leads directly to the next, these methods involve the evaluation of a much smaller number of equations than the 2^p required for the all possible regressions approach. Stepwise methods are a particularly useful option when the selection procedures discussed in Section 8.2 are not practically feasible, such as when a very large number of predictor variables are available for consideration. Their major limitation is that the end result is only a single subset that is not necessarily best for a given size and few alternative subsets are suggested.

The actual term stepwise procedure has come to apply to modifications of two basic approaches, the Forward Selection and the Backward Elimination methods. Both of these techniques and a further modification called the Stepwise procedure will be discussed in this section.

8.3.1   Forward Selection Method

The Forward Selection (FS) method adds predictor variables one at a time to the prediction equation. This process continues until either all predictors are included or until some selection criterion is satisfied. The variable considered for inclusion at each step is the one that produces the greatest decrease in SSE_1.

The actual steps involved in the FS technique are summarized as follows. If only one predictor variable is to be used in predicting the response, choose the variable that yields the smallest SSE_1. This will involve the calculation of at most p regressions, one for each predictor variable. If two variables are to be used in predicting Y, choose from the $p - 1$ remaining variables the one which, when used with the predictor chosen in step one, produces the smallest SSE_1. At step two at most $p - 1$ regressions will be needed. Continue this process, each time adding one more predictor variable to the previous set until the addition of the next best variable does not sufficiently reduce SSE_1.

The most popular selection criterion for assessing whether the reduction in SSE_1 is sufficiently small to terminate the forward selection process is an F statistic. At stage 1 the F statistics that are calculated are simply the SSR_1/MSE_1 ratios, where SSR_1 is the regression sum of squares for one of the individual predictor variables and MSE_1 is the corresponding residual mean square. At subsequent stages, the F statistic is slightly different.

Suppose at stage s ($s \leqslant 1$ predictor variables are already included in the model) we denote the residual sum of squares by SSE_s. At stage $(s + 1)$, $(p - s)$ regressions must be run, one with each of the remaining predictor variables and the s that are already selected. Denote the residual sums of squares for these $(p - s)$ regressions by $SSE_{s+1}(j)$, where the subscript $(s+1)$ refers to the $(s+1)$-th stage of forward selection and j refers to X_j as the additional predictor variable. An F-type statistic that can be used to assess whether X_j is a worthwhile addition to the prediction equation is (note that $s+2$ terms are now in the prediction equation, $s+1$ predictor variables and the constant term)

$$F_j = \frac{SSE_s - SSE_{s+1}(j)}{MSE_{s+1}(j)} \quad , \tag{8.3.1}$$

where $MSE_{s+1}(j) = SSE_{s+1}(j) / (n-s-2)$. The largest F_j statistic identifies the next predictor variable to consider for insertion into the prediction equation.

In essence, the F_j statistic in eqn. (8.3.1) tests whether $\beta_j = 0$ in the reduced model

$$Y = \alpha + \beta_1 X_1 + \beta_2 X_2 + \ldots + \beta_s X_s + \beta_j X_j + \varepsilon . \tag{8.3.2}$$

The t statistic that is usually advocated for this test is (see Table 6.4)

$$t_j = \frac{\hat{\beta}_j}{[C_{jj} \cdot MSE_{s+1}(j)]^{1/2}} .$$

or, equivalently (since the square of a t statistic with v degrees of freedom is an F statistic with 1 and v degrees of freedom), the F statistic

$$F_j = \frac{\hat{\beta}_j^2}{C_{jj} \cdot MSE_{s+1}(j)} . \tag{8.3.3}$$

One can show, however, (Appendix 8.A) that

$$\frac{\hat{\beta}_j^2}{C_{jj}} = SSE_s - SSE_{s+1}(j) \tag{8.3.4}$$

and consequently that F_j in eqn. (8.3.1) does indeed test the significance of β_j in the $(s+1)$-term model (8.3.2).

Once the largest F_j is calculated, its value is compared to an appropriate percentage point from an F distribution with 1 and $(n-s-2)$ degrees of freedom. If $F_j \geq F_{1-\gamma}(1, n-s-2)$, predictor variable X_j is added to the prediction equation and FS proceeds to the next stage with $(s+1)$ predictor variables included in the prediction equation. If $F_j < F_{1-\gamma}(1, n-s-2)$, the process terminates with a reduced prediction equation containing s predictor variables. Again, we suggest using $\gamma = 0.25$ in choosing the cutoff value for the F statistic.

Table 8.9. Forward Selection Method for Four-Variable Housing Data

Stage(s)	Variables in Subset	SSR	SSE	MSE	F_j
1	X_1	4,993.78	889,886.75	1964.43	2.54
1	X_2	20,890.21	873,990.31	1929.34	10.83
1	X_3	301,351.25	593,529.25	1310.22	230.00
1	X_4	313,342.31	581,538.19	1283.75	244.08
2	X_1, X_4	313,700.56	581,179.94	1285.80	.28
2	X_2, X_4	326,852.06	568,028.44	1256.70	10.75
2	X_3, X_4	561,096.31	333,784.19	738.46	335.52
3	X_1, X_3, X_4	561,152.50	333,728.00	739.97	.08
3	X_2, X_3, X_4	574,910.75	319,969.75	709.47	19.47
4	X_1, X_2, X_3, X_4	574,936.31	319,944.19	710.99	.04

Table 8.9 outlines the forward selection procedure for the four-variable housing rent data. At step one each of the four predictor variables is evaluated separately in single variable regression analyses. The F_j statistics are simply $MSR_1/MSE_1 = SSR_1/MSE_1$ for each of the analyses. If the largest F_j is greater than $F_{.75}(1,453) = 1.33$, the predictor variable corresponding to the largest F_j will be the first one included in the prediction equation. From Table 8.9 the stage one analysis indicates that X_4 should be the first predictor variable selected.

At stage two, three regressions must be run, one each with predictor variable subsets (X_1, X_4), (X_2, X_4), and (X_3, X_4). Using SSE from the prediction equation for X_4 in stage one as SSE_s in eqn. (8.3.1), the SSE and MSE values for each of the regressions in stage 2 is inserted in eqn. (8.3.1) for $SSE_{s+1}(j)$, respectively, to form F_j. The cutoff value is $F_{.75}(1,452) = 1.33$. From Table 8.9, X_3 is the next variable to be added.

At stage three the FS procedure terminates since the F_j statistic for stage four is less than $F_{.75}(1,450) = 1.33$. Thus, forward selection has chosen the best subset of size three and the best overall subset, (X_2, X_3, X_4). Ten regressions have been computed but most of them involved relatively simple

**Table 8.10. Forward Selection Procedure
for 17 Variable Housing Data**

Stage	Predictor Variable	F_j	R_1^2
1	X_{12}	244.08	.350
2	X_7	335.52	.627
3	X_8	66.85	.675
4	X_{10}	25.37	.693
5	X_1	15.90	.703
6	X_6	17.15	.714
7	X_{14}	16.23	.724
8	X_{11}	12.24	.731
9	X_{16}	8.74	.737
10	X_3	8.05	.741
11	X_2	12.08	.748
12	X_9	8.15	.753
13	X_4	6.37	.756
14	X_{15}	5.11	.759
15	X_{17}	5.40	.762
16	X_{13}	2.23	.763
17	X_5	1.68	.764

calculations (4 regressions with one predictor variable, 3 with two predictor variables, etc.).

Forward selection does not always lead to optimal subsets of a given size. For example, in the seventeen variable housing data displayed in Table 8.10 the first five stages yield subsets that are identical to the best subsets in Table 8.4. After stage five the subsets selected by FS are not the best subsets. By comparing the R_1^2 values for the two procedures, however, it is apparent that the subsets selected by FS are good competitors to the best subsets. Note too that all predictor variables are retained in the prediction equation with FS.

The above results illustrate one of the major problems associated with the Forward Selection method; i.e., the stepwise procedure reveals only one subset of a given size with no guarantee that it is best. Hamaker (1962) and Mantel (1970) both provide examples of situations where good predictor variable sets are overlooked by the FS method due to its limitation of only examining one variable at a time. A second difficulty is that the F_j statistics are not true F random variables since (i) they are correlated at any stage (e.g., all use the same SSE_s in the numerator of F_j) and (ii) the F_j actually compared to an F percentage point is the largest of $(p-s)$ F_j values. In addition, the early stages of FS utilize residual mean squares, MSE_s, that can be poor estimates of σ^2 since important predictor variables have not yet been included in the prediction equation.

The Forward Selection method, despite these criticisms, does have many advantages. It requires much less work than the all possible regressions and best subset procedures, it is a simple method to understand and apply, and it generally results in good, although perhaps not best, subsets. It is particularly useful in analyzing problems involving polynomial terms, such as X, X^2, X^3, ..., since polynomials of increasing order can be viewed in a step-by-step process where the next predictor variable to add to the first s is unique; i.e., X^{s+1} should be added after X, X^2, ..., X^s. Finally, FS can be effectively employed with much larger pools of predictor variables than the other techniques previously mentioned.

One final cautionary comment on FS concerns the interpretation of the results. Often data analysts mistakenly associate an order of importance to the predictor variables according to their order of entry into the prediction equation; i.e., the first predictor variable that enters is the most important to the prediction of the response, the next one to enter is the second most important, etc. As demonstrated by comparing Tables 8.4 and 8.10, this reasoning is incorrect. In fact a variable that enters first in FS can later prove of little predictive value when other predictor variables are added; X_7 in Tables 8.4 and 8.10 is a good example.

8.3.2 Backward Elimination Method

The Backward Elimination (BE) procedure, in contrast to the FS method, begins with all predictor variables in the prediction equation. The variables are then eliminated one at a time until either all predictors are excluded or some selection criterion is satisfied. The variable considered for exclusion at each step is the one that produces the smallest increase in SSE_1 from the previous step.

The actual steps involved in the BE technique can be summarized as follows. At stage one, discard the predictor variable that, when eliminated, yields the smallest SSE_1 for the remaining $(p-1)$ variables. From the equivalence noted in eqn. (8.3.4) (see also Appendix 8.A), the predictor variable that corresponds to the smallest SSE_1 is the one with the smallest t statistic:

$$t_j = \frac{\hat{\beta}_j}{[C_{jj} \cdot \text{MSE}]^{1/2}} \ .$$

Equivalently, one could select the predictor variable that has the smallest $F_j = t_j^2$ value, where F_j can be written as

$$F_j = \frac{\hat{\beta}_j^2}{C_{jj} \cdot \text{MSE}} = \frac{\text{SSE}_{p-1}(j) - \text{SSE}}{\text{MSE}} \ .$$

In this equation, $SSE_{p-1}(j)$ is the residual sum of squares from the $(p-1)$ variable prediction equation with X_j removed. If $F_j < F_{1-\gamma}(1, n-p-1)$ for the smallest F_j statistic, X_j is deleted from the prediction equation; if not, BE terminates and all predictor variables are retained.

At the $(s+1)$-th stage, s predictor variables have already been removed from the full prediction equation and $(p-s)$ remain as candidates for elimination. The F_j statistics for these $(p-s)$ candidates can be written as (carefully note the difference in the definition of $SSE_{s+1}(j)$ for FS and BE despite the common notation)

$$F_j = \frac{SSE_{s+1}(j) - SSE_s}{MSE}. \tag{8.3.5}$$

where MSE is the estimate of σ^2 from the full prediction equation. We retain MSE in eqn. (8.3.5) rather than using $MSE_{s+1}(j)$ since MSE is an unbiased estimator of σ^2 and if too many predictor variables are eliminated in BE, $MSE_{s+1}(j)$ can be biased. Again, if the smallest F_j is less than $F_{1-\gamma}(1, n-p-1)$, the corresponding predictor variable is removed; otherwise, BE terminates.

Unlike all the variable selection techniques mentioned thus far, backward elimination only requires that one regression analysis be performed at each stage. This is because all the numerators in eqn. (8.3.5) can be obtained from t statistics at each stage. Specifically, at stage $(s+1)$

$$SSE_{s+1}(j) - SSE_s = t_j^2 \cdot MSE_s,$$

where t_j is the t statistic for testing $\beta_j = 0$ in the reduced model of $(p-s)$ predictor variables and MSE_s is the mean squared error from the same analysis. In other words, in the regression analysis in which s predictor variables have already been eliminated, t_j is the t statistic corresponding to X_j and MSE_s is the residual mean square from the ANOVA table.

To illustrate the backward elimination procedure, consider again the four variable housing data. Stage one, as indicated in Table 8.11, consists of examining the t statistics from the full prediction equation. The smallest t statistic corresponds to X_1 and since F_1 is less than $F_{.75}(1,450) = 1.33$ the process continues with X_1 eliminated from the prediction equation. At stage two, the smallest t statistic corresponds to X_2, but when F_2 is calculated using eqn. (8.3.5) its value is larger than $F_{.75}(1,450) = 1.33$ so X_2 must be retained and BE terminates. With this data set, all the procedures discussed so far suggest the same subset, (X_2, X_3, X_4).

Table 8.11. Backward Elimination Method for
Four-Variable Housing Data

Stage(s)	Variable Deleted	t_j	MSE_s	F_j
1	X_1	.19	710.99	.04
1	X_2	-4.40	710.99	19.39
1	X_3	18.67	710.99	348.53
1	X_4	18.77	710.99	352.43
2	X_2	-4.41	709.47	19.47
2	X_3	18.70	709.47	349.64
2	X_4	18.89	709.47	356.64

In the seventeen variable housing data, backward elimination only removes X_7 (see Table 8.12) while forward selection included all the predictor variables, inserting X_7 at stage two. Best subset regression in Table 8.4 indicated that both X_7 and X_{13} could be removed and perhaps a few others. These are not major differences but they could be important in situations where one is attempting to evaluate a theoretical model that is proposed as an explanation of some physical phenomenon. In other data sets, moreover, the discrepancies among the suggested subsets from each of these variable selection procedures can be more disparate than in these examples.

Table 8.12. Backward Elimination Procedure for
17-Variable Housing Data

Stage(s)	Variable Deleted	F_j	R_1^2
1	X_7	.15	.764
2	X_{13}	2.58	.763

As with the FS procedure, the Backward Elimination method reveals only one subset, which is not necessarily best, and this is its major criticism. Nevertheless, the BE method begins with all predictors in the equation so one starts out with the best possible prediction equation (assuming all relevant predictor variables are included and appear in their correct functional form) and MSE, used in all the denominators of F_j [eqn. (8.3.5)] is the best estimate of σ^2 that is obtainable from the data set. For these reasons the BE method is recommended over the FS procedure unless there are so many potential predictor variables that FS is the only feasible alternative.

8.3.3 Stepwise Procedure

The Stepwise (SW) procedure is a method that combines the FS and BE approaches. It is essentially a forward selection procedure but at each step of the process the predictor variables in the chosen subset are reexamined for possible deletion, similar to the backward elimination method. Hence, after each predictor variable is added consideration is given to discarding one of the previously accepted variables.

At each step of the SW technique a predictor variable is added to the prediction equation if its F statistic as determined from eqn. (8.3.1) is the largest one calculated and exceeds $F_{1-\gamma}(1, n-s-2)$. Next each predictor variable already chosen is reconsidered and eliminated from the selected subset if its F statistic, again as calculated using eqn. (8.3.1), is the smallest one and does not exceed $F_{1-\gamma}(1, n-s-2)$. This process continues until all p predictor variables are in the equation or until the selection criteria are no longer met.

The advantages of the new SW procedure are straightforward. At any given step the method allows one to judge the contribution of each predictor in the subset as though it were the most recent variable to enter the prediction equation. The actual order of entry of the predictor is thus ignored and any variable that does not make a significant contribution, at the given step, is eliminated.

An example will help to illustrate this technique. Consider again the four variable housing data. Step one of the SW procedure is identical to the first step of the FS method so that the first predictor chosen is X_4. Step two is also identical to that of the FS procedure and both X_3 and X_4 are included in the subset. Hence, the first two steps summarized in Table 8.13 agree with those given in Table 8.9.

The SW procedure deviates from the Forward Selection process at the third step. Here consideration is given to excluding X_4 by determining the contribution X_4 would have made if X_3 had entered first and X_4 second. The F statistic exceeds $F_{.75}(1,452) = 1.33$ so that X_4 is retained in the prediction equation. The SW process then repeats itself in step 4 where, as in the FS method, X_2 is chosen as the third entering variable.

Next X_3 and X_4, in step 5, are again reconsidered for exclusion. The two resultant F statistics are both large and exceed the F tabled value; consequently all three variables are retained. The final step involves entering X_1 into the prediction equation but it is rejected, as in the FS method, due to its small F value ($F = 0.04$). The SW method thus selects a three predictor variable set consisting of X_2, X_3, and X_4.

A quick review of the above example clearly demonstrates that more effort is involved in using the SW procedure as compared to the other stepwise methods. In spite of this shortcoming the Stepwise procedure is recommended as the best available stepwise technique by a number of authors (e.g., Draper and Smith, 1966). This is chiefly a result of its ability to reconsider the importance of each predictor variable in the subset at each step of its procedure.

Table 8.13. Stepwise Procedure for Four-Variable Housing Data

Step	In	Added	Deleted	F_j
		Variables		
1	None	X_1		2.54
1	None	X_2		10.83
1	None	X_3		230.00
1	None	X_4		244.08
2	X_4	X_1		.28
2	X_4	X_2		10.75
2	X_4	X_3		335.50
3	X_3, X_4		X_4	351.75
4	X_3, X_4	X_1		.08
4	X_3, X_4	X_2		19.47
5	X_2, X_3, X_4		X_3	349.64
5	X_2, X_3, X_4		X_4	356.64
6	X_2, X_3, X_4	X_1		.04

APPENDIX

8.A   Derivation of Equation (8.3.4)

Let X^* denote the $n \times (s+2)$ matrix of observations on the constant term and $(s+1)$ predictor variables in a reduced prediction equation. In addition, let $\underline{X}_j$ be the last column of X^*; i.e., $X^* = [\underline{1}:D_j:\underline{X}_j]$ where D_j contains the observations of s predictor variable other than X_j. As in Appendix 5.B, we can let $\underline{r}_{s+1}(X_j)$ denote the residuals of X_j from its regression on the constant term and s predictor variables:

$$\underline{r}_{s+1}(X_j) = (I - D_j(D_j' D_j)^{-1} D_j')\underline{X}_j.$$

Then from the reduced prediction equation,

$$\hat{\beta}_j = \frac{\underline{r}_{s+1}(X_j)' \underline{Y}}{\underline{r}_{s+1}(X_j)' \underline{r}_{s+1}(X_j)} \tag{8.A.1}$$

and the regression sum of squares for X_j adjusted for all other predictor variables is

$$\underline{Y}'\underline{r}_{s+1}(X_j)'\Big[\underline{r}_{s+1}(X_j)' \ \underline{r}_{s+1}(X_j)\Big]^{-1} r_{s+1}(X_j)' \ \underline{Y}$$

$$= \hat{\beta}_j^2\Big[\underline{r}_{s+1}(X_j)'\underline{r}_{s+1}(X_j)\Big] . \tag{8.A.2}$$

Equation (8.3.4) can be written as

$$\frac{\hat{\beta}_j^2}{C_{jj}} = \text{SSE}_s - \text{SSE}_{s+1}(j)$$

$$= (\text{TSS} - \text{SSR}_s) - [\text{TSS} - \text{SSR}_{s+1}(j)]$$

$$= \text{SSR}_{s+1}(j) - \text{SSR}_s , \tag{8.A.3}$$

where $\text{SSR}_{s+1}(j)$ is the regression sum of squares for a prediction equation
with $(s+1)$ predictor variables, including X_j, and SSR_s is the regression sum
of squares for the s predictor variables when X_j is excluded. Now define the
$(s+1) \times (s+1)$ nonsingular matrix T by

$$T = \begin{bmatrix} I_s & -(D_j'D_j)^{-1}D_j'\underline{X}_j \\ \underline{0}' & 1 \end{bmatrix} .$$

Then

$$\text{SSR}_{s+1}(j) = \underline{Y}'X^*(X^{*'}X^*)^{-1}X^{*'} \ \underline{Y}$$

$$= \underline{Y}'X^*(TT^{-1})(X^{*'}X^*)^{-1}(TT^{-1})X^{*'} \ \underline{Y}$$

$$= \underline{Y}'(X^*T)([X^*T]'[X^*T])^{-1}(X^*T)' \ \underline{Y}$$

$$= \underline{Y}'[D_j:\underline{r}_j(s+1)] \begin{bmatrix} D_j'D_j & \underline{0} \\ \underline{0}' & \underline{r}_j(s+1)'\underline{r}_j(s+1) \end{bmatrix}^{-1} \begin{pmatrix} D_j' \\ \underline{r}_j(s+1)' \end{pmatrix} \underline{Y}$$

$$= \text{SSR}_s + \hat{\beta}_j^2\Big[\underline{r}_j(s+1)'\underline{r}_j(s+1)\Big] .$$

Thus

$$\text{SSR}_{s+1}(j) - \text{SSR}_s = \hat{\beta}_j^2[\underline{r}_j(s+1)' \ \underline{r}_j(s+1)]$$

$$= \frac{\hat{\beta}_j^2}{C_{jj}}$$

by comparison with eqn. (8.A.3).

EXERCISES

1. Using eqn. (6.2.5) estimate the expectation for the coefficients of LTOTRMS and LTOTARE in a reduced model in which AREAPR was excluded. For estimation purposes use the least squares coefficients of these three variables from the full model in Table 8.1 in place of the true coefficients in eqn. (6.2.5). Note that the sum of squares and cross products for these three predictor variables can be obtained from the standard deviations and correlations in Data Set A.7 (ignore the other 14 predictor variables in determining the alias matrix). How do these estimated expectations compare with the estimated coefficients for LTOTRMS and LTOTARE in the eight variable prediction equation for FS in Table 8.1?

2. Fit a cubic polynomial (X, X^2 and X^3) to the first year college grades for high school 1 in Data Set A.2. Simultaneously test whether the quadratic and cubic functions of the grade 13 scores contribute significantly to the prediction equation.

3. Calculate all four selection criteria mentioned in Section 8.1.2 for the three variable emissions data in Data Set A.5. Construct a C_k plot and identify subsets which appear to be good alternatives to the full prediction equation.

4. Consider fitting a prediction equation for nitrous oxide using humidity, barometric pressure, and the interaction of humidity and barometric pressure. Compare the results of a FS procedure, a BE procedure, and all possible regressions. Which subset appears to be most useful in predicting NOX?

5. Perform a forward selection, backward elimination and t-directed search on the homicide data, Data Set A.3, using only the first nine predictor variables (ACC to WM). What do the results of these analyses suggest about the effect of the predictor variables on the homicide rate?

6. Analyze the prediction of ln(GNP) using subset prediction equations with forward selection, backward elimination, and stepwise selection procedures. What does a comparison of the results of these three analyses suggest about variable selection procedures when predictor variables are not multicollinear?

7.   How would the results of Exercises 5 and 6 be altered if a significance level of .05 was used instead of .25?

8.   Show that for a reduced model in which r predictor variables are eliminated,

$$C_k = r(F - 1) + k,$$

where F is the F statistic used to test the significance of the deleted variable [i.e., eqn. (8.1.4)]. What can one conclude about the sensitivity of C_k relative to the F statistic for assessing reduced prediction equations?

CHAPTER 9

MULTICOLLINEARITY EFFECTS

Throughout earlier portions of this text we have been concerned with relationships among the predictor variables. The problem has been referred to alternately as one of covariation, interrelationships, or multicollinearities among the predictor variables. All these terms suggest that the predictor variables are providing some redundant information that is to be used to predict the response variable. This redundancy confounds attempts to identify the individual contributions and relative merits of the predictor variables.

Least squares parameter estimation and its associated regression methodology is highly susceptible to the effects of multicollinearities. Numerical values of coefficient estimates, variances and covariances of the estimators, test statistics, and predicted responses can all be adversely affected by strong multicollinearities among the predictor variables. Not only the ability to detect multicollinearities but also knowledge of their ramifications is important to the proper utilization of fitted prediction equations.

Data Sets A.5 and A.7 present two different types of multicollinearities and illustrate the complications associated with them. Consider first Data Set A.7, the housing rent study. Moderately large pairwise correlations exist between two pairs of predictor variables including AREAPR(X_5) and LTOTARE(X_7) with $r = 0.729$, and LTOTRMS(X_6) and LTOTARE(X_7) with $r = 0.833$, although the pairwise correlation between AREAPR and LTOTRMS is not large ($r = 0.243$). Ordinarily, these first two pairwise correlations would not cause great concern but in this instance they are symptomatic of a more severe problem.

The common occurrence of a single variable, such as LTOTARE, in two pairwise correlations is sometimes indicative of a three-variable multicollinearity. Indeed, this is the case with Data Set A.7. Table 9.1 reveals that the smallest latent root of $W'W$, the matrix of correlations among the predictor variables, is 0.00632 and it identifies a three-variable multicollinearity among X_5, X_6, and X_7. This is the only multicollinearity among the predictor variables, moreover, since the second smallest latent root of $W'W$ is 0.27961.

Table 9.1. Multicollinearity in the Housing Rent Data

			$\ell_1 = 0.00632$			
Predictor Variable	X_1	X_2	X_3	X_4	X_5	X_6
V_{i1}	-.0018	.0032	-.0004	.0064	-.4183	-.5234
Predictor Variable	X_7	X_8	X_9	X_{10}	X_{11}	X_{12}
V_{i1}	.7419	.0093	-.0070	-.0007	.0027	-.0161
Predictor Variable	X_{13}	X_{14}	X_{15}	X_{16}	X_{17}	
V_{i1}	-.0075	-.0028	.0102	.0000	.0033	

If one calculates the variances of the least squares estimators of the housing rent standardized (unit length) coefficients (recall $Var[\hat{\beta}_j^*] = q_{jj}\sigma^2$ where $Q = (W'W)^{-1}$) the 14 predictor variables that are not multicollinear all have variances between $1.03\sigma^2$ and $2.54\sigma^2$. The variances of the coefficient estimators for the three multicollinear predictor variables, however, are $28.62\sigma^2$, $44.13\sigma^2$, and $87.34\sigma^2$. The variances for the estimators of the multicollinear predictor variable coefficients are much greater than the nonmulticollinear ones. As we shall see in Section 9.1.1, this property is characteristic of estimators associated with multicollinear predictor variables.

Another noticeable effect of the multicollinearity occurs in the stepwise regression procedures that were discussed in the last chapter. In the forward selection procedure summarized in Table 8.10, LTOTARE(X_7) was one of the first predictor variables to enter and AREAPR(X_5) was the last one. In the backward elimination procedure which is partially summarized in Table 8.12, however, the value of these two predictor variables appears to be reversed: X_7 is the first variable to be deleted and X_5 would be one of the last if one continued the backward elimination procedure until all variables are removed. The erratic behavior of these predictor variables is also apparent in the best subset listing, Table 8.4, wherein X_7 is included in the smaller "best" subsets but is replaced by X_5 and X_6 in the larger ones. In contrast, note the relative consistency of the entry of the nonmulticollinear predictor variables in these tables.

Consider next the emissions study, Data Set A.5. The three predictor variables HUMID, TEMP, and BPRES are not extremely highly correlated (pairwise correlations of -0.816 between HUMID and BPRES, 0.505 between HUMID and TEMP, and -0.254 between BPRES and TEMP) and the smallest latent root of $W'W$ is only 0.13851. Multicollinearities in this example are induced among the predictor variables by adding interaction and quadratic terms. As Table 9.2 clearly shows, some of the induced correlations indicate that the additional terms are virtually identical, apart from location and scale changes, with the original ones; e.g., BPRES and $(BPRES)^2$.

Table 9.2. Pairwise Correlations for Emissions Data Predictor Variables*

	TEMP	BPRES	HT	HB	TB	HH	TT	BB
HUMID	.505	-.816	.992	1.000	.442	.968	.509	-.816
TEMP		-.254	.593	.505	.996	.546	.999	-.253
BPRES			-.780	-.813	-.170	-.703	-.257	1.000
HT				.992	.534	.977	.597	-.780
HB					.443	.968	.509	-.813
TB						.493	.995	-.170
HH							.551	-.702
TT								-.256

*HT = HUMID × TEMP, HH = $(HUMID)^2$, etc.

The extreme covariation in the nine-term quadratic model causes wide variation in parameter estimates, depending heavily on which terms are included in the prediction equation. As Table 9.3 reveals, the sequential

Table 9.3. Coefficient Estimates for Subsets of Emissions Data

Variables Added	Standardized (Unit Length) Estimates								
	HUMID	TEMP	BPRES	TB	HB	HT	BB	TT	HH
H, T, B	-.2397	.0550	.3942						
TB	-.2907	18.4848	1.9461	-18.0765					
HB	17.4443	17.6003	2.0139	-17.2332	-17.6063				
HT	6.8412	-29.2152	-2.0901	28.0896	-12.0095	5.3087			
BB	62.6109	-9.9237	80.0837	9.5054	-64.7444	2.4078	-80.1073		
TT	61.3015	-8.9121	78.1342	7.8843	-63.2031	2.1474	-78.0224	.6736	
HH	44.7380	.7343	69.7362	-1.7097	-46.0186	.7968	-69.0049	.9706	.4577

addition of the interaction and quadratic terms can cause an estimated coefficient to change in magnitude or in sign, sometimes drastically. Yet if one has no a priori knowledge of which is the correct model, any one of these estimates could be some researcher's preferred choice. If the researcher is unaware of the presence of multicollinearities among these predictor variables, moreover, his choice might be considered a "theoretical" model by other researchers.

Sign changes, distorted magnitudes, and large variances are just some of the properties typically exhibited by least squares coefficient estimates of multicollinear predictor variables. Knowledge of multicollinearities and their attendant problems is the first step in correcting the deleterious effects of the multicollinearities. In this chapter we address the effects of multicollinearities on least squares estimators and test statistics. Once the effects are noted, some remedial suggestions are offered to make effective use of the test statistics. Alternative coefficient estimators are discussed in the next chapter as another solution to the problem.

Numerical comparison of the magnitudes of estimated regression coefficients and their variances and covariances must be made with standardized predictor variables in order to remove the distortions due to different scales and locations for the predictors. For this reason, and to eliminate the need to reexpress all results in terms of the various transformations of the predictor variables, we will discuss only the unit length standardization in this chapter. All conclusions remain valid for the raw, centered, and standardized prediction equations although specific expressions may differ notationally.

9.1 COEFFICIENT ESTIMATORS

In this section the effects of multicollinearities on least squares estimators are examined. We emphasize the properties of the estimators in this section and treat test statistics and selection criteria in the next. The reason for the large variances of the coefficient estimators for the multicollinear predictor variables in the housing rent data will be explained. Likewise, the sign and magnitude changes noted in Table 9.3 with the emissions data will be related to the multicollinearities.

9.1.1 Variances and Covariances

Write the multiple linear regression model as

$$\underline{Y} = \alpha^*\underline{1} + W\beta^* + \underline{\varepsilon},$$

where the elements of W are $(X_{ij} - \overline{X}_j)/d_j$. The least squares estimators of the model coefficients are

$$\hat{\alpha}^* = \overline{Y} \quad \text{and} \quad \hat{\underline{\beta}}^* = (W'W)^{-1}W'\underline{Y},$$

with moments

$$E[\hat{\alpha}^*] = \alpha^* \qquad \mathrm{Var}[\alpha^*] = \frac{\sigma^2}{n}$$

$$E[\hat{\underline{\beta}}^*] = \underline{\beta}^* \qquad \mathrm{Var}[\hat{\underline{\beta}}^*] = \sigma^2(W'W)^{-1}$$

under Assumptions 1-3 of Table 6.2. Let $Q = (W'W)^{-1}$. It can be shown* that the jth diagonal element of Q is

$$q_{jj} = [\underline{W}_j'\underline{W}_j - \underline{W}_j'D_j(D_j'D_j)^{-1}D_j'\underline{W}_j]^{-1}, \qquad (9.1.1)$$

where $\underline{W}_j$ is the jth column of W and D_j is an $n \times (p-1)$ matrix obtained by removing the jth column, $\underline{W}_j$, from W. Since all the columns of W are of unit length, $\underline{W}_j'\underline{W}_j = 1$.

In order to further simplify expression (9.1.1), consider a regression of the jth standardized predictor variable, $\underline{W}_j$, on the $(p-1)$ predictor variables remaining in D_j. Since all the predictor variables are centered (including $\underline{W}_j$) the regression necessarily passes through the origin; hence a "model" for this regression is

$$\underline{W}_j = D_j\underline{\eta} + \underline{\varepsilon}.$$

The least squares "estimator" of $\underline{\eta}$ is

$$\hat{\underline{\eta}} = (D_j'D_j)^{-1}D_j'\underline{W}_j.$$

Now the total variability of $\underline{W}_j$ is given by

$$\mathrm{TSS} = \mathrm{TSS(adj)} = \underline{W}_j'\underline{W}_j = 1.0$$

and the regression sum of squares is

$$\mathrm{SSR} = \underline{\eta}'D_j'\underline{W}_j = \underline{W}_j'D_j(D_j'D_j)^{-1}D_j'\underline{W}_j.$$

---

*Using formulas for the inverse of a partitioned matrix, this result can be shown as in Searle (1971, p. 46) or Graybill (1976, p. 19).

Since TSS(adj) = 1.0, moreover, R_j^2 = SSR, where R_j^2 denotes the coefficient of determination obtained from the regression of the jth predictor variable on the other $(p-1)$ predictor variables. We can now reexpress eqn. (9.1.1) as

$$q_{jj} = (1.0 - R_j^2)^{-1} . \qquad (9.1.2)$$

The implications of eqn. (9.1.2) are extremely important for least squares parameter estimation. Highly multicollinear predictor variables will have R_j^2 values close to 1.0 since the jth predictor variable (if it is involved in the multicollinearity) can be approximated well by some of the other predictor variables. If R_j^2 is close to 1.0, q_{jj} will be large in magnitude as will be the variance of $\hat{\beta}_j^*$:

$$\mathrm{Var}[\hat{\beta}_j^*] = q_{jj}\sigma^2 .$$

If $\underline{W}_j$ is orthogonal to all the other columns of W, $\underline{W}_j'D_j = \underline{0}'$ and $R_j^2 = 0$. Then $\mathrm{Var}[\hat{\beta}_j^*] = \sigma^2$; otherwise, $\mathrm{Var}[\hat{\beta}_j^*] = q_{jj}\sigma^2 > \sigma^2$ since $q_{jj} > 1$ when $R_j^2 > 0$. Consequently, the q_{jj} measure the "inflation" of the variances of the least squares estimators due to multicollinearities among the predictor variables. The diagonal elements of $(W'W)^{-1}$ are referred to as variance inflation factors* (VIF). Any VIF larger than 4 implies that the associated $R_j^2 > 0.75$, while $q_{jj} > 10$ implies $R_j^2 > 0.90$. If either the variance inflation factors or the R_j^2 are known, the other quantity can be calculated from eqn. (9.1.2).

Not only are the variances of least squares estimators increased by multicollinearities among the predictor variables, but the covariances and correlations are also greater than those for orthogonal predictor variables. In order to better understand this point recall that one can write $(W'W)^{-1}$ in terms of the latent roots and latent vectors of $W'W$ as in eqn. (4.1.11):

$$(W'W)^{-1} = \sum_{r=1}^{p} \ell_r^{-1} \underline{V}_j \underline{V}_r' .$$

From this expression observe that

$$q_{jj} = \sum_{r=1}^{p} \ell_r^{-1} V_{jr}^2 . \qquad (9.1.3)$$

---

*The term "variance inflation factor" is due to Marquardt (1970). Note that the calculation of the VIF as the diagonal elements of $(W'W)^{-1}$ requires the unit length standardization.

Equation (9.1.3) shows the reason for the inflation of the variances of the least squares estimators from another point of view. If multicollinearities exist in W, some of the latent roots are near zero and so some of the ℓ_r^{-1} are large in magnitude. Large elements in latent vectors corresponding to latent roots near zero identify predictor variables that are involved in the multicollinearities. For these same predictor variables, q_{jj} is large because of the magnitude of the $\ell_r^{-1} V_{jr}^2$ terms. Predictor variables that are not involved in multicollinearities do not suffer from large VIF because the large magnitude of ℓ_r^{-1} is compensated for by small values of V_{jr}^2.

Off-diagonal terms of $(W'W)^{-1}$ can be written as

$$q_{jk} = \sum_{r=1}^{p} \ell_r^{-1} V_{jr} V_{kr}.$$

These terms will be large for any pair of variables that are involved in a multicollinearity; i.e., if neither V_{jr} nor V_{kr} is small enough to cancel the effect of a large ℓ_r^{-1}, q_{jk} will be large in magnitude. But neither V_{jr} nor V_{kr} will be near zero when both are involved in a multicollinearity that is associated with a small latent root ℓ_r. Thus if both X_j and X_k (or W_j and W_k) are multicollinear,

$$\mathrm{Cov}(\hat{\beta}_j^*, \hat{\beta}_k^*) = q_{jk}\sigma^2$$

will tend to be large, as will

$$\mathrm{Corr}(\hat{\beta}_j^*, \hat{\beta}_k^*) = \frac{q_{jk}}{(q_{jj} \cdot q_{kk})^{1/2}}. \tag{9.1.4}$$

Let us now reexamine several of the previous data sets and assess the effects of multicollinearities on the variances and correlations of the least squares estimators. We have already seen that the multicollinearity in the housing data causes the VIF for the three predictor variables to be 28.62, 44.13, and 87.34. Those of the nonmulticollinear predictor variables were all between 1.03 and 2.54.

Next consider the police applicant data, Data Set A.6. Table 9.4 gives the three smallest latent roots of $W'W$ and the corresponding latent vectors. Also shown are the variance inflation factors, R_j^2 values, and a portion of the correlation matrix, $W'W$. Examine first the VIF and R_j^2 values. Highly multicollinear predictor variables are WEIGHT, CHEST, THIGH, and FAT; but the VIF and R_j^2 values can only tell us which variables are

Table 9.4. Detection of Multicollinearities for the Police Applicant Data, Data Set A.6

VARIABLE	$\underline{V}_1$	$\underline{V}_2$	$\underline{V}_3$	VIF(q_{jj})	R_j^2	$\hat{\beta}_j^*$
	$\ell_1 = 0.0425$	$\ell_2 = 0.0432$	$\ell_3 = 0.1377$			
HEIGHT	-.046	.140	.060	3.35	.70	.1894
WEIGHT	-.122	-.844	.034	17.06	.94	-.0852
SHLDR	.245	.077	-.363	3.92	.74	.0261
PELVIC	-.131	.092	.327	3.31	.70	-.1664
CHEST	-.346	.465	.030	8.92	.89	-.1469
THIGH	-.391	.078	-.507	6.30	.84	-.0933
PULSE	-.092	.035	.087	1.93	.48	.0278
DIAST	.015	-.012	-.043	1.44	.31	.0831
CHNUP	.144	-.081	-.462	3.48	.71	.0461
BREATH	.072	.001	-.213	2.07	.52	.0497
RECVR	-.046	-.070	.325	2.30	.57	-.1504
SPEED	-.114	.027	.326	2.41	.58	-.1255
ENDUR	.030	.007	-.058	1.40	.29	-.0140
FAT	.763	.135	.125	14.63	.93	.2998

Condensed Correlation Matrix

	WEIGHT	CHEST	THIGH	FAT
WEIGHT	1.000	.889	.554	.810
CHEST		1.000	.398	.725
THIGH			1.000	.844
FAT				1.000

multicollinear and not the nature of the multicollinearities. The latent roots and latent vectors of $W'W$ can provide us with the latter information.

The latent vector corresponding to the smallest latent root of $W'W$, $\ell_1 = 0.0425$, contains three large elements, those associated with CHEST, THIGH, and FAT (the element corresponding to SHLDR, 0.245, could be considered large if one decides that its VIF, 3.92, indicates that SHLDR is multicollinear). The latent vector corresponding to the next smallest latent root, $\ell_2 = 0.0432$, identifies two multicollinear predictor variables, WEIGHT and CHEST. Together these two latent vectors identify all the predictor variables with VIF over 4.0. The third latent vector, corresponding to $\ell_3 = 0.1377$, has a latent root over three times larger than the previous two. Its large elements, moreover, include SHLDR, PELVIC, CHNUP, RECVR, and SPEED, variables that do not appear to be strongly multicollinear on the basis of the VIF and R_j^2 values. Consequently, we conclude that there are two multicollinearities in W.

A three-variable multicollinearity is indicated by $\underline{V}_1$. That this is not identifying strong pairwise correlations among the three variables can be seen by examining $W'W$, a portion of which is shown in Table 9.4. CHEST and

THIGH have a small pairwise correlation, CHEST and FAT as well as THIGH and FAT have moderately large pairwise correlations.

From $\underline{V}_1$ it appears that the relationship between CHEST (W_5), THIGH (W_6), and FAT (W_{14}) is

$$W_{14} = 0.45\,W_5 + 0.51\,W_6 .$$

Interestingly, total body fat was not directly measured on each subject in this data set but was computed from well-established formulas that relate body fat to other anthropometric measurements taken on these individuals, including skinfold thickness and body density. This multicollinearity, then, has a natural physical interpretation as does that between WEIGHT and CHEST, both of which are indirect measurements of the size of the subjects.

Table 9.5. Calculation of Estimator Correlations, Police Applicant Data, Selected Predictor Variables: Elements of $(W'W)^{-1}$ on and Below the Diagonal, Correlations Above

	WEIGHT	CHEST	THIGH	PULSE	DIAST	ENDUR	FAT
WEIGHT	17.059	-.622	-.063	-.044	.035	-.023	-.294
CHEST	-7.670	8.917	.461	.266	-.046	-.008	-.398
THIGH	-.655	3.455	6.302	.170	-.054	.041	-.721
PULSE	-.254	1.102	.591	1.927	-.239	.094	-.245
DIAST	.176	-.164	-.162	-.398	1.444	-.153	.015
ENDUR	-.114	-.028	.123	.154	-.218	1.401	.159
FAT	-4.647	-4.551	-6.928	-1.299	.068	.718	14.634

Turning now to the estimator correlations, Table 9.5 provides representative calculations for seven of the fourteen variables in Data Set A.6. The numerical values on and below the diagonal of the matrix in Table 9.5 are the elements of $(W'W)^{-1}$ for the seven variables listed (i.e., $Q = (W'W)^{-1}$ was calculated for the entire fourteen variable data set and the elements, q_{ij}, corresponding to the seven predictor variables in Table 9.5 were obtained from Q; since Q is symmetric, moreover, only the elements on and below the diagonal need be shown). The correlations shown in the upper portion of the table are then obtainable as expressed in eqn. (9.1.4).

As indicated by eqn. (9.1.3), the largest q_{ij} in the lower portion of Table 9.5 correspond to pairs of multicollinear predictor variables. Also, the largest correlations in the upper half of the table belong to multicollinear predictor variables. It is just as important to note that coefficient estimators corresponding to nonmulticollinear predictor variables do not have large correlations with any of the other estimators; for example, PULSE,

DIAST, and ENDUR do not have correlations larger than 0.3 in magnitude in Table 9.5.

As a final illustration of the variable inflation attributable to multicollinearities, Table 9.6 lists the VIF for three fits to the emissions data. The first fit uses only the three linear terms, the second one adds the interaction terms, and the third one includes all nine terms of the full quadratic model. The changes in the VIF as terms are added to the linear fit are dramatic.

Table 9.6. Variance Inflation Factors for Three Fits to the Emissions Data

Predictor Variable	Linear Fit	Linear and Interactions	Full Quadratic
HUMID	4.17	40,962.48	562,602.19
BPRES	3.32	798.80	455,999.18
TEMP	1.49	105,494.42	247,571.79
HB		37,460.88	570,400.12
HT		1,030.07	3,988.18
TB		99,497.08	242,697.67
HH			270.23
BB			452,788.25
TT			1,197.92

From these three examples it should be apparent that multicollinearities result in much larger variances and correlations for the least squares estimators of the regression coefficients over corresponding quantities for orthogonal predictor variables. Large variances imply imprecision in the estimation of model parameters. Imprecision is the first direct effect of multicollinearities.

9.1.2 Estimator Effects

The effect of multicollinearities on least squares estimates can be as dramatic as the effect on the estimator variances. While the following derivation will reveal that the parameter estimates are not always detrimentally affected, researchers frequently observe signs and magnitudes of estimates that disagree with well-established theory or previous empirical studies. One estimator property is certain: the estimators associated with multicollinear predictor variables have much larger variances than nonmulticollinear ones and, therefore, a much greater likelihood of being a poor estimate of their respective parameters.

Just as we reexpressed $(W'W)^{-1}$ in terms of its latent roots and vectors in the last subsection, let us now rewrite the least squares estimator as a function of these same latent roots and vectors:

$$\hat{\beta}^* = (W'W)^{-1}W'\underline{Y}$$

$$= \sum_{r=1}^{p} \ell_r^{-1} \underline{V}_r \underline{V}_r' W' \underline{Y}$$

$$= \sum_{r=1}^{p} \ell_r^{-1} c_r \underline{V}_r, \tag{9.1.5}$$

where $c_r = \underline{V}_r' W' \underline{Y}$. Observe from eqn. (9.1.5) that unless c_r is near zero, small latent roots of $W'W$ produce large multipliers (ℓ_r) on the corresponding latent vectors. Consequently, multicollinear predictor variables, those with large elements in the $\underline{V}_r$ that have small roots, will receive large weights in eqn. (9.1.5). Also, the signs on these coefficient estimates tend to have the same pattern as the signs in the $\underline{V}_r$.

Let us now look at a special case. Suppose the latent roots of $W'W$ are ordered so that $\ell_1 \leqslant \ell_2 \leqslant \ldots \leqslant \ell_p$, and that ℓ_1 is several orders of magnitude smaller than ℓ_2, implying that there is a single very strong multicollinearity in W. Then ℓ_1^{-1} will be several orders of magnitude larger than ℓ_2^{-1}, ℓ_3^{-1}, ..., ℓ_p^{-1} and the first term in eqn. (9.1.5) should dominate the entire summation for elements of $\underline{V}_1$ that are not near zero; i.e., for the estimators of the multicollinear predictor variables. Thus when there is a single strong multicollinearity in W, from eqn. (9.1.5)

$$\hat{\beta}_j^* \approx \ell_1^{-1} c_1 V_{j1} \tag{9.1.6}$$

for the estimated coefficients of multicollinear predictor variables.

Expression (9.1.6) points out two characteristics of least squares estimates when a single strong multicollinearity occurs in W:

(i)   the estimates tend to be large in magnitude due to the multiplier ℓ_1^{-1}, unless ℓ_1^{-1} is cancelled by a small value of c_1, and

(ii) the signs on the estimates tend to be determined more by the multicollinearity than by relationships of the predictor variables with the response; i.e., if $c_1 > 0$, the sign of $\hat{\beta}_j^*$ (for multicollinear predictor variables) is the same as that of V_{j1}, while if $c_1 < 0$, the sign of $\hat{\beta}_j^*$ is opposite that of V_{j1}.

Often in the physical sciences or economics, theoretical derivations suggest that certain coefficients should have particular signs and the magnitudes should not be large. With highly multicollinear predictor variables, the above two characteristics can dominate the theoretical relationships among response and predictor variables to such an extent that the least squares estimates end up with incorrect signs and magnitudes that are much larger than expected. With several strong multicollinearities in W, expression (9.1.5) can be used to draw the same general conclusions as above.

Comparison of the signs of the standardized coefficient estimates of the police applicant data in Table 9.4 with the signs of the elements of $\underline{V}_1$ reveals that the estimated coefficients for CHEST, THIGH, and FAT, the three multicollinear predictor variables identified by $\underline{V}_1$, have the same signs as the corresponding elements in $\underline{V}_1$. Moreover, FAT has the largest estimated regression coefficient and CHEST has the fifth largest one. Whether these estimated coefficients actually reflect the true relative influences of the individual predictor variables on the response is speculative.

WEIGHT and CHEST are the two multicollinear predictor variables that are identified in $\underline{V}_2$. The estimated regression coefficient for WEIGHT in Table 9.4 has the same sign as the corresponding element in $\underline{V}_2$. CHEST does not have the same sign as the corresponding element in $\underline{V}_2$ but its magnitude is large in both $\underline{V}_1$ and $\underline{V}_2$. Apparently the magnitude of the element corresponding to CHEST in $\ell_1^{-1} c_1 \underline{V}_1$ is outweighing that for CHEST in $\ell_2^{-1} c_2 \underline{V}_2$ resulting in a cumulative estimate [using eqn. (9.1.5)] that is negative.

The single multicollinearity in the housing rent data does not produce coefficient estimates that have the same sign patterns as the three large elements in $\underline{V}_1$. From Table 9.1, the large elements of $\underline{V}_1$ are

$$V_{51} = -.4183 , \qquad V_{61} = -.5234 , \qquad V_{71} = .7419$$

while the corresponding standardized coefficients are (multiplying each by 10^{-3})

$$\hat{\beta}_5^* = .1525 , \qquad \hat{\beta}_6^* = .3400 , \qquad \hat{\beta}_7^* = .0785 .$$

These coefficient estimates are also comparable in magnitude to many of the estimates for nonmulticollinear predictor variables. Apparently the value of c_1 is small enough to cancel most of the effect of ℓ_1^{-1} on $\underline{V}_1$ in eqn. (9.1.5). Despite the lack of obvious sign and magnitude problems with this data set, the frequent occurrence of such problems and the illustration with the police applicant data (another example will be given in the next chapter) should lead one to carefully study the effects of multicollinearities on coefficient estimates.

9.2 INFERENCE PROCEDURES

Because of the discussion in the last section, one should anticipate that inferences drawn on the regression coefficients might be suspect when predictor variables are multicollinear. The large variances of the estimators suggest that test statistics will tend to be smaller than they should be and that nonrejection of hypotheses would occur more frequently than if the predictor variables were not multicollinear. In this section we examine some of the test procedures that were proposed in Chapter 6 and 8 to determine the impact of multicollinearities on them.

9.2.1 t Statistics

Individual t statistics used to test $H_0:\beta_j = 0$ vs. $H_a:\beta_j \neq 0$ or, equivalently, $H_0:\beta_j^* = 0$ vs. $H_a:\beta_j^* = 0$ can be expressed as (see Section 6.3.1):

$$t_j = \frac{\hat{\beta}_j^*}{\hat{\sigma} q_{jj}^{1/2}}. \tag{9.2.1}$$

Since $q_{jj} = (1 - R_j^2)^{-1}$, an equivalent expression is

$$t_j = \frac{(1 - R_j^2)^{1/2} \hat{\beta}_j^*}{\hat{\sigma}}. \tag{9.2.2}$$

Either of two characteristics can be observed from t statistics for multicollinear data, depending on the relative magnitudes of $(1 - R_j^2)^{1/2}$ and $\hat{\beta}_j^*$. If the multicollinearities are extremely strong, $(1 - R_j^2)^{1/2}$ can be very close to zero, causing the t statistics to be small even though β_j might be nonzero. Alternatively, if R_j^2 is not extremely close to 1.0 but $\hat{\beta}_j^*$ has nevertheless been inflated by the multicollinearities, the t statistics can be inflated as well. Frequently all the t statistics corresponding to multicollinear predictor variables will be large or all will be small.

Of the four multicollinear predictor variables in the police applicant data, two of the four, WEIGHT and THIGH, have t statistics (see Table 6.6, Section 6.3.1) that are less than 1.0 in magnitude. The other two, CHEST and FAT, have t statistics between 1.0 and 2.0. At the 5% significance level $[t_{.975}(35) = 2.030]$, none of the coefficients for the multicollinear predictor variables would be judged nonzero in an individual t test. This lack of significance could be attributed either to the minor influence of these variables on the response or it could be due to the multicollinearities.

If one's sole purpose in analyzing a data set is to form a prediction equation, the effects of multicollinearities on t statistics may not prove serious. If model specification or parameter estimation is the goal, inferences from t

statistics can be devastatingly inaccurate. The tendency to remove one or more of the multicollinear predictor variables from a prediction equation because of the multicollinearities themselves and not because the variables are necessarily unimportant predictors of the response is one of the major defects of least squares methodology.

Several remedial steps can be taken to lessen the likelihood of erroneously deleting important multicollinear predictor variables. The first and primary solution is to be aware of the problem and seek to detect multicollinearities in regression data bases. One may then opt to ignore the t statistics and leave all multicollinear predictor variables in a prediction equation, preferring to make no inferences on the true coefficients rather than risking erroneous ones. Of course if more data can be collected and the multicollinearities thereby removed from the data base, adequate inferences may be possible on all model parameters.

Many regression analyses call for the elimination of ineffective predictor variables regardless of whether multicollinearities are present. In order to guard against wholesale elimination of multicollinear predictor variables in such circumstances, we recommend raising the significance probability from $p = 0.05$ to a higher level. Our preference is to use $p = 0.25$ and be conservative: it is better to leave in some relatively uninfluential predictor variables than to remove important ones. The use of a significance probability larger than 0.05 has been supported by several studies* and although the particular choice of $p = 0.25$ is a subjective one, it has also found support in the statistical literature.

Adopting a significance probability of $p = 0.25$ now leads one to conclude the regression coefficient for FAT is nonzero $[t_{.875}(35) = 1.18]$ and that for CHEST is bordering on significance if individual t tests are performed on either of these predictor variables. WEIGHT and THIGH would be judged nonsignificant although one must still question whether WEIGHT truly has little influence on reaction times. Again being conservative, one might reconsider whether WEIGHT should indeed be deleted in light of its large R_j^2 value. This is a subjective decision, but one that may be merited by the analysis of the multicollinearities.

One might question at this juncture why FAT with its large R_j^2 value does not suffer from the same problems as WEIGHT. The answer is, it does. If $q_{jj} = 1$ for FAT its t statistic (given the same numerical values for $\hat{\beta}_j^*$ and $\hat{\sigma}$ as in the present analysis) would rise from 1.77 to 6.77. Also if a significance probability of 0.05 is used in conjunction with the t value of 1.77, a test of the hypothesis $H_0 : \beta_{14} = 0$ would not be rejected while with a significance probability of 0.25 it would be. On the other hand, if $t = 6.77$ both

---

*See Kennedy and Bancroft (1971), Pope and Webster (1972), and Bendel and Afifi (1977).

tests would reject the hypothesis. The behavior of these t statistics can again be attributed to either of two explanations: (i) β_{14} is nonzero but only large enough to overcome the magnitude of $q_{14,14}$ at the 0.25 significance level, or (ii) β_{14} actually is zero and the t statistic attained its numerical value randomly (with a probability of at least 0.25).

One additional solution is available for making inferences with multicollinear predictor variables. Rather than estimating the regression coefficients with the unbiased least squares estimators and using t statistics with large significance probabilities (which can still result in the elimination of important predictor variables), employ biased regression estimators. Biased estimation techniques will be studied in Chapter 10.

9.2.2   Other Selection Criteria

In Section 8.1.2 three criteria were introduced for selecting predictor variable subsets. These are the coefficient of determination, R_1^2, the error mean square, MSE_1, and the C_k statistic. All three criteria are a function of the degree of change in SSE_1 induced by deleting a subset of predictor variables. Since this change in SSE_1 can be shown to be directly dependent on linear relationships among the predictor variables, each criteria is affected by the problem of multicollinearity.

In a reduced prediction equation of size s, the error sum of squares SSE_s measures the accuracy of prediction of the observed responses. If X_j is also removed from the model, SSE_{s+1} can be close to SSE_s for either of two reasons:

(i) X_j is not useful as a predictor of the response variable, or

(ii) X_j is multicollinear with other predictor variables.

When X_j is multicollinear the information provided by X_j is redundant and no essential loss of predictive ability of the *observed response* occurs when X_j is deleted. This characteristic of SSE_s can occur not only in the full prediction equation but also in any subset predictor for which some of the X_j are multicollinear.

Small differences in SSE_s and SSE_{s+1} also result in small changes in R_1^2 (provided the degrees of freedom are not extremely small). Hence, a small decrease in R_1^2 or trivial change in MSE_1 could indicate the presence of a multicollinearity. In fact, examination of changes in MSE_1 or R_1^2 has been recommended as a means of detecting if a multicollinearity exists (e.g., Farrar and Glauber, 1967). Unfortunately such a change might also reflect the worthlessness of X_j as a predictor. For these reasons, the detection methods discussed in Chapter 4 are preferred over examination of R_1^2 and MSE_1 variations.

Overreliance on R_1^2 or MSE_1 can easily lead to the removal of one or more of the multicollinear predictor variables from a prediction equation. The problem, however, is not necessarily the removal of the predictors but the choice of which of the multicollinear variables to delete. Neither the R_1^2 nor the MSE_1 criteria are able to distinguish between important ($\beta_j = 0$) and unimportant ($\beta_j \neq 0$) multicollinear predictor variables; they can only point to the redundancy of at least one of these variables.

It is evident that any subset selection criteria that depends on examination of changes in SSE_1 from one equation to the next is hampered in its ability to select useful subsets when multicollinearities are present. Since C_k is also a function of SSE_1 it is similarly affected, although due to the influence of the sample size [see eqn. (8.2.2)] minor changes in SSE_1 can appear as nontrivial changes in C_k.

The seventeen-variable housing data contains a strong multicollinearity among X_5, X_6, and X_7, all measures of the size of the dwelling unit. The best subset of size 16 deleted X_7 and this subset was selected by several of the variable selection techniques discussed earlier. The C_k statistics for subsets of size 16 in which X_5, X_6, or X_7 are the only variables deleted are, respectively, 17.68, 21.42, and 16.15. The corresponding R_1^2 values are 0.763, 0.761, and 0.764. Both C_k and R_1^2 change with SSE_1 yet the C_k values appear to indicate clear preferences among the predictor variables whereas the R_1^2 values do not. As mentioned in Section 8.1.2, C_k is usually a better indicator of the importance of predictor variables than is R_1^2, but when the variables are multicollinear caution needs to be exercized, as we now demonstrate.

Recall that in the discussion of the t-directed search, the t statistics for X_5, X_6, and X_7 were all less than 3.0 (see Table 8.7, Section 8.2.4). One application of a t-directed search recommended that a basic set of predictor variables be formed from all t statistics larger than 3.0 and a search be made of all regressions of the basic set and subsets of the remaining predictor variables. We did not draw any conclusions about predictor variables with small t statistics even though it is tempting to say that all t statistics with small t values should be eliminated.

Suppose we now consider t statistics for a reduced model in which X_5 is removed from the full prediction equation. The t statistics for X_6 and X_7 in this reduced model are 3.81 and 8.14, respectively. Surprisingly, the t statistic for X_7 is now the largest one in the collection of sixteen predictor variables. Jointly X_6 and X_7 are exhibiting the predictive ability of all three predictor variables. In fact, dramatic increases in the t statistics for any two of these three predictor variables occur when the third one is eliminated from the full prediction equation. Thus the multicollinearity among X_5, X_6, and X_7 is distorting the true influence of these predictor variables by deflating the values of the selection criteria [recall eqn. (9.2.2)]. It is almost arbitrary which predictor variable is eliminated; its predictive power, if any, will be retained by the other two predictors. The other selection criteria do not, moreover, reflect the relative importance of the multicollinear predictor variables as the C_k statistics appeared to do. Note carefully that these comments refer to prediction of the *observed* responses and do not necessarily reflect the importance of the predictor variables on *all possible* responses; i.e., the relative magnitudes of the true coefficients in the theoretical model. Further discussion on this point is found in Section 9.3.2.

Stepwise selection methods suffer from multicollinearities since they use these same criteria to determine whether to add or remove predictor variables. Differences in the final subsets for FS and BE with the seventeen-variable housing data are due to the multicollinearity among X_5, X_6, and X_7. Likewise, the differences occurring in both the chosen subsets and the estimated coefficients for the FS and BE eight-variable subsets in Table 8.1 are attributable to the multicollinearity.

9.2.3 Effects on Subset Selection Methods

In the all possible regressions procedure, R_1^2 is used as the criteria for comparing different subsets of predictor variables. As was shown in the previous subsection, this selection criterion can be misleading when multicollinearity is present. Nevertheless, since all possible regressions are performed and each prediction equation is examined, multicollinearity is more evident. If it is properly diagnosed and isolated before variable selection begins, one should be able to distinguish whether a small change in R_1^2 is indicating a useless predictor variable or a redundant variable. In such a situation one may decide not to eliminate any of the multicollinear predictor variables since their coefficient estimates and corresponding t values are possibly distorted. It might be better to retain some uninfluential predictor variables rather than risk removing important ones.

The best possible subset approach is affected in a similar manner. Since its basis is change in SSE_1, it follows from the discussion in the previous subsection that important subsets could be ignored. Furthermore, all multicollinear predictor-variable coefficient estimates could be distorted if the multicollinearity is severe. Hence, deciding which, if any, of the multicollinear predictors are unimportant will be most difficult.

As was previously mentioned there exists a three-variable multicollinearity in the housing data between X_5, X_6, and X_7. Its influence is particularly prominent in the sixth row of Table 8.4 (Section 8.2.2) where the best subset approach has replaced X_7 with X_5 and X_6 to end up with the best subset of size 6. Notice that X_7 was initially chosen over the other two multicollinear predictors as contributing most to explaining the response. Now it is removed and does not enter into the prediction equation again until the last step.

Table 9.7. Alternative Solution to Best Possible Subset Regression for 17 Variable Housing Data

Variables Added	p	R_1^2	Cumulative C_k
X_{12}	1	.350	752.46
X_7	2	.627	241.73
X_8	3	.675	154.57
X_{10}	4	.693	124.46
X_1	5	.703	106.98
X_6	6	.715	88.71
X_{14}	7	.725	72.15
X_{11}	8	.732	60.49
X_{16}	9	.737	52.91
$-X_{16}, X_2, X_3$	10	.744	43.13
X_9	11	.749	34.55
X_{16}	12	.753	29.05
X_4	13	.757	24.53
X_{15}	14	.759	21.34
X_{17}	15	.762	17.92
X_{13}	16	.764	17.68
X_5	17	.764	18.00

Table 9.7, in contrast to the results summarized in Table 8.4, does not remove X_7 at the sixth step of the best subset procedure but forces in X_6 and holds out X_5 until the last step. Other than these two changes, all variables are added to the prediction equation in the same order in both tables. The resultant changes in R_1^2 and C_7 are slight. With X_5 and X_6 in the prediction equation, $R_1^2 = 0.715$ and $C_7 = 87.38$, while with X_6 and X_7, $R_1^2 = 0.714$ and $C_7 = 88.71$. If X_7 had been retained and X_5 added instead of X_6, R_1^2 would be 0.712 and $C_7 = 93.28$. Hence, regardless of which two multicollinear predictors are retained the outcome would be essentially the same. Apart from the change at step 6, Tables 8.4 and 9.7 are identical in their subset selection and little change is noted in the corresponding values of R_1^2 and C_k.

This outcome is expected. Since X_5, X_6, and X_7 are multicollinear any two of these predictors contain the essential information of the third and make it needless. Choosing which pair to include is no easy task and, in the we not known of this three-way multicollinearity the results of Table 8.4 might have been taken as the best subset. Now there is some doubt as to which multicollinear predictor variable to exclude.

9.3 POPULATION-INHERENT MULTICOLLINEARITIES

Beginning with Section 4.4, we have tried to distinguish sample and population-inherent multicollinearities. Sample-based multicollinearities are a property of the particular data base collected and are not truly characteristic of the population under study. Knowing that the true regression coefficients might be poorly estimated and that inference procedures could be ambiguous at best, erroneous at worst, places severe restrictions on the usefulness of least squares methods when multicollinearities are sample-based. The key point to remember is that we are desiring to estimate model parameters for a population model from data that is not characteristic of the population.

On the other hand, if the multicollinearities are inherent to the population or if one merely wishes to construct adequate prediction equations for other data values that possess the same multicollinearities as are exhibited in the data base, the effects of multicollinearities are not as damaging. In this section we elaborate on remedies for population-inherent multicollinearities.

9.3.1 Variable Selection

Erroneous removal of important predictor variables from an estimated model can have dramatic consequences when one's goal is model specification or parameter estimation. Magnitudes and signs of estimated coefficients can change when a multicollinear predictor variable is removed since the multicollinearity no longer exists in the reduced data base. If the variable that is eliminated is actually a good predictor of the response, not only is an erroneous inference being made by deleting that predictor variable but other misleading conclusions might also be drawn. Uninfluential predictor variables might suddenly appear to be important ones because of their association with the deleted one. For example, if WEIGHT is removed from the full prediction equation for reaction time, the t statistic for CHEST rises from -1.11 to -1.80, significant at the .25 significance level. Does this change occur because CHEST actually is an important predictor variable or because it is acting as a surrogate for WEIGHT in the reduced prediction equation?

If the multicollinearities are actually inherent to the population, variable selection procedures are not only desirable, they may be essential for proper inferences. In the emissions study, for example, only one humidity variable was reported, although in the original study three humidity readings were taken. After the experiment was concluded the three humidity readings were found to be highly correlated. Pairwise correlations taken between the three possible pairs of humidity measurements were 0.965, 0.925, and 0.893. It was then decided to discard two of the humidity readings and concentrate on analyzing the third.

Elimination of the two alternative humidity readings was called for precisely because they would be highly correlated in all measurements of humidity; i.e., not just because they highly correlated in this data base. Since they can be expected to be highly correlated in all humidity measurements similar to those taken in this experiment, the correlations among the humidity readings constitute population-inherent multicollinearities. Failure to eliminate the other two humidity readings could have produced confusion in the subsequent analysis of NOX. The relative importance of TEMP and BPRES when compared to the three humidity variables could have been greatly distorted because of the correlations between the humidity readings. Estimated coefficients for the humidity variables could have had incorrect signs, enlarged variances, and t statistics that were too large or too small—all because of the multicollinearities.

A similar argument can be raised for the exclusion of many of the interaction and quadratic terms of the full nine-variable model. Over the range of humidity, temperature, and barometric pressure values in this experiment, many of the interaction and quadratic terms will always be highly correlated with the linear ones. Two cases dictate which terms should be deleted in situations like this: (i) if it is known theoretically, say, that NOX is a function of the square of barometric pressure and not of the linear or interaction terms, these latter terms should be eliminated and $(BPRES)^2$ should be retained in the prediction equation; or (ii) if no theoretical or past empirical studies dictate the functional form of the model, the principle of model parsimony generally suggests that the simpler linear terms be retained and the more complex interaction and quadratic terms be candidates for deletion.

Another characteristic of multicollinearities and model specification is important to recognize at this point. Interactions and quadratic terms in the raw predictor variables have been analyzed in the emissions data. These functions of the raw predictor variables are used because of the tendency of data analysts to do so. Actually, it is more beneficial to standardize or at least center the predictor variables prior to forming interactions and quadratic terms. In doing so, the constant effect, $\overline{X}_j$, is removed from all model terms and reduces the covariation among them. The interaction and quadratic terms are then less multicollinear with the linear ones and each other; their individual effects can be assessed with less confusion (see Exercise 8).

Questions relating to whether multicollinearities are population-inherent are sometimes subjective and inextricably tied to the goals of a study. If population-inherent multicollinearities can be identified, one of the redundant predictor variables often can be eliminated without sacrificing any of the objectives of the regression analysis. The subsequent lessening of ambiguities in the interpretation of coefficient estimates and selection criteria are the main benefits of such an approach.

9.3.2 Prediction

Multicollinearities do not necessarily have as deleterious effects on predicted responses as they do on estimated coefficients. Even if one erroneously removes redundant predictor variables from the full prediction equation, predictions may remain as accurate with reduced prediction equations as with full ones. The key to the successful utilization of prediction equations formed from multicollinear predictor variables is in an examination of where prediction is desired.

In the first three chapters of this text we belabored the dangers of extrapolation, especially with single-variable prediction equations. If extrapolation beyond the range of the data base is desired, the true theoretical model for the response must be correctly specified and good parameter estimates must be obtained; otherwise, there are regions of the predictor variables for which prediction of the response will be highly inaccurate.

Adequate prediction of a response can occur with multicollinear predictor variables so long as the predictor variables are restricted to (i) be within the limits of the predictor variables in the data base used to formulate the prediction equation, and (ii) have the same multicollinearities as occur in the data base. The second requirement is often overlooked when prediction equations are utilized. Its violation, however, is tantamount to extrapolation.

Predicting the observed responses in a data base satisfies both (i) and (ii) since the same predictor-variable values are used to formulate the prediction equation and predict the responses. It is not surprising, therefore, that several subset models from one data base can predict the observed responses as well as the full prediction equation. None of the fitted models need be properly specified in order to achieve accurate prediction of the observed responses, but poor prediction of future responses may result from all the fits.

An indication of the potential trouble that can arise due to extrapolation by ignoring the second requirement mentioned above is apparent from an appraisal of two prediction equations fit to the police applicant data. The full prediction equation using the unit length standardization is

$$\hat{Y} = 0.316 + 0.189\,W_1 - 0.085\,W_2 + 0.026\,W_3 - 0.166\,W_4 - 0.147\,W_5$$
$$- 0.093\,W_6 + 0.028\,W_7 + 0.083\,W_8 + 0.046\,W_9 + 0.050\,W_{10}$$
$$- 0.150\,W_{11} - 0.126\,W_{12} - 0.014\,W_{13} + 0.300\,W_{14}. \tag{9.3.1}$$

Eliminating WEIGHT (W_2) yields the reduced prediction equation

$$\hat{Y} = 0.316 + 0.177\,W_1 + 0.014\,W_3 - 0.174\,W_4 - 0.185\,W_5 - 0.097\,W_6$$
$$+ 0.027\,W_7 + 0.084\,W_8 + 0.052\,W_9 + 0.048\,W_{10} - 0.143\,W_{11}$$
$$- 0.126\,W_{12} - 0.015\,W_{13} + 0.277\,W_{14}. \tag{9.3.2}$$

Some of the coefficients in the above two equations are virtually identical: PULSE(W_7), DIAST(W_8), BREATH(W_{10}), SPEED(W_{12}), ENDUR(W_{13}). Others have undergone substantive changes; e.g., CHEST(W_5) and FAT(W_{14}) have both changed by more than 0.02 units. It is important to note that the variable eliminated in eqn. (9.3.2), WEIGHT, is highly multicollinear with CHEST and, through CHEST, with FAT (see Table 9.4). The five predictor variables listed above whose coefficients changed the least are five of the six variables listed in Table 9.4 that have variance inflation factors less than 3.0. Elimination of WEIGHT appears to have affected some of the multicollinear coefficient estimates the most while leaving several of the nonmulticollinear ones untouched.

To see whether changing from the full to the reduced prediction equation seriously affects the predicted responses, let us use both to predict the observed responses for the first five subjects in Data Set A.6. The upper portion of Table 9.8 records the predicted responses using eqns. (9.3.1) and (9.3.2). Overall the predictions are very close as one might expect.

Let us now make a very minor change in the values for one of the multicollinear predictor variables, FAT. In the second half of Table 9.8 only the signs, not the magnitudes of the values for FAT are altered. The predicted responses now differ more for the two prediction equations. Using the predicted responses for the full prediction equation for comparison, the average difference in predicted responses for the upper portion of Table 9.8 is about 0.9% of the predicted responses while for the lower portion it is 2.4%. Increasing the error from 0.9% to 2.4% may not seem like too great a change in the similarity of predictions for eqns. (9.3.1) and (9.3.2), but the only alteration in the predictor variables was a sign change in one of the fourteen variables.

These differences can be magnified even further by choosing values of the predictor variables that do not extend beyond the ranges of the individual variables but do violate requirement (ii) above. For example if

Table 9.8. Comparison of Predictions From Full and Reduced Prediction Equations for Reaction Times

| | Interpolation | | | | |
| | Total Body Fat | | Observed | Predicted Response | |
Subject	Raw	Standardized	Response	Eqn. (9.3.1)	Eqn. (9.3.2)
1	11.91	-.03640	.310	.344	.335
2	3.13	-.20711	.345	.320	.320
3	16.89	.06042	.293	.326	.327
4	19.59	.11291	.254	.274	.274
5	7.74	-.11748	.384	.351	.345

| | Extrapolation | | |
| | Standardized | Predicted Response | |
Subject	Total Body Fat	Eqn. (9.3.1)	Eqn. (9.3.2)
1	.03640	.367	.355
2	.20711	.444	.435
3	-.06042	.290	.294
4	-.11291	.206	.212
5	.11748	.421	.410

$X_1 = 165$, $X_2 = 100$, $X_3 = 38$, $X_4 = 25$, $X_5 = 81$, $X_6 = 6$, $X_7 = 50$,

$X_8 = 65$, $X_9 = 0$, $X_{10} = 150$, $X_{11} = 91$, $X_{12} = 4$, $X_{13} = 3$, $X_{14} = 4$,

the predicted value from the full prediction equation is $\hat{Y} = 0.369$ while that from the reduced one is $\hat{Y} = 0.412$, a difference of about 11.7%. (Note: the entire range of reaction times is from 0.221 to 0.427, so the difference in these predicted values is substantial.) Virtually the entire difference in these predicted responses is due to the extremely large value of WEIGHT coupled with the small value for CHEST. In Data Set A.6 these two variables are positively correlated, implying that large values of WEIGHT are accompanied by large values of CHEST, and similarly for small values. Requirement (ii) is thus violated by the preceeding choice of X_2 and X_5. Further magnification of the differences in eqns. (9.3.1) and (9.3.2) can be obtained by not adhering to requirement (i).

Thus, if one only requires that adequate prediction of a response variable be attained by the final fitted model, it is sometimes of relatively minor importance which of the multicollinear predictor variables are eliminated from the prediction equation. Several reduced models with sufficiently large R^2 values might be able to predict as accurately as the full prediction equation. One must be careful, however, not to extrapolate. This requires that one adhere to both requirements (i) and (ii) mentioned earlier in this subsection.

EXERCISES

1. Make two-variable graphs of NOX versus HUMID, NOX versus HUMID × TEMP and NOX versus HUMID × BPRES. What do these graphs reveal concerning the dominance of humidity in the latter two interactions? Do the correlations in Table 9.2 confirm the visual impression left by the plots?

2. How would one calculate variance inflation factors if computer output only provides standard errors of the raw coefficient estimates (estimates of the standard deviation of $\hat{\beta}_j$)?

3. Construct prediction equations for first-year-of-college scores using (i) linear, (ii) linear and quadratic and (iii) linear, quadratic and cubic functions of the grade 13 scores for high school 1. How does the variance inflation factor of the linear coefficient estimator change as the quadratic and cubic terms are added? Calculate the correlations among the estimators in (ii) and (iii).

4. Convert the variance inflation factors of the previous problem to standard errors for the raw, centered, normal deviate and unit length transformations of the predictor variables. Compare the magnitudes of these standard errors (i) among the various transformations for each prediction equation and (ii) for each transformation among the three prediction equations.

5. Consider a two-variable prediction equation wherein both predictor variables are (unit length) standardized. Suppose $X_1' Y \approx X_2' Y$, as might occur if X_1 and X_2 are very highly correlated. If r denotes the correlation coefficient between the two predictor variables, evaluate $\hat{\beta}_1^*$ and $\hat{\beta}_2^*$ using eqn. (5.1.17) as r approaches 1 in value.

6. Construct a prediction equation for NOX using HUMID, TEMP, BPRES, and $BPRES^2$. Compare the standardized coefficient estimates of BPRES and $BPRES^2$ with the corresponding elements of the latent vector of $W'W$ associated with the smallest latent root. Are the signs on the coefficient estimates consistent with what one would expect from observing the correlation coefficient between these two predictor variables? Also compare these coefficient estimates with those in the last three rows of Table 9.3. Calculate the correlation between the coefficient estimators and the variance inflation factors.

7. Are the results of Exercises 4, 5, and 6 in Chapter 8 consistent with the effects of multicollinearity in Sections 9.2.2 and 9.2.3?

8. Standardize the three predictor variables in the emissions data and then form interaction and quadratic terms. Calculate pairwise correlations, VIFs, and the latent roots and latent vectors of $W'W$. How has standardization affected these quantities? What are the effects on the regression estimates for the linear, linear and interaction, and full quadratic fits (compare with Table 9.4).

CHAPTER 10

BIASED REGRESSION ESTIMATORS

Least squares methodology provides a wide variety of model-fitting techniques for constructing prediction equations. As discussed in Chapter 9, however, multicollinear data bases can destroy the effectiveness of least squares estimation in spite of the highly desirable estimator properties of unbiasedness, minimum variance (among unbiased estimators of the regression coefficients), and known probability distributions. In previous chapters we stressed the identification of multicollinearities in a data set and the elimination of them through variable selection techniques when the multicollinearities are inherent to the population under study. If the multicollinearities are merely a function of the sample data set and are not inherent to the population under study such a strategy could result in important predictor variables being deleted from the prediction equation. Biases thereby induced on the remaining coefficient estimators may render interpretations of the fitted model erroneous.

In order to stress these points further, consider an apparently simple task of predicting height from the information in Data Set A.8. Accurate body measurements and the characterization of modelling of body features are of great importance to physicians, health science personnel, anthropologists, and many others. Physicians studying disease or aging, for example, might wish to investigate the deterioration of body muscle as a function of other physical measurements. Medical examiners could be called upon to determine the height, weight, sex, and age of a deceased person based on only partial remains of a body. Biological and anthropological scientists are often interested in growth rates of civilizations for which only incomplete

315

skeletal remains are available for analysis. Depending on the precise purpose of the investigation, studies such as these could require that (i) the fitted model predict well for a wide range of values of the predictor variables, or (ii) accurate coefficient estimates be obtained. In either situation multicollinearities in the data base can hamper the effectiveness of least squares methodology.

Data Set A.8 contains measurements on height (HEIGHT) for 33 black female police department applicants along with values for nine predictor variables: sitting height (SITHT), upper arm length (UARM), forearm length (FORE), hand length (HAND), upper leg length (ULEG), lower leg length (LLEG), foot length (FOOT), and two indices which will be discussed below (BRACH and TIBIO). Preliminary examination of the original variables and residuals from fitted models revealed no apparent outliers nor model inadequacies. As is indicated in Table 10.1, almost 90% of the response variability is accounted for by the nine predictor variables.

Table 10.1. Analysis of Variance Table for Police Height Data

ANOVA					
Source	d.f.	S.S.	M.S.	F	R^2
Mean	1	893,679.28	893,679.28	250,201.43	.893
Regression	9	683.82	75.98	21.27	
Error	23	82.15	3.57		
Total	33	894,445.26			

Least squares coefficient estimates for the full model are displayed in the second column of Table 10.2. Unexpectedly, several of the estimates are negative, suggesting that *increases* in these variables are associated with decreases in height. If one attempts to make such interpretations inconsistencies abound; for example, increases in height are associated with *decreases* in upper leg length and increases in lower leg length. We know from Section 5.2 that this is not the proper interpretation for the estimated coefficients and suspect from these sign inconsistencies that multicollinearities exist in the data.

Columns 3-5 of Table 10.2 confirm that two strong multicollinearities occur among the predictor variables. Each is a three variable multicollinearity: $\underline{V}_1$ identifies a multicollinearity among ULEG, LLEG, and TIBIO, while $\underline{V}_2$ identifies one among UARM, FORE, and BRACH. Observe that the coefficient estimates with negative signs also have negative signs and large magnitudes in either $\underline{V}_1$ or $\underline{V}_2$. Thus these estimates are being forced to be negative by the multicollinearities. Note too that the coefficient estimate for LLEG, by far the largest in magnitude, also has a large value in $\underline{V}_1$.

Table 10.2. Least Squares Coefficient Estimates For (Standardized) Height Data; First Three Latent Roots and Latent Vectors of $W'W$

Variable	Standardized Coefficient Estimate	$\ell_1 = .00047$ V_1	$\ell_2 = .00087$ V_2	$\ell_3 = .23145$ V_3
Constant	164.56			
SITHT	11.91	-.010	-.000	.174
UARM	4.36	.141	.585	-.403
FORE	-3.39	-.139	-.521	-.500
HAND	4.26	-.004	.001	.542
ULEG	-9.64	-.611	.148	.038
LLEG	25.44	.604	-.154	.290
FOOT	3.37	.002	.001	-.243
BRACH	6.48	.135	.574	-.040
TIBIO	-9.52	-.452	.109	.342

The dilemma that would now face the data analyst is whether to report the fitted model or attempt to remedy the negative signs. If one chooses to report the least squares coefficient estimates there is great potential for misinterpretation or lack of understanding by other users. On the other hand, if one attempts to delete some of the multicollinear predictor variables one risks deletion of variables that may be important for accurate predictions of future heights; i.e., heights of black females who are not included in Data Set A.8. One thus jeopardizes the attainment of goals (i) and (ii) mentioned above.

Circumstances similar to these have prompted research into biased regression estimators. These estimators attempt to introduce a small bias into the regression estimator while greatly reducing the variance from that of least squares. Three biased estimators will be examined in this chapter: principal component, latent root, and ridge regression.

10.1 PRINCIPAL COMPONENT REGRESSION

Predictor variables contribute to prediction of a response variable only insofar as variation in the predictor variables helps to account for or explain variation in the response variable. In single-variable prediction equations, for example, prediction is generally deemed adequate when increases in the predictor variable are strongly associated with increases or decreases in the response variable. If the predictor variable remains relatively constant as the response variable changes, the predictor variable is generally of little value in predicting the response.

Multicollinearities are linear combinations of predictor variables that are essentially constant (zero, when expressed in terms of standardized variables, see Section 4.4) for all responses. As with single-variable least squares, when the response variable varies but the multicollinearities remain fairly constant, they are generally of little value in predicting the response. Their presence in a data base greatly inflates least squares estimator variances and introduces other problems with coefficient estimates (see Section 9.1). Principal component coefficient estimators eliminate multicollinearities from the least squares estimator, thereby greatly reducing estimator variances while attempting to introduce only a small amount of bias. In the following subsections the principal component estimator will be more precisely defined and its use illustrated.*

10.1.1   Motivation

Using the unit length standardization of the predictor variables the least squares coefficient estimators are

$$\hat{\alpha}^* = \overline{Y} \qquad \text{and} \qquad \hat{\underline{\beta}}^* = (W'W)^{-1}W'\underline{Y}$$

and their respective variances are given by

$$\text{Var}[\hat{\alpha}^*] = \frac{\sigma^2}{n} \qquad \text{and} \qquad \Sigma_{\beta^*} = (W'W)^{-1}\sigma^2.$$

The estimator of α^* is unaffected by the predictor variables (although the estimator of α is since $\hat{\alpha} = \overline{Y} - \Sigma\overline{X}_j d_j^{-1}\hat{\beta}_j^*$) so we will concentrate attention on $\hat{\underline{\beta}}^*$.

Observe that both $\hat{\underline{\beta}}^*$ and Σ_β are functions of $(W'W)^{-1}$. Expressing this latter matrix in terms of its latent roots, $\ell_1 \leqslant \ell_2 \leqslant \ldots \leqslant \ell_p$, and its latent vectors, $\underline{V}_1, \underline{V}_2, \ldots, \underline{V}_p$, yields

$$(W'W)^{-1} = \sum_{j=1}^{p} \ell_j^{-1} \underline{V}_j \underline{V}_j' \tag{10.1.1}$$

---

*For a comprehensive treatment of principal component regression, the interested reader is referred to Massy (1965), Toro-Vizarrondo and Wallace (1968), Marquardt (1970), and Bock, Yancey, and Judge (1973).

as in eqn. (4.1.11). Suppose now that $\underline{V}_1$, $\underline{V}_2$, ..., $\underline{V}_s$ identify s strong multi-collinearities among the predictor variables. Deletion of the first s terms from eqn. (10.1.1) will greatly reduce the magnitudes of the diagonal and off-diagonal elements of

$$(W'W)^+ = \sum_{j=s+1}^{p} \ell_j^{-1} \underline{V}_j \underline{V}'_j \tag{10.1.2}$$

for multicollinear predictor variables when compared to the corresponding elements of $(W'W)^{-1}$. Note, however, that very little of the original variation among the predictor variables is lost by eliminating $\underline{V}_1$, $\underline{V}_2$, ..., $\underline{V}_s$ from $W'W$ since if ℓ_1, ℓ_2, ..., ℓ_s are suitably small

$$W'W = \sum_{j=1}^{p} \ell_j \underline{V}_j \underline{V}'_j \approx \sum_{j=s+1}^{p} \ell_j \underline{V}_j \underline{V}'_j . \tag{10.1.3}$$

Consider now the following estimator

$$\hat{\underline{\beta}}^*_{PC} = (W'W)^+ W' \underline{Y} \tag{10.1.4}$$

where $(W'W)^+$ eliminates only those terms from eqn. (10.1.1) that correspond to "nonpredictive" multicollinearities (as will be seen in Section 10.1.3, multicollinearities can have predictive value and, if so, should be retained in $\hat{\underline{\beta}}^*_{PC}$). These might be the first s terms in eqn. (10.1.1) or some subset of them.

The estimator in eqn. (10.1.4) is referred to as the "principal component" estimator of $\underline{\beta}^*$ since only the principal (nonnegligible) elements of $W'W$ have been retained in eqn. (10.1.3). In other words, since

$$W'W = \sum_{j=1}^{s} \ell_j \underline{V}_j \underline{V}'_j + \sum_{j=s+1}^{p} \ell_j \underline{V}_j \underline{V}'_j$$

and because $\ell_j \approx 0$ for $j = 1, 2, ..., s$,

$$\sum_{j=1}^{s} \ell_j \underline{V}_j \underline{V}'_j \approx \Phi ,$$

the first s terms (or some subset of them) can be eliminated from $W'W$ without loss of essential information on the variation of the predictor variables. As we now demonstrate, the principal component estimator is biased but the variances of the individual estimators are much smaller than the least squares estimators.

Since $\underline{V}_1, \underline{V}_2, \ldots, \underline{V}_p$ are mutually orthogonal, it is easily verified that

$$(W'W)^+ W'W = \sum_{j=s+1}^{p} \underline{V}_j\underline{V}_j' = I - \sum_{j=1}^{s} \underline{V}_j\underline{V}_j'$$

and

$$(W'W)^+ W'W(W'W)^+ = (W'W)^+.$$

Then under Assumptions 1-3 of Table 6.2

$$E[\hat{\underline{\beta}}^*_{PC}] = (W'W)^+ W'E[\underline{Y}]$$

$$= (W'W)^+ W'W\underline{\beta}^*$$

$$= \underline{\beta}^* - \sum_{j=1}^{s} \underline{V}_j\underline{V}_j' \ \underline{\beta}^* \qquad (10.1.5)$$

and the covariance matrix for the principal component estimator [using eqn. (6.B.1) of Appendix 6.B] is

$$\Sigma_{\beta^*} = (W'W)^+ W'\mathrm{Var}[\underline{Y}] W(W'W)^+$$

$$= (W'W)^+ \sigma^2. \qquad (10.1.6)$$

Hence for an individual estimator, $\hat{\beta}^*_{PC}(k)$, from eqns. (10.1.5) and (10.1.6)

$$E[\hat{\beta}^*_{PC}(k)] = \beta^*_k - \sum_{j=1}^{s} V_{kj}(\underline{V}_j'\underline{\beta}^*)$$

and

$$\mathrm{Var}[\hat{\beta}^*_{PC}(k)] = \sigma^2 \sum_{j=s+1}^{p} \ell_j^{-1} V_{kj}^2$$

which can be compared with the corresponding expectation and variance for least squares estimators

$$E[\hat{\beta}^*_k] = \beta^*_k \quad \text{and} \quad \mathrm{Var}[\hat{\beta}^*_k] = \sigma^2 \sum_{j=1}^{p} \ell_j^{-1} V_{kj}^2.$$

By comparing these last two expressions for the mean and variance of principal component and least squares estimators, it should be clear that the principal component estimator greatly reduces the variances of the coefficient estimators since the largest ℓ_j^{-1} terms are removed. The tradeoff is the bias that is introduced,

$$E[\hat{\beta}_k^*] - \beta_k^* = -\sum_{j=1}^{s} V_{kj}(\underline{V}_j'\beta^*) \, .$$

A comparison of least squares and principal component estimates of the regression coefficients of the police height data is given in Table 10.3. The principal component estimator uses eqn. (10.1.4) and eliminates the two multicollinear latent vectors ($s = 2$) of $W'W$. Also displayed in Table 10.3 are the variance inflation factors (VIF) for the two estimators: the diagonal elements of $Q = (W'W)^{-1}$ and $Q^+ = (W'W)^+$ for least squares and principal components, respectively.

Table 10.3. Comparison of Least Squares and Principal Component (Deleting $\underline{V}_1$ and $\underline{V}_2$) Estimates For Height Data

	Standardized Coefficient Estimates		Variance Inflation Factors	
Variable	Least Squares	Principal Component	q_{jj}	q_{jj}^+
SITHT	11.91	12.19	1.52	1.29
UARM	4.36	-.48	436.42	1.28
FORE	-3.39	1.30	354.00	1.61
HAND	4.26	4.37	2.43	2.39
ULEG	-9.64	6.84	817.57	.96
LLEG	25.44	9.16	802.17	1.09
FOOT	3.37	3.31	1.77	1.76
BRACH	6.48	1.83	417.37	.63
TIBIO	-9.52	2.69	448.58	1.18
$\hat{\sigma}^2$	3.57	3.30		
R^2	.893	.892		

Note first in Table 10.3 that least squares and principal component estimates differ very little for all predictor variables that have small VIF for least squares. These predictor variables are not multicollinear and are affected only slightly by the deletion of $\underline{V}_1$ and $\underline{V}_2$. The multicollinear predictor variables, on the other hand, have estimates that are greatly altered by the elimination of $\underline{V}_1$ and $\underline{V}_2$. All three estimates that are negative with least squares are positive with principal components. Likewise, the magnitudes of all the multicollinear coefficient estimates are reduced, including the largest one, LLEG. Finally, the VIF of the multicollinear coefficient estimators are reduced by over two orders of magnitude with the principal component estimator.

Overall, the principal component estimates for the police height data appear much more reasonable than do those for least squares. The accuracy of predicting the response variable as measured* by R^2 is virtually unchanged and the estimates of the error variance are quite close (calculation of these quantities is detailed in the next subsection). Even the negative sign on the principal component coefficient estimate for UARM is not disturbing since it does not differ significantly from zero (see Section 10.1.2) and can be deleted from the prediction equation. While the extent of the bias which may have been introduced with principal components is unknown (since $\underline{\beta}^*$ is unknown), the reasonableness of the magnitudes and signs of the estimates, the close agreement of $\hat{\sigma}^2$ and R^2 with the corresponding values for least squares, and the great reductions in the variances of the estimators [from eqn. (10.1.6), $\mathrm{Var}[\hat{\beta}^*_{PC}(j)] = q^+_{jj}\sigma^2$] over least squares lead one to prefer the principal component estimates for the police height data.

10.1.2   Analysis of Variance

Variation in the response variable, Y, can be partitioned algebraically for principal component regression just as it was for least squares in Section 5.3.1. Recall that for least squares estimation

$$\mathrm{TSS} = \sum_{i=1}^{n} Y_i^2 = \mathrm{SSM} + \mathrm{SSR} + \mathrm{SSE},$$

where $\mathrm{SSM} = n\overline{Y}^2$, $\mathrm{SSR} = \underline{Y}' W(W'W)^{-1} W'\underline{Y}$, and $\mathrm{SSE} = \mathrm{TSS} - \mathrm{SSM} - \mathrm{SSR}$. Principal component regression alters the least squares estimator, $\hat{\beta}^*$, by adjusting for multicollinearities among the predictor variables. The effect of this adjustment on the partitioning of TSS is that SSR can be partitioned into two terms: one for the coefficient estimator, $\hat{\beta}^*_{PC}$, and one for the components that are deleted from $\hat{\beta}^*$ to form $\hat{\beta}^*_{PC}$.

Consider a transformation of W to a matrix $U = [U_{CD} : U_{PC}]$ as follows:

---

*A general definition for R^2 that is applicable to biased estimators as well as least squares is $R^2 = 1 - [\mathrm{SSE}/\mathrm{TSS(adj)}]$. This definition does not require that R^2 be bounded by 0 and 1 nor can R^2 for biased estimators necessarily be written as the square of Pearson's r between Y_i and Y_j, eqn. (3.3.1).

$$U = [\underline{U}_1, \underline{U}_2, ..., \underline{U}_p] = W\underline{V} = [W\underline{V}_1, W\underline{V}_2, ..., W\underline{V}_p] .$$

The columns of U are often referred to as the principal components of W, with $U_{PC} = [\underline{U}_{s+1}, \underline{U}_{s+2}, ..., \underline{U}_p]$ containing the components of W associated with the latent vectors retained in $\hat{\beta}_{PC}^*$ (the "principal components") and $U_{CD} = [\underline{U}_1, \underline{U}_2, ..., \underline{U}_s]$ containing those components that are deleted from $\hat{\beta}_{PC}^*$ (the "components deleted"). We can then write the regression model as

$$\underline{Y} = \alpha^*\underline{1} + W\underline{\beta}^* + \underline{\varepsilon}$$

$$= \alpha^*\underline{1} + WVV'\underline{\beta}^* + \underline{\varepsilon},$$

$$= \alpha^*\underline{1} + U_{CD}\underline{\gamma}_{CD} + U_{PC}\underline{\gamma}_{PC} + \underline{\varepsilon} . \qquad (10.1.7)$$

where $\gamma' = [\gamma_1, \gamma_2, ..., \gamma_p] = [\underline{\gamma}'_{CD}, \underline{\gamma}'_{PC}]$ and $\underline{\gamma}'_j = \underline{V}'_j\underline{\beta}^*$. Since $U'_{CD}U_{PC} = \Phi$, the least squares estimators of $\underline{\gamma}_{CD}$ and $\underline{\gamma}_{PC}$ are

$$\hat{\underline{\gamma}}_{CD} = (U'_{CD}U_{CD})^{-1}U'_{CD}\underline{Y}$$

and

$$\hat{\underline{\gamma}}_{PC} = (U'_{PC}U_{PC})^{-1}U'_{PC}\underline{Y} . \qquad (10.1.8)$$

The principal component estimator of $\underline{\beta}^*$, eqn. (10.1.4), can be obtained from $\hat{\underline{\gamma}}_{PC}$ by noting that

$$V'_{PC}W'WV_{PC} = L_{PC} \quad \text{and} \quad (V'_{PC}W'WV_{PC})^{-1} = L_{PC}^{-1} \qquad (10.1.9)$$

where $L_{PC} = \text{diag}(\ell_{s+1}, \ell_{s+2}, ..., \ell_p)$. Thus,

$$\hat{\underline{\beta}}_{PC}^* = V_{PC}\hat{\underline{\gamma}}_{PC}$$

$$= V_{PC}(V'_{PC}W'WV_{PC})^{-1}V'_{PC}W'\underline{Y}$$

$$= \sum_{j=s+1}^{p} \ell_j^{-1}\underline{V}_j\underline{V}'_j W'\underline{Y}$$

$$= (W'W)^+ W'\underline{Y} . \qquad (10.1.10)$$

Similarly, $\hat{\underline{\beta}}_{CD}^* = V_{CD}\hat{\underline{\gamma}}_{CD}$ is the principal component estimator of $\underline{\beta}^*$ using only the components associated with multicollinearities in W.

Partitioning SSR can now be accomplished in a straightforward fashion:

$$\text{SSR} = \underline{Y}' W (W' W)^{-1} W' Y$$

$$= \underline{Y}' W \left(\sum_{j=1}^{s} \ell_j^{-1} \underline{V}_j \underline{V}_j' \right) W' \underline{Y} + \underline{Y}' W \left(\sum_{j=s+1}^{p} \ell_j^{-1} \underline{V}_j \underline{V}_j' \right) W' \underline{Y}$$

$$= \underline{Y}' W \left(\sum_{j=1}^{s} \ell_j^{-1} \underline{V}_j \underline{V}_j' \right) W' \underline{Y} + \underline{Y}' W (W' W)^+ W' \underline{Y}$$

$$= \text{SSR}_{\text{CD}} + \text{SSR}_{\text{PC}}. \tag{10.1.11}$$

Using the definitions of U_{CD} and U_{PC}, it is readily verified that

$$\text{SSR}_{\text{PC}} = \hat{\underline{\gamma}}_{\text{PC}}' W' W \hat{\underline{\gamma}}_{\text{PC}} = \hat{\underline{\beta}}_{\text{PC}}^{*\prime} V_{\text{PC}} W' W V_{\text{PC}} \hat{\underline{\beta}}_{\text{PC}}^*$$

and

$$\text{SSR}_{\text{CD}} = \hat{\underline{\gamma}}_{\text{CD}}' W' W \hat{\underline{\gamma}}_{\text{CD}} = \hat{\underline{\beta}}_{\text{CD}}^{*\prime} V_{\text{CD}} W' W V_{\text{CD}} \hat{\underline{\beta}}_{\text{CD}}^*,$$

where $V_{\text{CD}} = [\underline{V}_1, \underline{V}_2, ..., \underline{V}_s]$ and $V_{\text{PC}} = [\underline{V}_{s+1}, \underline{V}_{s+2}, ..., \underline{V}_p]$. Thus, if the principal components associated with multicollinearities in W have little predictive value (i.e., $\hat{\underline{\beta}}_{\text{CD}}^* \approx \underline{0}$ or equivalently, $\underline{\gamma}_{\text{CD}} \approx \underline{0}$), SSR_{CD} should be small relative to SSR_{PC}. In this partitioning of SSR, SSR_{PC} is the sum of squares attributable to the regression of the response variable on the $(p-s)$ components associated with the largest latent roots of $W' W$ and SSR_{CD} is the sum of squares attributable to the regression of the response variable on the multicollinear components of W.

An analysis of variance table can be written for principal component regression similar to Table 5.4 for least squares. Table 10.4 exhibits two ANOVA tables for principal component regression. In the upper ANOVA table each of the sums of squares derived above is shown separately. Under Assumptions 1-4 of Table 6.2, each of the F statistics listed follows an F distribution. The F statistics for the regression sums of squares test, respectively, that $\underline{\gamma}_{\text{PC}} = \underline{0}$ and that $\underline{\gamma}_{CD} = \underline{0}$.

If the test that $\underline{\gamma}_{CD} = \underline{0}$ is not rejected, one can feel relatively confident that the components deleted from the least squares estimator, $\hat{\beta}^*$, to obtain the principal component estimator, $\hat{\underline{\beta}}_{\text{PC}}^*$, were of little predictive value. This in turn implies that the bias term in eqn. (10.1.5) is small relative to the elements of β^*. Coupled with the reduced variances of the principal component coefficient estimators, small bias in $\hat{\underline{\beta}}_{\text{PC}}^*$ suggests that the principal component estimator (for a given data set) is more accurate an estimator of $\underline{\beta}^*$ than is the least squares estimator.

**Table 10.4. Symbolic Analysis of Variance Tables for
Principal Component Regression**

(a) Without Pooling Sums of Squares for Error and Components Deleted
ANOVA

Source	d.f.	S.S.	M.S.	F	R^2
Mean	1	SSM	MSM	MSM/MSE	$SSR_{PC}/TSS(adj)$
Regression					
Principal Components	$p-s$	SSR_{PC}	MSR_{PC}	MSR_{PC}/MSE	
Components Deleted	s	SSR_{CD}	MSR_{CD}	MSR_{CD}/MSE	
Error	$n-p-1$	SSE	MSE		
Total	n	TSS			

(b) Pooling Sums of Squares for Error and Components Deleted
ANOVA

Source	d.f.	S.S.	M.S.	F	R^2
Mean	1	SSM	MSM	MSM/MSE_{PC}	$SSR_{PC}/TSS(adj)$
Principal Components	$p-s$	SSR_{PC}	MSR_{PC}	MSR_{PC}/MSE_{PC}	
Error	$n-p+s-1$	SSE_{PC}	MSE_{PC}		
Total	n	TSS			

Note: $SSE_{PC} = SSE + SSR_{CD}$

Frequently nonrejection of $H_0:\gamma_{CD} = \underline{0}$ leads researchers to pool SSE and SSR_{CD} into a common error term $SSE_{PC} = SSE + SSR_{CD}$. This pooling is advocated since nonrejection of the above hypothesis implies that SSR_{CD} is essentially a sum of squares attributable to random fluctuation, just as is SSE. If few degrees of freedom are available with SSE, a better estimator of σ^2 can be obtained by such a pooling. The lower ANOVA table in Table 10.4 indicates the pooling of the two sums of squares, their degrees of freedom, and the resulting F statistics. It should be stressed that the F statistics in Table 10.4(b) do not possess exact F distributions unless γ_{CD} is identically $\underline{0}$ (thus SSR_{CD} only measures random fluctuation); nevertheless, the F statistics are generally treated as though they are true F random variables.

Analysis of variance tables for the police height data are presented in Table 10.5 wherein the first two components are deleted from the principal component estimator. The sum of squares attributable to the two deleted components in Table 10.5(a) is quite small, indicating that those two components add little to the prediction of height. The corresponding F statistic is also quite small, supporting the belief that the principal component estimates in Table 10.3, due to their smaller variances and no indication of substantial bias, are more accurate than those of least squares. Whether one

Table 10.5. Principal Component Regression Anova Tables
For Police Height Data

	(a) Without Pooling ANOVA				
Source	d.f.	S.S.	M.S.	F	R^2
Mean	1	893,679.28	893,679.28	250,201.43	.892
Regression					
Principal Components	7	683.47	97.64	27.35	
Components Deleted	2	.36	.18	.05	
Error	23	82.15	3.57		
Total	33	894,445.26			

	(b) With Pooling ANOVA				
Source	d.f.	S.S.	M.S.	F	R^2
Mean	1	893,679.28	893,679.28	270,784.94	.892
Principal Components	7	683.47	97.64	29.58	
Error	25	82.51	3.30		
Total	33	894,445.26			

chooses to estimate σ^2 with the least squares MSE as in Table 10.5(a) or with the pooled estimate as in Table 10.5(b), the overall analysis of the prediction equation changes very little as the similarity of the F statistics in Tables 10.1 and 10.5 reveal.

10.1.3 Inference Techniques

In addition to the techniques presented in the previous subsection for assessing the adequacy of the overall principal component prediction equation, it is possible to test for the significance of individual components and predictor variables. The analysis of individual components follows the same development as the F test for several components that was shown in Table 10.4.

Expressing the regression model as in eqn. (10.1.7) actually enables the response variable to be written as a function of all p components:

$$\underline{Y} = \alpha^*\underline{1} + \sum_{j=1}^{p} \underline{U}_j\gamma_j + \underline{\varepsilon} \; . \tag{10.1.12}$$

A test for the significance of the jth component, $\underline{U}_j$, is then a test of the hypothesis $H_0\colon \gamma_j = 0$ vs. $H_a\colon\gamma_j \neq 0$. Since the $\underline{U}_j$ are mutually orthogonal, moreover, the least squares estimators of the γ_j are

$$\hat{\gamma}_j = (\underline{U}_j'\underline{U}_j)^{-1}\underline{U}_j'\underline{Y}$$

$$= \ell_j^{-1}\underline{U}_j'\underline{Y} \qquad j = 1, 2, ..., p \qquad (10.1.13)$$

since $\underline{U}_j'\underline{U}_j = \underline{V}_j'W'W\underline{V}_j = \ell_j$. The model (10.1.7) or (10.1.12) is simply a multiple regression model in the *components* of W instead of the original variables. As such, all the inferential techniques of Chapters 6 and 8 can be utilized to analyze the regression of Y on the components. Our interest in these components is concentrated on the first few since they are associated with multicollinearities in W; hence, we will formulate tests for the first, say, s components only.

Individual components can be assessed through the use of t statistics. Under Assumptions 1-4 of Table 6.2, from eqn. (10.1.13) one can show that

$$\hat{\gamma}_j \sim \text{NID} (\gamma_j, \sigma^2/\ell_j) \qquad j = 1, 2, ..., p.$$

Also since

$$\underline{V}_j'(W'W)^{-1}W = \ell_j^{-1}\underline{V}_j'W = \ell_j^{-1}\underline{U}_j',$$

$$\hat{\gamma}_j = \underline{V}_j'(W'W)^{-1}W'\underline{Y} = \underline{V}_j'\hat{\underline{\beta}}^*,$$

and SSE$/\sigma^2 \sim \chi^2(n-p-1)$ independently of $\hat{\underline{\beta}}^*$, SSE$/\sigma^2$ is distributed independently of the $\hat{\gamma}_j$. Thus, the following t statistic can be used to test the significance of individual components:

$$t = \frac{\hat{\gamma}_j}{(\hat{\sigma}^2/\ell_j)^{1/2}} \qquad (10.1.14)$$

where $\hat{\sigma}^2 = $ MSE. Once one determines which components are to be retained in $\hat{\underline{\beta}}_{PC}$, note from the second expression in eqn. (10.1.11),

$$\text{SSR}_{PC} = \sum_{PC}\ell_j\hat{\gamma}_j^2 \qquad \text{and} \qquad \text{SSR}_{CD} = \sum_{CD}\ell_j\hat{\gamma}_j^2,$$

where the two summations are over the components retained and deleted, respectively, from the principal component estimator of $\underline{\beta}^*$.

Two strategies are now available for deleting components. The first strategy deletes all components associated with small latent roots of $W'W$ (i.e., all components associated with multicollinearities in W). The second strategy deletes only those multicollinear components of W for which the t statistic (10.1.14) is not significant. Our experience has been that the first strategy often works better in practice than the second, although the individual t

tests can be more effective if a very small significance level is used (say α = .001). The rationale behind this suggestion is that the decrease in variance associated with deletion of multicollinear components generally is much greater than the bias incurred by doing so. If the bias does exceed the variance and a component, despite being associated with a multicollinearity, should be retained, the t statistic eqn. (10.1.14) will be significant even at small significance levels.

Estimated values for the coefficients for the first two principal components of the police height data are

$$\hat{\gamma}_1 = 27.39 \quad \text{and} \quad \hat{\gamma}_2 = -1.66$$

$$t_1 = 0.31 \quad \text{and} \quad t_2 = -0.03.$$

Neither of these two t statistics would be judged significant at any reasonable significance level so both components should be deleted and the principal component estimates of the regression coefficients are again those displayed in Table 10.3.

Once components have been deleted one is often still concerned about assessing individual predictor variables. Mansfield et al. (1977) derive a technique for evaluating the elimination of any number of predictor variables from a principal component estimator. Here we will only discuss the construction of a t statistic for examining individual predictor variables.

Suppose that the principal component estimator deletes some of the principal components when forming $\hat{\beta}_{PC}$. If one now desires to test the hypothesis $H_0: \beta_j^* = 0$ vs. $H_a: \beta_j^* \neq 0$ with the principal component estimator, Mansfield et al. (1977) showed that the appropriate statistic to use is

$$t = \frac{\hat{\beta}_{PC}^*(j)}{[MSE \cdot (\Sigma \, \ell_r^{-1} \, V_{jr}^2)]^{1/2}}, \qquad (10.1.15)$$

where $\hat{\beta}_{PC}^*(j)$ is the principal component estimate of β_j^* in $\hat{\beta}_{PC}^*$ and the summation in eqn. (10.1.15) is again taken over only those components retained by the principal component estimator. The statistic in eqn. (10.1.15) follows Student's t distribution with $(n-p-1)$ degrees of freedom under H_0 provided that $\gamma_1 = \gamma_2 = \ldots = \gamma_s = 0$; i.e., if the true coefficients of the components that are deleted from $\hat{\beta}_{PC}^*$ are zero. If the deleted component coefficients are not identically zero, then the t statistics in eqn. (10.1.15) can be considered approximate Student t random variables (the smaller the bias in $\hat{\beta}_{PC}^*(j)$, the better the approximation).

Following deletion of the first two principal components for the police height data, individual t statistics were calculated for the predictor variables and are displayed in Table 10.6. The smallest t statistic corresponds to

Table 10.6. Individual t Statistics For Principal Component Regression of Police Height Data ($\underline{V}_1$ and $\underline{V}_2$ Deleted)

Variable	SITHT	UARM	FORE	HAND	ULEG
$\hat{\beta}^*_{PC}(j)$	12.19	-.48	1.30	4.37	6.84
t	5.68	-.22	.54	1.50	3.68

Variable	LLEG	FOOT	BRACH	TIBIO
$\hat{\beta}^*_{PC}(j)$	9.16	3.31	1.83	2.69
t	4.64	1.32	1.22	1.31

UARM, the only principal component coefficient estimate having a negative sign. Its small magnitude reveals that this variable can be eliminated from the prediction equation without seriously affecting the prediction of height. If one deletes UARM from the prediction equation, one of the multicollinearities in W no longer exists. At this point we would recommend reassessing the prediction equation with the 8 remaining predictor variables. If at some later stage of the analysis the other multicollinearity is broken up we would again recommend reanalysis of the prediction equation using least squares with only the variables remaining at that time; otherwise the principal component procedures presented in this section can be utilized with the multicollinearity still present.

10.2   LATENT ROOT REGRESSION ANALYSIS

Latent root regression analysis* is similar to principal component regression. As the name implies, latent root regression utilizes latent roots and latent vectors to construct regression coefficient estimators. Like principal component regression, this methodology adjusts the least squares estimator for multicollinearities among the predictor variables. Two major differences exist between principal component regression and latent root regression: (i) the matrices from which latent roots and vectors are extracted differ, and (ii) the techniques for deciding whether to delete multicollinear latent vectors from the respective estimators differ.

---

*Latent root regression is discussed more completely in Hawkins (1973), Webster, Gunst, and Mason (1974), Gunst, Webster, and Mason (1976), and Gunst and Mason (1977a, b).

10.2.1 Motivation

Latent root regression extends principal component regression by utilizing
an assessment of whether multicollinearities have predictive value and
should be retained in the resulting estimator. An assessment of the predictivity of multicollinearities is only partially successful with principal components using t statistics and MSR_{CD}/MSE. For example, the t statistic in eqn.
(10.1.14) for testing $H_0:\gamma_j = 0$ can be written as

$$t = \ell_j^{1/2} \hat{\gamma}_j / \hat{\sigma} .$$

The numerator of this statistic is $\ell_j^{1/2} \hat{\gamma}_j$ which is normally distributed with
mean $\ell_j^{1/2} \gamma_j$ and variance σ^2. If ℓ_j is small this t statistic will tend to have a
small value even if $\gamma_j \neq 0$ since the multiplier $\ell_j^{1/2}$ will force both $\ell_j^{1/2} \gamma_j$ and
$\ell_j^{1/2} \hat{\gamma}_j$ to be closer to zero than are γ_j and $\hat{\gamma}_j$. Thus inferences made with
eqn. (10.1.14) will tend to eliminate components which do have predictive
value.

As argued in Section 10.1.3, use of a small significance level with tests on
the components serves to insure that a multicollinear component will only
be retained when there is clear evidence that substantial bias would otherwise result. With this type of a strategy full advantage of the variance-
reducing properties of principal component regression can be realized. As
we have just seen, however, if γ_j is nonzero but not extremely large in magnitude and ℓ_j is very close to zero, these tests tend to remove the corresponding latent vector from $\hat{\beta}_{PC}$ and could thereby produce large biases. Thus
when $H_0: \gamma_j = 0$ is rejected, one can feel relatively certain that $\underline{V}_j$ should be
retained in $\hat{\beta}_{PC}$, but when the hypothesis is not rejected the component
could still have predictive value.

An alternative means of assessing the predictivity of multicollinearities is
to examine the latent roots and latent vectors of a matrix consisting of the
correlations among both predictor and response variables. Specifically, let

$$A = [\underline{Y}^*, \ W], \tag{10.2.1}$$

where $\underline{Y}^*$ is a vector containing the unit length standardization of the response variables; i.e., $Y_i^* = (Y_i - \overline{Y})/[\Sigma(Y_i - \overline{Y})^2]^{1/2}$. The matrix $A'A$ is then
a matrix of the correlations among the $(p + 1)$ response and predictor varia-
bles, just as $W'W$ contains the correlations among the p predictor variables.
Denote the latent roots of $A'A$ by

$$\lambda_0 \leqslant \lambda_1 \leqslant \lambda_2 ... \leqslant \lambda_p$$

and the corresponding latent vectors by

$$\underline{\gamma}_0, \underline{\gamma}_1, \underline{\gamma}_2, \cdots, \underline{\gamma}_p.$$

As with the latent roots and latent vectors of $W'W$, multicollinearities among the variables in A are indicated by small latent roots of $A'A$. Likewise, the variables involved in multicollinearities are determined by the relative magnitudes of the elements of $\underline{\gamma}_j$. Suppose now that $\lambda_j \approx 0$, indicating a multicollinearity. Then

$$\gamma_{0j}\underline{Y}^* + \gamma_{1j}\underline{W}_1 + \cdots + \gamma_{pj}\underline{W}_p \approx \underline{0}. \tag{10.2.2}$$

Note that the key difference between detecting multicollinearities in $W'W$ and $A'A$ is that the response variable, Y, can affect the multicollinearities in $A'A$. If the weight on $\underline{Y}^*$, γ_{0j}, is not trivially close to zero, the response variable is involved in the multicollinearity. We then conclude that the multicollinearity has predictive value since, from eqn. (10.2.2),

$$\underline{Y}^* \approx -\gamma_{0j}^{-1}(\gamma_{1j}\underline{W}_1 + \gamma_{2j}\underline{W}_2 + \cdots + \gamma_{pj}\underline{W}_p). \tag{10.2.3}$$

When $\ell_j \approx 0$ and $\gamma_{0j} \approx 0$, the multicollinearity is referred to as a *nonpredictive multicollinearity*; i.e., a multicollinearity among the predictor variables that is of little value in predicting the response variable.

Latent root regression utilizes the latent roots and latent vectors of $A'A$ to construct a biased estimator of β^* by eliminating all nonpredictive multicollinearities from $\hat{\underline{\beta}}^*$. To accomplish this the latent root estimator is written as

$$\hat{\underline{\beta}}^*_{LR} = \sum_{j=0}^{p} f_j \underline{\delta}_j,$$

where the f_j are variables whose values will be determined shortly and $\underline{\delta}_j$ is the p-dimensional vector consisting of the last p elements of $\underline{\gamma}_j$; i.e.,

$$\underline{\gamma}_j = \begin{pmatrix} \gamma_{0j} \\ \underline{\delta}_j \end{pmatrix}, \qquad j = 0, 1, 2, \ldots, p. \tag{10.2.4}$$

The f_j are determined by first setting $f_j = 0$ if $\underline{\gamma}_j$ identifies a nonpredictive multicollinearity and then selecting the remaining values to minimize the residual sum of squares, $\Sigma(Y_i - \hat{Y}_i)^2$. The resulting latent root regression estimator* is

---

*A derivation of this result can be found in Webster, Gunst, and Mason (1974).

$$\hat{\beta}_{LR}^{*} = \sum_{j=0}^{p} f_j \underline{\delta}_j ,$$

$$f_j = \begin{cases} 0 & \lambda_j \approx 0 \text{ and } \gamma_{0j} \approx 0 \\ -\eta\gamma_{0j}\lambda_j^{-1} \Big/ \sum_{q} \gamma_{0q}^2 \lambda_q^{-1} & \text{otherwise} \end{cases} \qquad (10.2.5)$$

where $\eta^2 = \Sigma(Y_i - \bar{Y})^2$ and where the summation sign with q under it extends only over those latent roots and latent vectors which do not identify nonpredictive multicollinearities. If none of the f_j equal zero, the latent root estimator and the least squares estimator are identical since both minimize SSE without imposing any restrictions on the estimators.

When determining how small λ_j and γ_{0j} must be in order to label a multicollinearity as nonpredictive, no precise rules have been formulated since the exact distributional properties of $\hat{\beta}_{LR}^{*}$ are unknown. Several properties of $W'W$ and $A'A$ can aid in such a determination. First, if all the columns of A are orthogonal, all the latent roots of $A'A$ will be equal to 1.0. Consequently, any latent root that is much smaller than 1.0, say less than 0.1, should be investigated further. If the elements of the corresponding latent vector are very close in magnitude to similar ones in a latent vector of $W'W$ and γ_{0j} is sufficiently close to zero, perhaps less than 0.1 or 0.05 in magnitude, a nonpredictive multicollinearity has been identified.

Table 10.7 lists the smallest three latent roots of $A'A$ for the police height data along with their corresponding latent vectors. The first two latent roots of $A'A$ are virtually identical with those of $W'W$ and the elements of $\underline{\delta}_0$ and $\underline{\delta}_1$ are almost the same as those of $\underline{V}_1$ and $\underline{V}_2$. The magnitudes of γ_{01} and γ_{02} are so close to zero that is clear these are nonpredictive multicollinearities. The last column of Table 10.7 shows the latent root estimates of the regression coefficients. For this example the latent root estimator yields coefficient estimates that are quite similar to the principal component estimator. An example for which the two estimators do not provide similar results is provided next.

10.2.2 Predictive Multicollinearities

Principal component regression and latent root regression need not produce similar parameter estimates. Each methodology utilizes different techniques for deleting terms from the respective estimators and the latent vectors of $W'W$ and $A'A$ are not identical. To illustrate how the two estimators can arrive at different results, consider the regression of the Detroit homicide

Table 10.7. Latent Root Estimation for the Police Height Data (γ_0 and γ_1 Deleted)

Variable	$\lambda_0 = .00047$ γ_0	$\lambda_1 = .00087$ γ_1	$\lambda_2 = .07328$ γ_2	Standardized Estimates
HEIGHT	-.0044	-.0005	.8069	
SITHT	-.0085	-.0002	-.3902	12.20
UARM	.1414	.5849	.0614	-.51
FORE	-.1395	-.5205	.0141	1.33
HAND	-.0034	.0011	-.1596	4.37
ULEG	-.6096	.1488	-.2255	6.94
LLEG	.6054	-.1538	-.3076	9.06
FOOT	.0027	.0009	-.0855	3.31
BRACH	.1355	.5739	-.0522	1.80
TIBIO	-.4517	.1090	-.1084	2.76

rate (HOM) on the full-time police rate (FTP), handgun license rate (LIC), and handgun registration rate (GR). Table 10.8 summarizes a least squares fit to the data.

Table 10.8. Least Squares Fit to the Regression of HOM on FTP, LIC, and GR, Detroit Homicide Data

		ANOVA			
Source	d.f.	S.S.	M.S.	F	R^2
Mean	1	8,207.71	8,207.71	981.56	.977
Regression	3	3,146.53	1,048.84	125.43	
Error	9	75.26	8.63		
Total	13	11,429.50			

Variable	$\ell_1 = .0776$ V_1	$\ell_2 = .4636$ V_1	$\hat{\beta}_j^*$	t_j
FTP	.1925	:8288	45.46	10.92
LIC	.6237	-.5167	12.90	1.86
GR	-.7576	-.2147	2.75	.34

Although the least squares prediction equation provides an excellent fit to the data, V_1 reveals that a multicollinearity exists between LIC and GR (from $W'W$ the pairwise correlation between these two predictor variables is 0.904). The t statistics for both predictor variables are small, leading to the deletion of GR at the 0.25 significance level. It is interesting to note that the coefficient estimates for LIC and GR do not have opposite signs as do the corresponding elements in V_1. This observation suggests that the multicollinearity might not be strong enough to critically distort inferences on the true regression parameters.

The two strategies for deleting components from the principal component estimator both result in the deletion of $\underline{V}_1$. For the second strategy mentioned in Section 10.1.3,

$$\hat{\gamma}_1 = 14.71 \quad \text{and} \quad t = 1.42.$$

The t statistic for testing $H_0:\gamma_1 = 0$ is only significant at the 0.20 significance level, suggesting that the component should be deleted. Table 10.9 summarizes the principal component analysis of the data.

Table 10.9. Principal Component Fit to the Regression of HOM on FTP, LIC, and GR, Detroit Homicide Data

	ANOVA				
Source	d.f.	S.S.	M.S.	F	R^2
Mean	1	8,207.71	8,207.71	981.56	.977
Regression					
Principal Components	2	3,129.75	1,564.88	187.19	
Component Deleted	1	16.78	16.78	2.01	
Error	9	75.26	8.36		
Total	13	11,429.50			

Variable:	FTP	LIC	GR
$\hat{\beta}^*_{PC}$:	42.63	3.72	13.89
t:	11.68	1.52	9.53

Examination of the summary information on the principal component fit to HOM shows that the overall fit is just as good as the least squares fit but the estimates of the coefficients for LIC and GR are greatly altered. All three coefficients would be judged significantly different from zero in the principal component analysis (using a 0.25 significance level).

A latent root regression analysis of this data sheds additional light on the previous analyses. Table 10.10 contains the latent roots and latent vectors of $A'A$ for this data. Substantive differences exist between the latent roots and latent vectors of $A'A$ and those of $W'W$. The smallest latent root of $A'A$ is not approximately equal to $\ell_1 = .0776$; rather, $\lambda_0 = .0136$. The elements of $\underline{d}_0$, moreover, are not similar to those in $\underline{V}_1$. Scanning the second smallest latent root of $A'A$, λ_1, and its associated latent vector, one observes that $\lambda_1 \approx \ell_1$ and $\underline{d}_1 \approx \underline{V}_1$. It is the second latent vector of $A'A$ that identifies the multicollinearity between LIC and GR.

Neither of the first two latent vectors of $A'A$ have small first elements. In particular, the multicollinearity between LIC and GR is seen to be predictive since γ_{01} is so large. Unlike the values of γ_{00} and γ_{01} that are displayed in

Table 10.10. Latent Roots and Latent Vectors of $A'A$ For Prediction of HOM from FTP, LIC, and GR, Detroit Homicide Data

Variable	$\lambda_0 = .0136$ γ_0	$\lambda_1 = .0812$ γ_1	$\lambda_2 = .5595$ γ_2	$\lambda_3 = 3.3456$ γ_3
HOM	-.7563	.1781	.3474	.5250
FTP	.6251	.0345	.6109	.4846
LIC	.1929	.6025	-.6100	.4772
GR	.0040	-.7773	-.3661	.5117

Table 10.7, the value of γ_{01} in Table 10.10 is larger than the value suggested in the last subsection as the cutoff for nonpredictive multicollinearities; viz, 0.10. Since neither of the multicollinearities in $A'A$ are judged nonpredictive, the latent root estimator is identical with least squares. Apparently the smallest latent root of $W'W$, $\ell_1 = 0.0776$, is sufficiently small to mask the nonzero value of the coefficient of the first principal component when the t-statistic, eqn. (10.1.14), is used to test $H_0: \gamma_1 = 0$. Considering the dramatic differences in the coefficient estimates between principal components and least squares, the latent root analysis has provided valuable information concerning which set of estimates should be used.

Although the latent root analysis substantiates the least squares analysis with this data set and agrees with the principal component analysis of the police height data, it need not agree with either analysis. If several multicollinearities occur in a data set, the latent root analysis could indicate that some components should be deleted but the components selected and their exact numerical values could differ from principal components.

10.2.3 Analysis of Variance

A slightly different approach must be taken from that of least squares and principal components in order to partition the total sum of squares. First, the sum of squares attributable to the mean is extracted, leaving the total adjusted sum of squares:

$$TSS = SSM + TSS \text{ (adj)}.$$

Instead of calculating SSR and SSE and then partitioning SSR into components SSR_{LR} and SSR_{CD} as was done for principal components, it is easier to compute $SSE_{LR} = SSR_{CD} + SSE$ directly and obtain the other sums of squares by subtraction.

The pooled error sum of squares, SSE_{LR}, is calculable as

$$\text{SSE}_{LR} = \sum_{i=1}^{n} (Y_i - \hat{Y}_i)^2$$

$$= \sum_{i=1}^{n} (Y_i - \overline{Y} - \underline{u}_i'\hat{\beta}^*_{LR})^2$$

$$= \eta^2 \left(\sum_q \gamma_{0q}^2 \lambda_q^{-1} \right)^{-1}, \tag{10.2.6}$$

where $\underline{u}_i'$ is the ith row of W and η^2 is again $\Sigma(Y_i - \overline{Y})^2$. As in the expression for $\hat{\beta}^*_{LR}$, eqn. (10.2.5), the summation in eqn. (10.2.6) only extends over those latent vectors of $A'A$ which are not deleted from the estimator. Once SSE_{LR} is calculated the other quantities needed for the partitioning of TSS are readily obtained:

$$\text{SSR}_{LR} = \text{TSS(adj)} - \text{SSE}_{LR}$$
$$\tag{10.2.7}$$
$$\text{SSR}_{CD} = \text{SSE}_{LR} - \text{SSE}.$$

One caution needs to be pointed out concerning this process. If predictive multicollinearities are inadvertently removed from $\hat{\beta}^*_{LR}$, prediction of the response variable can be so poor that SSE_{LR} exceeds TSS(adj). A negative value for SSR_{LR} then results when SSE_{LR} is subtracted from TSS(adj).

Table 10.11 shows symbolic analysis of variance tables for latent root regression that are similar to those in Table 10.4 for principal components. By analogy with the principal component analysis, the degrees of freedom and F statistics are determined as though exact distributional properties of the various statistics were known. Unfortunately this is not the case; exact distributional theory has not been derived for the latent root estimator. In practice, however, one treats the F statistics in these tables as though they are approximately distributed as F random variables. Note too that the number of components deleted by latent root regression, d, can differ from the number of components deleted by principal component regression, s.

Analysis of the police height data in Table 10.12 is essentially identical with the principal component analysis in Table 10.5. The exact agreement is unusual and is due to the extremely small first elements in γ_0 and γ_1 in Table 10.7 and the closeness of λ_0 and λ_1 to ℓ_1 and ℓ_2, respectively. Generally the first elements of the latent vectors are not this close to zero even for nonpredictive multicollinearities nor are the small latent roots so similar. Naturally if predictive multicollinearities are found the latent root and principal component ANOVA tables can be markedly different.

Table 10.11. Symbolic Analysis of Variance Tables for Latent Root Regression

(a) Without Pooling
ANOVA

Source	d.f.	S.S.	M.S.	F	R^2
Mean	1	SSM	MSM	MSM/MSE	SSR_{LR}/TSS(adj)
Regression					
Components Retained	$p-d$	SSR_{LR}	MSR_{LR}	MSR_{LR}/MSE	
Components Deleted	d	SSR_{CD}			
Error	$n-p-1$	SSE	MSE		
Total	n	TSS			

(b) With Pooling
ANOVA

Source	d.f.	S.S.	M.S.	F	R^2
Mean	1	SSM	MSM	MSM/MSE_{LR}	SSR_{LR}/TSS(adj)
Components Retained	$p-d$	SSR_{LR}	MSR_{LR}	MSR_{LR}/MSE_{LR}	
Error	$n-p+d-1$	SSE_{LR}	MSE_{LR}		
Total	n	TSS			

Table 10.12. Latent Root Regression Anova Tables for Police Height Data

(a) Without Pooling
ANOVA

Source	d.f.	S.S.	M.S.	F	R^2
Mean	1	893,679.28	893,679.28	250,201.43	.892
Regression					
Components Retained	7	683.46	97.64	27.35	
Components Deleted	2	.36			
Error	23	82.15	3.57		
Total	33	894,445.26			

(b) With Pooling
ANOVA

Source	d.f.	S.S.	M.S.	F	R^2
Mean	1	893,679.28	893,679.28	270,770.20	.892
Components Retained	7	683.46	97.64	29.58	
Error	25	82.51	3.30		
Total	33	894,445.26			

One further difference exists between the latent root and principal component analyses. We do not recommend testing for the significance of the deleted components by comparing $\text{MSR}_{CD}/\text{MSE}$ to a critical value from F tables. This statistic, due to its similarity to the corresponding F statistic for principal component regression, suffers from the same defects as the F and t statistics discussed in Section 10.1.3. Rather than employing these statistics, we recommend using the procedure described above for determining non-predictive multicollinearities. To emphasize this point we have excluded the mean square and F statistic for SSR_{CD} in Tables 10.11 and 10.12.

10.2.4 Inference Techniques

Although exact distributional properties of the latent root estimator are unknown, we argued in the previous subsection that ratios of mean squares in ANOVA tables could be treated as F random variables. We did so for several reasons. The sums of squares attributable to the respective sources of variability were derived [see Webster et al. (1974)] using the same approach as least squares and principal components: sums of squares for all three procedures can be derived as differences in the residual sum of squares for fitted models with and without the individual sources of variability.* For ease of exposition we have not elected to utilize this approach to formulate the ANOVA tables in this chapter; nevertheless, all the sums of squares for least squares, principal components, and latent root regression can be found by this method.

Another reason for approximating ratios of mean squares in latent root regression as F random variables is that (i) when no components are deleted, latent root regression produces the same F ratios as least squares, and (ii) when small latent roots of $A'A$ equal those of $W'W$ and the first elements of the corresponding latent vectors of $A'A$ are identically zero, latent root regression and principal component regression produce the same F statistics. Thus when either of these two sets of conditions hold the F statistics in latent root regression are theoretically correct F random variables.

---

*For example, SSR_{PC} can be derived as the difference in the residual sum of squares, SSE, for fitting the two models

$$\underline{Y} = \alpha^* \underline{1} + \sum_{j=1}^{s} \underline{U}_j \gamma_j + \underline{\varepsilon}$$

and

$$\underline{Y} = \alpha^* \underline{1} + \sum_{j=1}^{s} \underline{U}_j \gamma_j + \sum_{j=s+1}^{p} \underline{U}_j \gamma_j + \underline{\varepsilon}.$$

Not only are approximate F statistics available for testing the significance of the regression coefficient vector, β^*, but approximate t statistics can also be derived for testing the significance of individual regression coefficients, β_j^*. For latent root regression the t statistics can be expressed as

$$t = \frac{\hat{\beta}_{LR}^*(j)}{[MSE \cdot (g_{jj} - g_{00}^{-1}g_{0j}^2)]^{1/2}} \qquad (10.2.7)$$

where $\hat{\beta}_{LR}^*(j)$ is the latent root estimate of β_j^* and

$$G = \begin{bmatrix} g_{00} & g_{01} & \cdots & g_{0p} \\ g_{10} & g_{11} & \cdots & g_{1p} \\ \vdots & \vdots & & \vdots \\ g_{p0} & g_{p1} & \cdots & g_{pp} \end{bmatrix} = \sum_q \lambda_q^{-1} \underline{\gamma}_q \underline{\gamma}_q' ; \qquad (10.2.8)$$

i.e., $g_{ij} = \Sigma \lambda_q^{-1} \gamma_{iq} \gamma_{jq}$. Again the summation in eqn. (10.2.8) includes only the latent roots and latent vectors of $A'A$ that do not identify nonpredictive multicollinearities.

The equivalence of eqn. (10.2.7) with the t statistics for least squares and principal components under conditions (i) and (ii) is readily established. Recall, for example, that the t statistic for least squares is

$$t = \frac{\hat{\beta}_j^*}{(MSE \cdot q_{jj})^{1/2}} , \qquad (10.2.9)$$

where q_{jj} is the jth diagonal element of $Q = (W'W)^{-1}$. Now if no components are deleted from the latent root estimator, eqn. (10.2.8) becomes

$$G = \sum_{j=0}^{p} \lambda_j^{-1} \underline{\gamma}_j \underline{\gamma}_j' = (A'A)^{-1} .$$

Partition $A'A$ and G as follows:

$$A'A = \begin{bmatrix} 1 & \underline{a}' \\ \underline{a} & W'W \end{bmatrix} \quad \text{and} \quad G = \begin{bmatrix} g_{00} & \underline{h}' \\ \underline{h} & G_{11} \end{bmatrix}$$

where the jth element of $\underline{a}$ is $a_j = \Sigma\, Y_i^* W_{ij}$, the correlation coefficient between Y and X_j, and the jth element of $\underline{h}$ is g_{0j}. A well-known relationship* between $A'A$ and $G = (A'A)^{-1}$ allows $(W'W)^{-1}$ to be obtained as follows

$$Q = (W'W)^{-1} = G_{11} - g_{00}^{-1}\underline{h}\underline{h}'$$

and, hence,

$$q_{jj} = g_{jj} - g_{00}^{-1}g_{0j}^2\,.$$

Since $\hat{\beta}_{LR}^*(j) = \hat{\beta}_j^*$ when no components are deleted, insertion of this expression for q_{jj} into eqn. (10.2.9) establishes its equivalence to eqn. (10.2.7). A similar derivation establishes the equivalence between eqn. (10.2.7) and eqn. (10.1.15) under condition (*ii*) above.

After deleting γ_0 and γ_1 from the latent root estimator for the police height data, individual t statistics were calculated using eqn. (10.2.7) and eqn. (10.2.8). The t values are shown in Table 10.13. As one might expect from the similarity of $\hat{\underline{\beta}}_{LR}^*$ and $\hat{\underline{\beta}}_{PC}$ for this data set, the t statistics in Table 10.13 are numerically close to those for principal components (see Table 10.6), although minor differences are apparent.

Table 10.13. Individual t Statistics for Latent Root Regression
of Police Height Data
(γ_0 and γ_1 Deleted)

Variable	SITHT	UARM	FORE	HAND	ULEG
$\hat{\beta}_{LR}^*(j)$	12.20	-.51	1.33	4.37	6.94
t	5.68	-.24	.56	1.50	3.74
Variable	LLEG	FOOT	BRACH	TIBIO	
$\hat{\beta}_{LR}^*(j)$	9.06	3.31	1.80	2.76	
t	4.59	1.32	1.20	1.35	

10.3   RIDGE REGRESSION

Ridge regression† adjusts the least squares estimator by reducing the influence of multicollinearities on the parameter estimates. Unlike principal component and latent root regression, no latent vectors defining multicollinearities are eliminated from the resulting estimator but their effect is greatly diminished. Ridge estimators are defined as a function of a ridge parameter, k, whose value can be selected by the data analyst. Proper

---

*See, for example, Searle (1971, p. 46) and Graybill (1976, p. 19).
†For derivations and a thorough discussion of ridge regression see Hoerl and Kennard (1970 a, b), Marquardt (1970), and Marquardt and Snee (1975).

choice of the ridge parameter can guarantee that the ridge estimator of the regression coefficients is more accurate than the least squares estimator.

10.3.1 Motivation

Least squares parameter estimators can be written in two equivalent forms [see eqn. (9.1.5)]:

$$\hat{\beta}^* = (W'W)^{-1}W'\underline{Y}$$

$$= \sum_{j=1}^{p} \ell_j^{-1} c_j \underline{V}_j, \qquad c_j = \underline{V}_j'W'\underline{Y}. \qquad (10.3.1)$$

The second expression for $\hat{\beta}^*$ stresses the major influence of multicollinearities since strong multicollinearities result in small latent roots of $W'W$ and, therefore, large values of ℓ_j^{-1}. The dominance of latent vectors associated with small latent roots of $W'W$ on the least squares estimates of regression coefficients is evident in the analysis of the police height data, Table 10.2.

Principal component regression and latent root regression, perhaps after an assessment of the predictivity of multicollinearities, completely remove latent vectors of $W'W$ or $A'A$, respectively, from the estimators while leaving the relative influence of all other latent vectors unaltered. Ridge regression, on the other hand, alters the influence of all the latent vectors by reducing the coefficients in (10.3.1). This is accomplished by adding a small constant, k, to the diagonal elements of $W'W$ before inverting. The ridge estimator is thus defined as

$$\hat{\beta}_R^* = (W'W + kI)^{-1}W'\underline{Y}$$

$$= \sum_{j=1}^{p} (\ell_j + k)^{-1} c_j \underline{V}_j \qquad (10.3.2)$$

where $k > 0$.

Observe carefully the effect of k in eqn. (10.3.2). By adding even a small amount, k, to small latent roots, ℓ_j, the influence of latent vectors identifying multicollinearities can be greatly reduced over that of eqn. (10.3.1). If k is sufficiently small, moreover, only very small latent roots will be appreciably altered in eqn. (10.3.2); i.e., latent vectors with moderate or large latent roots will have approximately the same influence in $\hat{\beta}^*$ and $\hat{\beta}_R^*$ since for these roots $\ell_j^{-1} \approx (\ell_j + k)^{-1}$.

One of the most important reasons for the popularity of ridge regression is that if k is chosen suitably small not only will the effects of the multicollinearities be reduced but the ridge estimator will be closer* to β^* than the least squares estimator. A theoretical condition that will guarantee that the ridge estimator is closer to β^* than least squares is

$$0 < k < \sigma^2/\beta^{*\prime}\beta^* .\tag{10.3.3}$$

Unfortunately, this interval for k is of limited practical value since its upper bound is a function of the parameters we wish to estimate; nevertheless, it is of immense theoretical value to know that for every regression model there does exist a range of values for the ridge parameter that guarantees that the ridge estimator is more accurate than least squares.

Under Assumptions 1-3 of Table 6.2 the mean vector and covariance matrix of $\hat{\beta}_R^*$ are

$$
\begin{aligned}
E[\hat{\beta}_R^*] &= (W'W + kI)^{-1}W'E[\underline{Y}]\\[4pt]
&= (W'W + kI)^{-1}W'W\beta^*\\[4pt]
&= \beta^* - k(W'W + kI)^{-1}\beta^*\\[4pt]
&= \underline{\beta}^* - k\sum_{j=1}^{p}(\ell_j + k)^{-1}\underline{V}_j\underline{V}_j'\beta^* .
\end{aligned}\tag{10.3.4}
$$

and

$$
\begin{aligned}
\mathrm{Var}[\hat{\beta}_R^*] &= (W'W + kI)^{-1}W'\mathrm{Var}[\underline{Y}]W(W'W + kI)^{-1}\\[4pt]
&= (W'W + kI)^{-1}W'W(W'W + kI)^{-1}\sigma^2\\[4pt]
&= \sum_{j=1}^{p}\ell_j(\ell_j + k)^{-2}\underline{V}_j\underline{V}_j'\sigma^2
\end{aligned}\tag{10.3.5}
$$

---

*"Closeness" is defined in terms of the total squared error (tse) of a regression estimator. For any regression estimator, say $\tilde{\underline{\beta}}^*$,

$$\mathrm{tse}(\tilde{\underline{\beta}}^*) = E[(\tilde{\underline{\beta}}^* - \underline{\beta}^*)'(\tilde{\underline{\beta}}^* - \underline{\beta}^*)] = \sum_{j=1}^{p}\mathrm{Var}[\tilde{\beta}_j^*] + \sum_{j=1}^{p}\mathrm{Bias}(\tilde{\beta}_j^*)^2,$$

where $\mathrm{Bias}(\tilde{\beta}_j^*) = E[\tilde{\beta}_j^*] - \beta_j^*$. Hoerl and Kennard (1970a) proved that there always exists a range of k values [e.g., eqn. (10.3.3)] for which $\mathrm{tse}(\hat{\underline{\beta}}_R^*) < \mathrm{tse}(\hat{\beta}^*)$.

For an individual parameter estimator, $\hat{\beta}_R^*(r)$, from eqns. (10.3.4) and (10.3.5),

$$E[\hat{\beta}_R^*(r)] = \beta_r^* - k \sum_{j=1}^{p} (\ell_j + k)^{-1} V_{rj}(\underline{V}_j'\underline{\beta}^*)$$

and

$$\text{Var}[\hat{\beta}_R^*(r)] = \sigma^2 \sum_{j=1}^{p} \ell_j(\ell_j + k)^{-2} V_{rj}^2.$$

By comparing these means and variances with those for least squares, it is evident that the ridge estimators are biased but have smaller variances than least squares since

$$\ell_j(\ell_j + k)^{-2} < \ell_j^{-1} \qquad \text{for } k > 0.$$

So like principal components and latent root regression, ridge regression attempts to greatly reduce the least squares variances while only introducing small biases. The key to whether this trade-off is successful is the determination of the ridge parameter, k.

10.3.2 Selecting Ridge Parameters

Ideally one would like to always choose ridge parameter values in the interval (10.3.3). Since the upper bound to this interval is unknown, it is impossible to guarantee that k is so chosen. Because of this problem several other approaches have been developed to guide data analysts in the selection of ridge parameter values.

One of the first techniques proposed for selecting k is referred to as the *ridge trace*. Parameter estimates, $\hat{\beta}_R^*(r)$, are computed for several values of k, often equally spaced throughout the interval (0, 1), and plotted. Some coefficient estimates for multicollinear predictor variables will change very rapidly as k is increased from zero and may even change sign. These sign and magnitude changes, while dramatic initially, soon stabilize and change only gradually thereafter. Coefficient estimates for nonmulticollinear predictor variables do not change dramatically, only gradually, as k is increased from zero. A value for k is selected when the coefficient estimates for the multicollinear predictor variables first stabilize and change gradually like the nonmulticollinear ones.

The rationale behind use of the ridge trace for selectfng a value for k is that as k increases $\hat{\beta}_R^*(r)$ behaves more like it was calculated from an orthogonal matrix of predictor variables than a multicollinear one. Note that as k increases in eqn. (10.3.2), substantial changes in signs and magnitudes can occur among the coefficient estimates as the effects of the small latent roots are lessened. On the other hand, if $W'W = I$ eqn. (10.3.2) becomes

$$\hat{\underline{\beta}}_R^* = (1 + k)^{-1} \sum_{j=1}^{p} c_j \underline{V}_j$$

and increasing k serves only to gradually lessen the magnitudes of the individual coefficients.

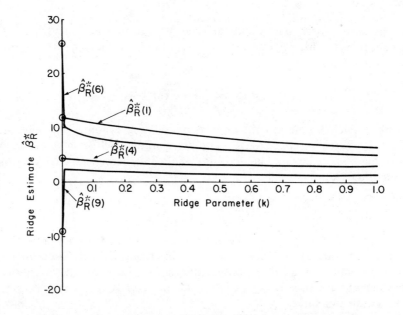

Figure 10.1. Ridge trace of four coefficient estimates, police height data.

Figure 10.1 exhibits a ridge trace for the police height data; for clarity of presentation only four of the coefficient estimates are plotted. Increasing k only slightly has the expected precipitous effect on the coefficient estimates for the two multicollinear predictor variables that are plotted, X_6(LLEG) and X_9(TIBIO). Both estimates drop in magnitude and the sign of $\hat{\beta}_R^*(9)$ quickly becomes positive. Coefficient estimates for the two nonmulticollinear predictor variables, X_1(SITHT) and X_4(HAND), change relatively little as k is increased.

Visual inspection of the ridge trace suggests that a small value of k, one in the interval $(0,0.1)$, produces a stabilization of the ridge trace. In order to more closely pinpoint which values of k stabilize the trace, an examination of differences in the coefficient estimates for successive choices of k can be helpful. Table 10.14 lists the ridge estimtes and their differences for several choices of k in the interval $(0,0.1)$. Note that until about $k = 0.07$ or $k = 0.08$ the change in $\hat{\beta}_R^*(6)$ is relatively larger than the changes in other coefficients. The difference for $\hat{\beta}_R^*(6)$ is comparable to that of $\hat{\beta}_R^*(1)$ as k changes from 0.07 to 0.08. We will use $k = 0.08$ as the value for the ridge parameter. All the ridge coefficient estimates for this value of k are shown in Table 10.15 along with the least squares estimates.

Table 10.14. Ridge Estimates and Their Differences For the Four Predictor Variables in Figure 10.1

Ridge Estimate	$k = .04$	$k = .05$	$k = .06$	$k = .07$	$k = .08$
$\hat{\beta}_R^*(1)$	11.684	11.568	11.455	11.346	11.240
$\hat{\beta}_R^*(4)$	4.068	4.010	3.957	3.909	3.865
$\hat{\beta}_R^*(6)$	8.815	8.665	8.534	8.415	8.306
$\hat{\beta}_R^*(9)$	2.258	2.227	2.192	2.156	2.119
		Differences			
	.05-.04	.06-.05	.07-.06	.08-.07	
$\hat{\beta}_R^*(1)$	-.116	-.112	-.109	-.106	
$\hat{\beta}_R^*(4)$	-.058	-.053	-.048	-.044	
$\hat{\beta}_R^*(6)$	-.150	-.132	-.119	-.109	
$\hat{\beta}_R^*(9)$	-.030	-.035	-.036	-.036	

The ridge estimates in Table 10.15 have positive signs and the magnitudes of all the coefficient estimates for the multicollinear predictor variables are reduced. These estimates are not very different from the principal component and latent root estimates. A direct comparison of Tables 10.3, 10.7, and 10.15 reveals that the ridge estimates of the nonmulticollinear predictor variable coefficients differ from those of least squares more than do those of principal components or latent root, but the three biased estimators do not yield estimates that differ substantially from one another for this data set.

Also displayed in Table 10.15 are the variance inflation factors, q_{jj}^R, of the ridge estimator for $k = 0.08$. By analogy with least squares the variance inflation factors are the diagonal elements of Q^R, where $\text{Var}[\underline{\hat{\beta}}_R^*] = Q^R \sigma^2$ and from eqn. (10.3.5),

Table 10.15. Least Squares and Ridge Estimation ($k = 0.08$) for Police Height Data

Predictor Variable	Coefficient Estimates		Variance Inflation Factors	
	$\hat{\beta}_j^*$	$\hat{\beta}_k^*(j)$	q_{jj}	q_{jj}^R
SITHT	11.91	11.24	1.52	1.01
UARM	4.36	.46	436.42	.86
FORE	-3.39	2.24	354.00	1.02
HAND	4.26	3.86	2.43	1.52
ULEG	-9.64	6.17	817.57	.69
LLEG	25.44	8.31	802.17	.75
FOOT	3.37	3.46	1.77	1.24
BRACH	6.48	1.82	417.37	.56
TIBIO	-9.52	2.12	448.58	.85
$\hat{\sigma}^2$	3.57	3.66		
R^2	0.893	0.890		

$$Q^R = (W'W + kI)^{-1} W'W(W'W + kI)^{-1}.$$

From Table 10.15 it is apparent that enormous reductions in the variances of individual ridge coefficient estimators are achieved.

Several other techniques have been proposed for selecting ridge parameter values. These techniques have been evaluated by simulation studies and many have been found to produce ridge estimators that are generally superior to least squares. Rather than detail the more effective selection rules for k and their advantages and disadvantages, the interested reader is referred to the following papers and the references contained therein: McDonald and Galarneau (1975), Hocking, Speed, and Lynn (1976), Dempster, Schatzoff, and Wermuth (1977), Gunst and Mason (1977b), and Wichern and Churchill (1978).

10.3.3   Analysis of Variance

Partitioning the total sum of squares proceeds along lines similar to that of latent root regression. First partition TSS as

$$TSS = SSM + TSS(adj).$$

Next, SSE_R is found by evaluating $\Sigma(Y_i - \hat{Y}_i)^2$, where

$$\hat{Y}_i = \bar{Y} + \underline{u}_i'\hat{\beta}_k^*$$

and $\underline{u}_i$ is the ith row of W. Without going through all the algebraic details, the result is

$$\text{SSE}_R = \text{TSS(adj)} - \underline{Y}' W\hat{\underline{\beta}}_R^* - k\hat{\underline{\beta}}_R^*\hat{\underline{\beta}}_R^* . \qquad (10.3.6)$$

Hence,

$$\text{SSR}_R = \text{TSS(adj)} - \text{SSE}_R$$

$$= \underline{Y}' W\hat{\underline{\beta}}_R^* + k\hat{\underline{\beta}}_R^{*'}\hat{\underline{\beta}}_R^* \qquad (10.3.7)$$

Table 10.16 is a symbolic analysis of variance table similar to the pooled analysis of variance tables for principal components and latent root regression, Tables 10.4(b) and 10.11(b). Caution must be exercised when interpreting Table 10.16. While SSE_R is similar to SSE_{PC} and SSE_{LR}, both of the latter sums of squares have more degrees of freedom associated with them. The residual sum of squares for ridge regression is a pooled sum of squares like SSE_{PC} and SSE_{LR} but there are no methods currently available for assigning additional degrees of freedom to it.

Table 10.16. Symbolic Analysis of Variance Table for Ridge Regression

			ANOVA		
Source	d.f.	S.S.	M.S.	F	R^2
Mean	1	SSM	MSM	MSM/MSE_R	$\text{SSR}_R/\text{TSS(adj)}$
Regression	p	SSR_R	MSR_R	$\text{MSR}_R/\text{MSE}_R$	
Error	$n-p-1$	SSE_R	MSE_R		
Total	n	TSS			

Ridge regression inference procedures suffer from the lack of exact distributional theory, as did the latent root estimator. The F ratios in Table 10.16 are not true F random variables for two reasons. First, the wrong degrees of freedom may be assigned to SSE_R. Second, the ridge estimator is biased and the sums of squares eqns. (10.3.6) and (10.3.7) do not (when divided by σ^2) have chi-square distributions unless the bias is zero. If the bias is negligible the F ratios still might not be approximate F random variables due to the difficulty of assigning proper degrees of freedom. Despite these problems the F ratios and R^2 do measure the significance of the fit to the data and provide meaningful insight into the fit of the prediction equation.

Table 10.17 provides the ANOVA table for the police height data. The values in the ANOVA table are very similar to those in Tables 10.5(b) and 10.12(b) and R^2 and MSE reveal that the ridge estimator does produce a good fit to the response variable.

Table 10.17. Anova Table for Ridge Regression of Police Height Data
($k = 0.08$)

		ANOVA			
Source	d.f.	S.S.	M.S.	F	R^2
Mean	1	893,679.28	893,679.28	243,932.90	.890
Regression	9	681.71	75.75	20.68	
Error	23	84.26	3.66		
Total	33	894,445.26			

The distributional problems described above prevent the derivation of other inferential techniques such as t statistics. While approximate t statistics could be derived, these distributional problems preclude having confidence that the approximations would be valid. The ability of the ridge estimator to adjust for multicollinearities among the predictor variables and the knowledge that ridge estimators, when k is selected appropriately, can always provide more accurate estimators than least squares are, nevertheless, powerful reasons for utilizing ridge estimators with multicollinear data.

10.4 FINAL REMARKS

Least squares parameter estimation is by far the most versatile and powerful regression methodology yet developed. Not only are its distributional properties well-established but there is also a rich collection of related model-fitting tools: variable selection techniques, procedures for assessing model specification, etc. Because of the diversity of the least squares methodology and the abundance of exact distributional theory surrounding it, least squares should be the preferred regression methodology for the data analyst.

Situations do arise, however, when biased regression estimation must be utilized instead of least squares. Severe multicollinearities in a data base can force least squares parameter estimates to be unrealistic. Biased estimators, despite the relative paucity of exact distributional theory, can provide more realistic parameter estimates and have been shown to generally be more accurate than least squares when predictor variables are severely multicollinear.

Even the exact distributional properties of the principal component and ridge regression estimators (expectations, variances, etc.) are only valid under very restrictive conditions. For principal components the distributional properties of $\hat{\beta}^*_{PC}$ are only valid if the components to be deleted are selected according to some procedure (such as only by examining the latent roots and latent vectors of $W'W$) that does not utilize the observed response variables. Hence, use of statistical tests for the significance of the components invalidates the subsequent tests for significance of the predictor variables and the expectations and variances of the resulting estimators. Nonstochastic rules (rules that do not involve the response variable) for deleting components do not invalidate these properties. Likewise, stochastic rules for selecting ridge parameter values, such as the ridge trace or estimating k from some function of the response variables, renders the theoretical properties of the ridge estimator incorrect. Again, despite these drawbacks, all three biased regression estimators discussed in this chapter have been shown to be beneficial with multicollinear data.

As a final illustration of the efficacy of biased regression estimators some additional remarks on the police height data are in order. The last two predictor variables, BRACH and TIBIO, are actually functions of the other multicollinear predictor variables. Specifically,

$$BRACH = FORE/UARM \qquad and \qquad TIBIO = LLEG/ULEG.$$

The measurements on the arms and legs for these female police department applicants are over such a narrow range that the nonlinear transformations given above are well approximated by linear ones, the two multicollinearities shown in Table 10.2. Knowing these relationships exist for all the observations in the data set as well as all future measurements insures that these multicollinearities are inherent to the population under study (see Section 9.3.1) and, therefore, that BRACH and TIBIO are redundant predictor variables. The analysis of this data set should proceed with both variables removed.

If BRACH and TIBIO are removed from Data Set A.8, the reduced $W'W$ matrix has a smallest latent root of 0.113 and the largest variance inflation factor is 4.62. No strong multicollinearities exist in the reduced data set. An analysis of variance table for the reduced data set and the least squares coefficient estimates for the 7 predictor variables are shown in Table 10.18. Observe how closely the values in the ANOVA table match those of the previous analyses using biased regression estimators. The individual coefficient estimates and t statistics are also in fairly close agreement.

In this case the nature of the multicollinearities was known and the proper analysis consists of least squares regression on the reduced data set. It is important to note, however, that if the multicollinearity in this data set could not be explained, e.g., if it was due to a sampling deficiency, and the biased estimators were utilized instead of least squares, each of the biased estimators discussed above would yield coefficient estimates that are quite

Table 10.18. Least Squares Analysis of Reduced Police Height Data Set

ANOVA

Source	d.f.	S.S.	M.S.	F	R^2
Mean	1	893,679.28	893,679.28	271,070.37	.892
Regression	7	683.56	97.65	29.62	
Error	25	82.42	3.30		
Total	33	894,445.26			

Predictor Variable	Coefficients Estimate	t Statistic	Variance Inflation Factor
SITHT	12.15	5.92	1.28
UARM	-2.31	-.79	2.58
FORE	2.84	.93	2.81
HAND	4.36	1.56	2.38
ULEG	3.16	.87	4.01
LLEG	12.83	3.29	4.62
FOOT	3.31	1.37	1.76

close to the ones in Table 10.18 for the reduced model, estimates that are much more reasonable than the least squares estimates for the complete data set.

EXERCISES

1. Calculate the latent roots and latent vectors of $W'W$ and $A'A$ for a quadratic fit to the high school grade data. What does an examination of these latent roots and latent vectors suggest concerning multicollinearities?

2. Obtain principal component estimates for the standardized regression coefficients of the variables in Exercise 1 (i) when latent vectors are deleted solely on the basis of the magnitude of the latent roots and (ii) following a test of hypothesis for the significance of the principal components.

3. Repeat Exercises 1 and 2 for a two-variable fit to the emissions data using humidity and barometric pressure.

4. Construct analysis of variance tables for the least squares and principal component fits of NOX using HUMID, TEMP, and HUMID × TEMP as predictor variables.

5.   Calculate t statistics for the deletion of the interaction term in the fit of NOX to HUMID, TEMP, and HUMID × TEMP using least squares and principal component regression. If parsimony is an important consideration in the fit to this data what conclusions do you draw regarding the need for the interaction term?

6.   Conduct a latent root regression analysis of the fit of NOX by HUMID, TEMP, and HUMID × TEMP. Calculate t statistics for the three predictor variables. Are the conclusions drawn from this analysis the same as those from least squares and principal component analysis?

7.   Ridge estimates for fixed k can be obtained from least squares computer programs in the following manner. First standardize the predictor variables using the unit length standardization. Next add p additional rows to W where the p additional rows constitute a diagonal matrix of the form $\sqrt{k}I$. Finally add p zeros to the response vector Y corresponding to the p new rows in W. Let $W'_R = [W':\sqrt{k}I]$ and $Y'_R = [\underline{Y}':\underline{0}']$. Show that $\hat{\beta}_R = (W'_R W_R)^{-1} W'_R \underline{Y}_R$.

8.   Construct a ridge trace for the fit of NOX by HUMID, TEMP, and HUMID × TEMP. Select a suitable value for the ridge parameter, k, and compare the ridge estimators and the resulting analysis of variance table with the previous analyses obtained used least squares, principal component and latent root regression.

APPENDIX A

DATA SETS ANALYZED IN THE TEXT

Tabulation of the data sets in this appendix is motivated primarily by the conviction that effective use of regression methodology requires experience with real data. By providing many data sets, the analyses that are partially undertaken in the text can illustrate a wide variety of problems that the researcher can expect to encounter. None of the data sets are completely analyzed in the text, leaving ample opportunity for the interested reader to experiment with techniques described in the chapters. Some of the data sets are intentionally small so that hand-plotting and computation with desk calculators are feasible; others are larger and require the aid of a computer.

No attempt is made to provide in-depth discussions of the subject matter of the data sets; however, the references often contain extensive commentaries. The subject matter of the data sets covers a wide range of applications and should cause little difficulty for the reader in terms of understanding possible goals of the analysis. The choice of the data sets is made on the basis of the characteristics of the data and the size of the data set. Our main objective in choosing these data sets, then, is to include ones that will be interesting to analyze but will not require of the reader training in any specific discipline.

A.1 SELECTED DEMOGRAPHIC CHARACTERISTICS

This data set is a compilation of entries in a much larger data base from Data Set 41 of Loether, McTavish and Voxland (1974). Seven variables that are listed below were chosen for inclusion in this data set. More complete descriptions of the variables, measurement process, and accuracy are available in the above reference. The 49 countries included in the data set are the ones with complete information on the seven variables. Brief descriptions of the seven variables are:

GNP — Gross National Product per Capita, 1957, U. S. Dollars
INFD — Infant Deaths per 1,000 Live Births
PHYS — Number of Inhabitants per Physician
DENS — Population per Square Kilometer
AGDS — Population per 1,000 Hectares of Agricultural Land
LIT — Percentage Literate of Population Aged 15 and Over
HIED — Students Enrolled in Higher Education per 100,000 Population.

A.2. HIGH SCHOOL AND COLLEGE AVERAGE GRADES

This article analyzes admission data (average grades during the final year of high school) and the average final grades after the first year for students from nine Canadian high schools who matriculated during the six year period from 1968-1973. Ruben and Stroud (1977) derive several statistical models for predicting the first year of college averages from the final year of high school average grades. By comparing equations for the six years and nine high schools, the authors are able to compare the schools and describe the changes in the fit over time.

A small portion of the original data base is recorded in this data set. The average grades for students matriculating in 1969 who attended the first three high schools in the original data base are included.

A.3 HOMICIDES IN DETROIT

One goal of this study is to investigate "the role of firearms in accounting for the rising homicide rate in one city." In order to achieve this goal, Fisher (1976) collected data on the homicide rate in Detroit for the years 1961-1973 and on the availability of firearms and several other variables believed to influence, account for, or be related to the rise in the homicide rate. The raw data includes the following variables:

HOM — Number of Homicides per 100,000 Population
ACC — Accidental Death Rate (Number per 100,000 Population)
ASR — Number of Assaults per 100,000 Population
FTP — Number of Full-time Police per 100,000 Population
UEMP — Percent of the Population Unemployed
MAN — Number of Manufacturing Workers (in Thousands)
LIC — Number of Handgun Licenses Issued per 100,000 Population
GR — Number of Handgun Registrations per 100,000 Population
CLEAR — Percent of Homicides Cleared by Arrest
WM — Number of White Males in the Population
NMAN — Number of Nonmanufacturing Workers (in Thousands)
GOV — Number of Government Workers (in Thousands)
HE — Average Hourly Earnings
WE — Average Weekly Earnings.

A.4 EDUCATIONAL STATUS ATTAINMENT

Causal models linking six predictor variables with educational attainment and all seven of these variables with occupational attainment are investigated in Sewell, Haller, and Ohlendorf (1970). All male high school seniors in Wisconsin were administered an initial questionnaire in 1957, and a random sample of one-third were given a follow-up questionnaire in 1964-65. Included in this data set are summary information on the six predictor variables and a measure of educational attainment for 1,094 subjects who resided in small cities during the 1957 initial survey. Due to the large sample size only the sample correlations, sample means, and sample standard deviations are displayed in the table. The seven variables are:

EDATT — A measure of educational attainment as of 1964-65
LOA — Level of occupational aspiration based on the subject's occupational goals as indicated in 1957
LEA — Level of educational aspiration based on the subject's educational goals as indicated in 1957
SOI — An indicator of the influence of "significant others" (parents, teachers, friends) on the educational plans of the subject in 1957
AP — Academic performance measure (centile rank in class in 1957)
SES — Socioeconomic status index of subject's family in 1957-60
MA — Centile rank on a statewide mental ability test administered during the junior year of high school.

A.5 NITROUS OXIDE EMISSIONS MODELLING

Light-duty, diesel-powered vehicle emissions are governed by Federal exhaust emissions standards. As newer vehicles with different engine configurations are introduced, standards must be modified. This study (Hare and Bradow, 1977) undertook the construction of prediction equations for exhaust emissions of light-duty diesel trucks. The data set consists of 44 observations on a single truck as three environmental conditions were varied: humidity, temperature, and atmospheric pressure. The purpose of the investigation was to construct prediction equations that could be used to transform observed emissions readings to fixed humidity, temperature, and pressure levels so that emissions readings, regardless of the environmental conditions under which they are observed, can be converted to "standard" environmental conditions. The variables are:

NOX — Nitrous Oxides, NO and NO_2 (grams/km)
HUMID — Humidity (grains H_2O/lb_m dry air)
TEMP — Temperature (°F)
BPRES — Barometric Pressure (in. Hg).

A.6 ANTHROPOMETRIC AND PHYSICAL FITNESS MEASUREMENTS

A broad study of the physical fitness of over 650 applicants to the police department of a major metropolitan city was conducted by Dr. Ladislav P. Novak, Department of Anthropology, Southern Methodist University. This data base contains 15 of over 60 variables measured on each of 50 white male police applicants. A purpose for investigating this portion of the original data base is to assess the influence of the anthropometric and fitness variables on a reaction time measurement taken on the applicants. The variables are:

REACT — Reaction Time to a Visual Stimulus (in sec.)
HEIGHT — Height of Applicant (cm.)
WEIGHT — Weight of Applicant (kg.)
SHLDR — Shoulder Width (cm.)
PELVIC — Pelvic Width (cm.)
CHEST — Minimum Chest Circumference (cm.)
THIGH — Thigh Skinfold Thickness (mm.)
PULSE — Pulse Rate (count per min.)
DIAST — Diastolic Blood Pressure

CHNUP   — Number of Chinups Applicant Can Complete
BREATH — Maximum Breathing Capacity (liters)
RECVR   — Pulse Rate After 5 Mins. Recovery from Treadmill Running
SPEED   — Maximum Treadmill Speed (Individually Set for Each Applicant)
ENDUR   — Treadmill Endurance Time: Maximum Time to Exhaust While Running Treadmill (min.)
FAT   — Total Body Fat Measurement.

A.7 HOUSING RENT STUDY

Information on over 3,000 households in Mariposa County (Phoenix), Arizona and Pittsburg, Pennsylvania was collected for a study (Merrill, 1977) sponsored by the U. S. Department of Housing and Urban Development. One purpose of the study was to assess how low-income families utilize direct cash allowances for housing. As part of this assessment a prediction equation for housing rent (a measure of housing quality) was constructed from housing characteristics such as tenure conditions, dwelling unit features, and neighborhood qualities. The portion of the original data base that is included in this data set consists of summary information on rent and seventeen housing characteristics for 455 households in Mariposa County, Arizona. The large sample size precludes listing the raw data, but sample means, sample standard deviations, and sample correlations are provided. From these summary statistics prediction equations can be formulated. The eighteen variables are:

ACR   — Monthly Rent for an Unfurnished Dwelling Unit (including utilities)
RELLL*   — Indicates Whether Household Members are Related to the Landlord
RES15   — Length of Residence in Dwelling Unit is 1 to 5 Years
RES510*   — Length of Residence in Dwelling Unit is 5 to 10 Years
RESG10*   — Length of Residence in Dwelling Unit is Greater than 10 Years
AREAPR   — Average Room Area
LTOTRMS — Logarithm of Total Number of Rooms in Dwelling Unit
LTOTARE — Logarithm of Total Floor Area of Dwelling Unit
CENH*   — Indicates Whether Dwelling Unit has Central Heating
BADH*   — Indicates Whether Heating Equipment is Inferior or Not Present

*Dichotomos variables: 1 = Yes, 0 = No

STOREF* — Indicates Whether Stove and Refrigerator are Provided

APPL* — Indicates Whether Dishwasher and/or Disposal are Provided

QUAL2 — Measures Average Surface Area and Structural Quality (0 = Worst, 3 = Best)

MULTI5* — Indicator of Large Multifamily Structure (5 or more dwelling units)

CENFO1 — Composite Measure of Quality Housing Units and Socioeconomic Level for the Census Tract where Dwelling Unit is Located

CENFO2 — Composite Measure of the Preponderance of Owner-Occupied, Single-Family Housing Units in the Census Tract where Dwelling Unit is Located

CENFO3 — Composite Measure of Poor Quality Housing Units in the Census Tract where Dwelling Unit is Located

CNHF13 — Composite Measure of School Quality and Lack of Transportation Facilities in the Census Tract where Dwelling Unit is Located.

*Dichotomous variables: 1 = Yes, 0 = No.

A.8   BODY MEASUREMENTS

This data set is another portion of the large data base mentioned in Data Set A.6. It contains body measurements on 33 black female applicants to the police department of a large metropolitan city. The goal of this portion of the investigation was to construct prediction equations for height, including any reduced predictors that were deemed sufficiently accurate. Variables in the data set are:

HEIGHT — Subject Height (cm)

SITHT — Sitting Height (cm)

UARM — Upper Arm Length (cm)

FORE — Forearm Length (cm)

HAND — Hand Length (cm)

ULEG — Upper Leg Length (cm)

LLEG — Lower Leg Length (cm)

FOOT — Foot Length (cm)

BRACH* — Brachial Index

TIBIO* — Tibio-Femoral Index.

*These two indices are described more fully in Section 10.4.

Data Set A.1. Selected Demographic Characteristics of Countries of the World

COUNTRY	GNP	INFD	PHYS	DENS	AGDS	LIT	HIED
Australia	1,316	19.5	806	1	21	98.5	856
Austria	670	37.5	695	84	1,720	98.5	546
Barbados	200	60.4	3,000	548	7,121	91.1	24
Belgium	1,196	35.4	819	301	5,257	96.7	536
Brit. Guiana	235	67.1	3,900	3	192	74.0	27
Bulgaria	365	45.1	740	72	1,380	85.0	456
Canada	1,947	27.3	900	2	257	97.5	645
Chile	379	127.9	1,700	11	1,164	80.1	257
Costa Rica	357	78.9	2,600	24	948	79.4	326
Cyprus	467	29.9	1,400	62	1,042	60.5	78
Czechoslovakia	680	31.0	620	108	1,821	97.5	398
Denmark	1,057	23.7	830	107	1,434	98.5	570
El Salvador	219	76.3	5,400	127	1,497	39.4	89
Finland	794	21.0	1,600	13	1,512	98.5	529
France	943	27.4	1,014	83	1,288	96.4	667
Guatemala	189	91.9	6,400	36	1,365	29.4	135
Hong Kong	272	41.5	3,300	3,082	98,143	57.5	176
Hungary	490	47.6	650	108	1,370	97.5	258
Iceland	572	22.4	840	2	79	98.5	445
India	73	225.0	5,200	138	2,279	19.3	220
Ireland	550	30.5	1,000	40	598	98.5	362
Italy	516	48.7	746	164	2,323	87.5	362
Jamaica	316	58.7	4,300	143	3,410	77.0	42
Japan	306	37.7	930	254	7,563	98.0	750
Luxembourg	1,388	31.5	910	123	2,286	96.5	36
Malaya	356	68.9	6,400	54	2,980	38.4	475
Malta	377	38.3	980	1,041	8,050	57.6	142
Mauritius	225	69.5	4,500	352	4,711	51.8	14
Mexico	262	77.7	1,700	18	296	50.0	258
Netherlands	836	16.5	900	346	4,855	98.5	923
New Zealand	1,310	22.8	700	9	170	98.5	839
Nicaragua	160	71.7	2,800	10	824	38.4	110
Norway	1,130	20.2	946	11	3,420	98.5	258
Panama	329	54.8	3,200	15	838	65.7	371
Poland	475	74.7	1,100	96	1,411	95.0	351
Portugal	224	77.5	1,394	100	1,087	55.9	272
Puerto Rico	563	52.4	2,200	271	4,030	81.0	1,192
Romania	360	75.7	788	78	1,248	89.0	226
Singapore	400	32.3	2,400	2,904	108,214	50.0	437
Spain	293	43.5	1,000	61	1,347	87.0	258
Sweden	1,380	16.6	1,089	17	1,705	98.5	401
Switzerland	1,428	21.1	765	133	2,320	98.5	398
Taiwan	161	30.5	1,500	305	10,446	54.0	329
Trinidad	423	45.4	2,300	168	4,383	73.8	61
United Kingdom	1,189	24.1	935	217	2,677	98.5	460
United States	2,577	26.4	780	20	399	98.0	1,983
USSR	600	35.0	578	10	339	95.0	539
West Germany	927	33.8	798	217	3,631	98.5	528
Yugoslavia	265	100.0	1,637	73	1,215	77.0	524

Source: Loether, McTavish, and Voxland (1974). © 1974 by Allyn and Bacon, Inc. Reprinted with permission.

Data Set A.2. Final High School Average Grades and First Year of College Average Grades

High School 1		High School 2		High School 3	
Grade 13 Average	First Year Average	Grade 13 Average	First Year Average	Grade 13 Average	First Year Average
60.6	49.6	63.0	52.4	61.8	65.0
60.6	54.6	64.0	45.8	66.6	45.0
62.0	47.4	65.5	40.8	67.0	59.0
66.0	51.8	66.0	51.8	67.6	52.2
66.4	51.2	66.2	80.0	69.0	46.0
69.0	60.2	66.3	63.2	69.2	42.0
69.0	64.2	66.8	57.5	70.8	50.0
69.2	57.4	67.0	54.6	71.5	61.2
72.0	50.0	68.8	52.8	73.2	54.3
72.3	64.8	71.0	52.4	73.4	60.4
72.4	60.8	71.2	62.6	74.0	63.2
73.8	67.8	74.0	64.2	75.2	72.4
75.0	69.4	74.6	66.6	76.7	67.0
75.4	68.6	76.0	69.4	79.3	73.6
76.0	65.8	80.4	69.0	83.7	69.5
77.2	70.0	80.6	74.8	84.5	63.2
77.5	73.4	81.0	68.0		
82.0	73.0	86.6	73.2		
		87.0	71.2		

Source: Ruben and Stroud (1977). ©1977 by American Educational Research Association, Washington, D. C. Reprinted with permission.

Data Set A.3. Homicides in Detroit, 1961-1973

Year	HOM	ACC	ASR	FTP	UEMP	MAN	LIC
1961	8.60	39.17	306.18	260.35	11.0	455.5	178.15
1962	8.90	40.27	315.16	269.80	7.0	480.2	156.41
1963	8.52	45.31	277.53	272.04	5.2	506.1	198.02
1964	8.89	49.51	234.07	272.96	4.3	535.8	222.10
1965	13.07	55.05	230.84	272.51	3.5	576.0	301.92
1966	14.57	53.90	217.99	261.34	3.2	601.7	391.22
1967	21.36	50.62	286.11	268.89	4.1	577.3	665.56
1968	28.03	51.47	291.59	295.99	3.9	596.9	1,131.21
1969	31.49	49.16	320.39	319.87	3.6	613.5	837.60
1970	37.39	45.80	323.03	341.43	7.1	569.3	794.90
1971	46.26	44.54	357.38	356.59	8.4	548.8	817.74
1972	47.24	41.03	422.07	376.69	7.7	563.4	583.17
1973	52.33	44.17	473.01	390.19	6.3	609.3	709.59

YEAR	GR	CLEAR	WM	NMAN	GOV	HE	WE
1961	215.98	93.4	558,724	538.1	133.9	2.98	117.18
1962	180.48	88.5	538,584	547.6	137.6	3.09	134.02
1963	209.57	94.4	519,171	562.8	143.6	3.23	141.68
1964	231.67	92.0	500,457	591.0	150.3	3.33	147.98
1965	297.65	91.0	482,418	626.1	164.3	3.46	159.85
1966	367.62	87.4	465,029	659.8	179.5	3.60	157.19
1967	616.54	88.3	448,267	686.2	187.5	3.73	155.29
1968	1,029.75	86.1	432,109	699.6	195.4	2.91	131.75
1969	786.23	79.0	416,533	729.9	210.3	4.25	178.74
1970	713.77	73.9	401,518	757.8	223.8	4.47	178.30
1971	750.43	63.4	387,046	755.3	227.7	5.04	209.54
1972	1,027.38	62.5	373,095	787.0	230.9	5.47	240.05
1973	666.50	58.9	359,647	819.8	230.2	5.76	258.05

Source: Fisher (1976). Reprinted with permission of Dr. J. C. Fisher.

Data Set A.4. Educational Status Attainment

	EDATT	LOA	LEA	SOI	AP	SES	MA
			SAMPLE CORRELATIONS				
EDATT	1.000	.629	.721	.615	.572	.458	.492
LOA		1.000	.770	.543	.497	.340	.451
LEA			1.000	.624	.504	.390	.429
SOI				1.000	.490	.383	.475
AP					1.000	.251	.602
SES						1.000	.329
MA							1.000

	EDATT	LOA	LEA	SOI	AP	SES	MA
Sample Mean	1.250	47.077	.861	6.976	44.077	17.790	52.433
Standard Deviation	1.222	25.586	.947	1.752	27.460	11.267	29.016
$n = 1,094$							

Source: Sewell, Haller, and Ohlendorf (1970). ©1970 by American Sociological Association, Washington, D. C. Reprinted with permission.

Data Set A.5. Nitrous Oxide Emissions Modelling

No.	NOX	HUMID	TEMP	BPRES
1	.81	74.9	78.4	29.08
2	.96	44.6	72.5	29.37
3	.96	34.3	75.2	29.28
4	.94	42.4	67.3	29.29
5	.99	10.1	76.4	29.78
6	1.11	13.2	67.1	29.40
7	1.09	17.1	77.8	29.51
8	.77	73.7	77.4	29.14
9	1.01	21.5	67.6	29.50
10	1.03	33.9	77.2	29.36
11	.96	47.8	86.6	29.35
12	1.12	21.9	78.1	29.63
13	1.01	13.1	86.5	29.65
14	1.10	11.1	88.1	29.65
15	.86	78.4	78.1	29.43
16	.85	69.1	76.7	29.28
17	.70	96.5	78.7	29.08
18	.79	108.7	87.9	28.98
19	.95	61.4	68.3	29.35
20	.85	91.3	70.6	29.03
21	.79	96.8	71.0	29.05
22	.77	95.9	76.1	29.04
23	.76	83.6	78.3	28.87
24	.79	75.9	69.4	29.07
25	.77	108.7	75.4	28.99
26	.82	78.6	85.7	29.02
27	1.01	33.9	77.3	29.43
28	.94	49.2	77.3	29.43
29	.86	75.7	86.4	29.06
30	.79	128.8	86.8	28.96
31	.81	82.4	87.1	29.12
32	.87	122.6	86.2	29.15
33	.86	124.7	87.2	29.09
34	.82	120.0	87.5	29.09
35	.91	139.5	87.7	28.99
36	.89	105.4	86.1	29.21
37	.87	90.7	87.0	29.16
38	.85	142.2	87.1	28.99
39	.85	136.5	85.0	29.09
40	.70	138.9	85.9	29.16
41	.82	89.7	86.7	29.15
42	.84	92.6	85.3	28.18
43	.84	147.6	87.3	29.10
44	.85	141.4	86.3	29.06

Source: Hare (1977). Reprinted with permission.

Data Set A.6. Anthropometric and Physical Fitness Measurements on Police Department Applicants

No.	REACT	HEIGHT	WEIGHT	SHLDR	PELVIC	CHEST	THIGH	PULSE	DIAST	CHNUP	BREATH	RECVR	SPEED	ENDUR	FAT
1	.310	179.6	74.20	41.7	27.3	82.4	19.0	64	64	2	158	108	5.5	4.0	11.91
2	.345	175.6	62.04	37.5	29.1	84.1	5.5	88	78	20	166	108	5.5	4.0	3.13
3	.293	166.2	72.96	39.4	26.8	88.1	22.0	100	88	7	167	116	5.5	4.0	16.89
4	.254	173.8	85.92	41.2	27.6	97.6	19.5	64	62	4	220	120	5.5	4.0	19.59
5	.384	184.8	65.88	39.8	26.1	88.2	14.5	80	68	9	210	120	5.5	5.0	7.74
6	.406	189.1	102.26	43.3	30.1	101.2	22.0	60	68	4	188	91	6.0	4.0	30.42
7	.344	191.5	84.04	42.8	28.4	91.0	18.0	64	48	1	272	110	6.0	3.0	13.70
8	.321	180.2	68.34	41.6	27.3	90.4	5.5	74	64	14	193	117	5.5	4.0	3.04
9	.425	183.8	95.14	42.3	30.1	100.2	13.5	80	78	4	199	105	5.5	4.0	20.26
10	.385	163.1	54.28	37.2	24.2	80.5	7.0	84	78	13	157	113	6.0	4.0	3.04
11	.317	169.6	75.92	39.4	27.2	92.0	16.5	65	78	6	180	110	5.0	5.0	12.83
12	.353	171.6	71.70	39.1	27.0	86.2	25.5	68	72	0	193	105	5.5	4.0	15.95
13	.413	180.0	80.68	40.8	28.3	87.4	17.5	73	88	4	218	109	5.0	4.2	11.86
14	.392	174.6	70.40	39.8	25.9	83.9	16.5	104	78	6	190	129	5.0	4.0	9.93
15	.312	181.8	91.40	40.6	29.5	95.1	32.0	92	88	1	206	139	5.0	3.5	32.63
16	.342	167.4	65.74	39.7	26.4	86.0	13.0	80	86	6	181	120	5.5	4.0	6.64
17	.293	173.0	79.28	41.2	26.9	96.1	11.5	72	68	6	184	111	5.5	3.9	11.57
18	.317	179.8	92.06	40.0	29.8	100.9	15.0	60	78	0	205	92	5.0	4.0	24.21
19	.333	176.8	87.96	41.2	28.4	100.8	20.5	76	90	1	228	147	4.0	3.5	22.39
20	.317	179.3	77.66	41.4	31.6	90.1	9.5	58	86	15	198	98	5.5	4.1	6.29
21	.427	193.5	98.44	41.6	29.2	95.7	21.0	54	74	0	254	110	5.5	3.8	23.63
22	.266	178.8	65.42	39.3	27.1	83.0	16.5	88	72	7	206	121	5.5	4.0	10.53
23	.311	179.6	97.04	43.8	30.1	100.8	22.0	100	74	3	194	124	5.0	4.0	20.62
24	.284	172.6	81.72	40.9	27.3	91.5	22.0	74	76	4	201	113	5.5	5.1	18.39
25	.259	171.5	69.60	40.4	27.8	87.7	15.5	70	72	10	175	110	5.5	3.0	11.14
26	.317	168.9	63.66	39.8	26.7	83.9	6.0	68	70	7	179	119	5.5	5.0	5.16
27	.263	183.1	87.24	43.2	28.3	95.7	11.0	88	74	7	245	115	5.5	4.0	9.60
28	.336	163.6	64.86	37.5	26.6	84.0	15.5	64	64	6	146	115	5.0	4.4	11.93
29	.267	184.3	84.68	40.3	29.0	93.2	8.5	64	76	2	213	109	5.5	5.0	8.55
30	.271	181.0	73.78	42.8	29.7	90.3	8.5	56	88	11	181	109	6.0	5.0	4.94
31	.264	180.2	75.84	41.4	28.7	88.1	13.5	76	76	9	192	114	5.5	3.6	10.62
32	.357	184.1	70.48	42.0	28.9	81.3	14.0	84	72	5	231	123	5.5	4.5	8.46

Data Set A.6. Anthropometric and Physical Fitness Measurements on Police Department Applicants (Cont'd)

No.	REACT	HEIGHT	WEIGHT	SHLDR	PELVIC	CHEST	THIGH	PULSE	DIAST	CHNUP	BREATH	RECVR	SPEED	ENDUR	FAT
33	.259	178.9	86.90	42.5	28.7	95.0	16.0	54	68	12	186	118	6.0	4.0	13.47
34	.221	170.0	76.68	39.7	27.7	93.6	15.0	50	72	4	178	108	5.5	4.5	12.81
35	.333	180.6	77.32	42.1	27.3	89.5	16.0	88	72	11	200	119	5.5	4.6	13.34
36	.359	179.0	79.90	40.8	28.2	90.3	26.5	80	80	3	201	124	5.5	3.7	24.57
37	.314	186.6	100.36	42.5	31.5	100.3	27.0	62	76	2	208	120	5.5	4.1	28.35
38	.295	181.4	91.66	41.9	28.9	96.6	25.5	68	78	2	211	125	6.0	3.0	26.12
39	.296	176.5	79.00	40.7	29.1	86.5	20.5	60	66	5	210	117	5.5	4.2	15.21
40	.308	174.0	69.10	40.9	27.0	88.1	18.0	92	74	5	161	140	5.0	5.5	12.51
41	.327	178.2	87.78	42.9	27.2	100.3	16.5	72	72	4	189	115	5.5	3.5	20.50
42	.303	177.1	70.18	39.4	27.6	85.5	16.0	72	74	14	201	110	6.0	4.8	10.67
43	.297	180.0	67.66	40.9	28.7	86.1	15.0	76	76	5	177	110	5.5	4.5	10.76
44	.244	176.8	86.12	41.3	28.2	92.7	12.5	76	68	7	181	110	5.5	4.0	14.55
45	.282	176.3	65.00	39.0	26.0	83.3	7.0	88	72	12	167	127	5.5	5.0	5.27
46	.285	192.4	99.14	43.7	28.7	96.1	20.5	64	68	4	174	105	6.0	4.0	17.94
47	.299	175.2	75.70	39.4	27.3	90.8	19.0	56	76	7	174	111	5.5	4.5	12.64
48	.280	175.9	78.62	43.4	29.3	90.7	18.0	64	72	7	170	117	5.5	3.7	10.81
49	.268	174.6	64.88	42.3	29.2	82.6	3.5	72	80	11	199	113	6.0	4.5	2.01
50	.362	179.0	71.00	41.2	27.3	85.6	16.0	68	90	5	150	108	5.5	5.0	10.00

Source: Dr. Ladislav P. Novak, Department of Anthropology, Southern Methodist University. Reprinted with permission.

Data Set A.7. Housing Rent Study
Sample Correlations

	ACR	RELLL	RES15	RES510	RESG10	AREAPR	LTOTRMS	LTOTARE	CENH
ACR	1.000								
RELLL	-.119	1.000							
RES15	-.066	.019	1.000						
RES510	-.153	.120	-.261	1.000					
RESG10	-.158	-.033	-.152	-.071	1.000				
AREAPR	.356	-.035	.070	.037	-.067	1.000			
LTOTRMS	.537	.057	.051	-.037	-.079	.243	1.000		
LTOTARE	.580	.022	.071	-.002	-.101	.729	.833	1.000	
CENH	.572	-.049	-.037	-.060	-.109	.198	.152	.217	1.000
BADH	-.540	.020	.002	.037	.100	-.091	-.178	-.180	-.411
STOREF	.237	-.046	-.037	-.178	-.139	-.164	-.037	-.189	.267
APPL	.368	-.029	-.010	-.063	-.020	.133	.079	.134	.353
QUAL2	.592	-.026	-.082	-.051	-.096	.030	.092	.096	.502
MULTI5	.241	-.070	-.126	-.048	.008	.062	-.132	-.047	.334
CENFO1	.416	-.056	-.010	-.065	-.013	.060	.009	.046	.415
CENFO2	.146	.034	-.050	-.041	.008	.034	.167	.121	.092
CENFO3	-.251	-.006	.018	.053	.090	.098	-.133	-.043	-.066
CENF13	.075	.051	-.031	-.025	-.035	.013	.046	.032	.188

	BADH	STOREF	APPL	QUAL2	MULTI5	CENFO1	CENFO2	CENFO3	CNHF13
BADH	1.000								
STOREF	-.289	1.000							
APPL	-.215	.096	1.000						
QUAL2	-.680	.340	.311	1.000					
MULTI5	-.181	.208	.383	.283	1.000				
CENFO1	-.353	.236	.283	.473	.217	1.000			
CENFO2	-.037	-.074	-.049	.015	-.083	.038	1.000		
CENFO3	.308	-.243	.012	-.323	-.036	.016	.004	1.000	
CNHF13	-.065	.086	-.027	.093	-.032	.162	.279	-.051	1.000

Data Set A.7. Housing Rent Study (Cont'd)
Sample Means and Standard Deviations

	ACR	RELLL	RES15	RES10	RESG10	AREAPR	LTOTRMS	LTOTARE	CENH
Mean	129.8718	.0264	.3604	.1077	.0396	109.9253	1.5990	6.2794	.3033
Std. Dev.	44.3974	.1604	.4807	.3103	.1951	20.5918	.2396	.3539	.4602
$n = 455$									

	BADH	STOREF	APPL	QUAL2	MULTI5	CENFO1	CENFO2	CENFO3	CNHF13
Mean	.2791	.7824	.1451	2.2973	.1517	.1705	.0390	.0013	.1171
Std. Dev.	.4491	.4131	.3525	.6364	.3591	.9923	.9878	.9945	1.0906
$n = 455$									

Source: Merrill (1977). Reprinted with permission of the Department of Housing and Urban Development, Washington, D. C.

Data Set A.8. Body Measurements on Police Department Applicants

No.	HEIGHT	SITHT	UARM	FORE	HAND	ULEG	LLEG	FOOT	BRACH	TIBIO
1	165.8	88.7	31.8	28.1	18.7	40.3	38.9	6.7	88.36	96.53
2	169.8	90.0	32.4	29.1	18.3	43.3	42.7	6.4	89.81	98.61
3	170.7	87.7	33.6	29.5	20.7	43.7	41.1	7.2	87.80	94.05
4	170.9	87.1	31.0	28.2	18.6	43.7	40.6	6.7	90.97	92.91
5	157.5	81.3	32.1	27.3	17.5	38.1	39.6	6.6	85.05	103.94
6	165.9	88.2	31.8	29.0	18.6	42.0	40.6	6.5	91.19	96.67
7	158.7	86.1	30.6	27.8	18.4	40.0	37.0	5.9	90.85	92.50
8	166.0	88.7	30.2	26.9	17.5	41.6	39.0	5.9	89.07	93.75
9	158.7	83.7	31.1	27.1	18.3	38.9	37.5	6.1	87.14	96.40
10	161.5	81.2	32.3	27.8	19.1	42.8	40.1	6.2	86.07	93.69
11	167.3	88.6	34.8	27.3	18.3	43.1	41.8	7.3	78.45	96.98
12	167.4	83.2	34.3	30.1	19.2	43.4	42.2	6.8	87.76	97.24
13	159.2	81.5	31.0	27.3	17.5	39.8	39.6	4.9	88.06	99.50
14	170.0	87.9	34.2	30.9	19.4	43.1	43.7	6.3	90.35	101.39
15	166.3	88.3	30.6	28.8	18.3	41.8	41.0	5.9	94.12	98.09
16	169.0	85.6	32.6	28.8	19.1	42.7	42.0	6.0	88.34	98.36
17	156.2	81.6	31.0	25.6	17.0	44.2	39.0	5.1	82.58	88.24
18	159.6	86.6	32.7	25.4	17.7	42.0	37.5	5.0	77.68	89.29
19	155.0	82.0	30.3	26.6	17.3	37.9	36.1	5.2	87.79	95.25
20	161.1	84.1	29.5	26.6	17.8	38.6	38.2	5.9	90.17	98.96
21	170.3	88.1	34.0	29.3	18.2	43.2	41.4	5.9	86.18	95.83
22	167.8	83.9	32.5	28.6	20.2	43.3	42.9	7.2	88.00	99.08
23	163.1	88.1	31.7	26.9	18.1	40.1	39.0	5.9	84.86	97.26
24	165.8	87.0	33.2	26.3	19.5	43.2	40.7	5.9	79.22	94.21
25	175.4	89.6	35.2	30.1	19.1	45.1	44.5	6.3	85.51	98.67
26	159.8	85.6	31.5	27.1	19.2	42.3	39.0	5.7	86.03	92.20
27	166.0	84.9	30.5	28.1	17.8	41.2	43.0	6.1	92.13	104.37
28	161.2	84.1	32.8	29.2	18.4	42.6	41.1	5.9	89.02	96.48
29	160.4	84.3	30.5	27.8	16.8	41.0	39.8	6.0	91.15	97.07
30	164.3	85.0	35.0	27.8	19.0	47.2	42.4	5.0	79.43	89.83
31	165.5	82.6	36.2	28.6	20.2	45.0	42.3	5.6	79.01	94.00
32	167.2	85.0	33.6	27.1	19.8	46.0	41.6	5.6	80.65	90.43
33	167.2	83.4	33.5	29.7	19.4	45.2	44.0	5.2	88.66	97.35

Source: Dr. Ladislav P. Novak, Department of Anthropology, Southern Methodist University. Reprinted with permission.

APPENDIX B

DATA SETS FOR FURTHER STUDY

B.1 MORTALITY AND POLLUTION STUDY

This study (McDonald and Schwing, 1973) examines the effect of air pollution, environmental, demographic, and socioeconomic variables on mortality rates. The data consist of observations on 15 predictor variables and the response, mortality rate, for sixty Standard Metropolitan Statistical Areas of the United States. One purpose of the study was to determine whether the pollution variables (HC, NOX, SO2) were influential in predicting the response once the other predictor variables were included in the prediction equation. The regression variables are:

MORT — Total Age Adjusted Mortality Rate (deaths per 100,000 population)

PREC — Average Annual Precipitation (inches)

JANT — Average January Temperature (degrees F.)

JULT — Average July Temperature (degrees F.)

OVR65 — Percent of 1960 SMSA Population 65 Years Old or Older

POPN — Population per Household, 1960

EDUC — Median School Years Completed for Those Over 22

HOUS — Percent of Housing Units Which are Sound with All Facilities

DENS — Population per Square Mile in Urbanized Areas, 1960

NONW — Percent of 1960 Urbanized Area Population Which is Non-white

WWDRK — Percent Employed in White Collar Occupations in 1960
POOR   — Percent of Families with Income Under $3000
HC   — Relative Pollution Potential of Hydrocarbons
NOX   — Relative Pollution Potential of Oxides of Nitrogen
SO2   — Relative Pollution of Sulfur Dioxide
HUMID   — Percent Relative Humidity, Annual Average at 1 p.m.

B.2   SOLID WASTE DATA

This report studies waste production and land use. The authors simulate waste production from economic models and compile information on a variety of variables affecting the amount of solid waste produced in a region. Measurements on several land use variables are included in the data set for 40 regions in a nine-county area of California. One goal in an analysis of this data set is to assess the predictive ability of the land use variables on the amount of solid waste that is generated from the economic models. Variables included in the data set are:

WASTE   — Solid Waste (millions of tons)
INDUS   — Industrial Land (acres)
METAL   — Fabricated Metals (acres)
WHOLE — Trucking and Wholesale Trade (acres)
RETAIL — Retail Trade (acres)
REST   — Restaurants and Hotels (acres)
FINAN   — Finance, Insurance, and Real Estate (acres)
MISC   — Miscellaneous Services (acres)
HOME   — Residential Land (acres).

Data Set B.1 Mortality and Pollution Study

No.	PREC	JANT	JULT	OVR65	POPN	EDUC	HOUS	DENS	NONW	WWDRK	POOR	HC	NOX	SO2	HUMID	MORT
1	36	27	71	8.1	3.34	11.4	81.5	3,243	8.8	42.6	11.7	21	15	59	59	921.870
2	35	23	72	11.1	3.14	11.0	78.8	4,281	3.5	50.7	14.4	8	10	39	57	997.875
3	44	29	74	10.4	3.21	9.8	81.6	4,260	.8	39.4	12.4	6	6	33	54	962.354
4	47	45	79	6.5	3.41	11.1	77.5	3,125	27.1	50.2	20.6	18	8	24	56	982.291
5	43	35	77	7.6	3.44	9.6	84.6	6,441	24.4	43.7	14.3	43	38	206	55	1,071.289
6	53	45	80	7.7	3.45	10.2	66.8	3,325	38.5	43.1	25.5	30	32	72	54	1,030.380
7	43	30	74	10.9	3.23	12.1	83.9	4,679	3.5	49.2	11.3	21	32	62	56	934.700
8	45	30	73	9.3	3.29	10.6	86.0	2,140	5.3	40.4	10.5	6	4	4	56	899.529
9	36	24	70	9.0	3.31	10.5	83.2	6,582	8.1	42.5	12.6	18	12	37	61	1,001.902
10	36	27	72	9.5	3.36	10.7	79.3	4,213	6.7	41.0	13.2	12	7	20	59	912.347
11	52	42	79	7.7	3.39	9.6	69.2	2,302	22.2	41.3	24.2	18	8	27	56	1,017.613
12	33	26	76	8.6	3.20	10.9	83.4	6,122	16.3	44.9	10.7	88	63	278	58	1,024.885
13	40	34	77	9.2	3.21	10.2	77.0	4,101	13.0	45.7	15.1	26	26	146	57	970.467
14	35	28	71	8.8	3.29	11.1	86.3	3,042	14.7	44.6	11.4	31	21	64	60	985.950
15	37	31	75	8.0	3.26	11.9	78.4	4,259	13.1	49.6	13.9	23	9	15	58	958.839
16	35	46	85	7.1	3.22	11.8	79.9	1,441	14.8	51.2	16.1	1	1	1	54	860.101
17	36	30	75	7.5	3.35	11.4	81.9	4,029	12.4	44.0	12.0	6	4	16	58	936.234
18	15	30	73	8.2	3.15	12.2	84.2	4,824	4.7	53.1	12.7	17	8	28	38	871.766
19	31	27	74	7.2	3.44	10.8	87.0	4,834	15.8	43.5	13.6	52	35	124	59	959.221
20	30	24	72	6.5	3.53	10.8	79.5	3,694	13.1	33.8	12.4	11	4	11	61	941.181
21	31	45	85	7.3	3.22	11.4	80.7	1,844	11.5	48.1	18.5	1	1	1	53	891.708
22	31	24	72	9.0	3.37	10.9	82.8	3,226	5.1	45.2	12.3	5	3	10	61	871.338
23	42	40	77	6.1	3.45	10.4	71.8	2,269	22.7	41.4	19.5	8	3	5	53	971.122
24	43	27	72	9.0	3.25	11.5	87.1	2,909	7.2	51.6	9.5	7	3	10	56	887.466
25	46	55	84	5.6	3.35	11.4	79.7	2,647	21.0	46.9	17.9	6	5	1	59	952.529
26	39	29	75	8.7	3.23	11.4	78.6	4,412	15.6	46.6	13.2	13	7	33	60	968.665
27	35	31	81	9.2	3.10	12.0	78.3	3,262	12.6	48.6	13.9	7	4	4	55	919.729
28	43	32	74	10.1	3.38	9.5	79.2	3,214	2.9	43.7	12.0	11	7	32	54	844.053
29	11	53	68	9.2	2.99	12.1	90.6	4,700	7.8	48.9	12.3	648	319	130	47	861.833
30	30	35	71	8.3	3.37	9.9	77.4	4,474	13.1	42.6	17.7	38	37	193	57	989.265
31	50	42	82	7.3	3.49	10.4	72.5	3,497	36.7	43.3	26.4	15	18	34	59	1,006.490

No.	PREC	JANT	JULT	OVR65	POPN	EDUC	HOUS	DENS	NONW	WWDRK	POOR	HC	NOX	SO2	HUMID	MORT
32	60	67	82	10.0	2.98	11.5	88.6	4,657	13.5	47.3	22.4	3	1	1	60	861.439
33	30	20	69	8.8	3.26	11.1	85.4	2,934	5.8	44.0	9.4	33	23	125	64	929.150
34	25	12	73	9.2	3.28	12.1	83.1	2,095	2.0	51.9	9.8	20	11	26	58	857.622
35	45	40	80	8.3	3.32	10.1	70.3	2,682	21.0	46.1	24.1	17	14	78	56	961.009
36	46	30	72	10.2	3.16	11.3	83.2	3,327	8.8	45.3	12.2	4	3	8	58	923.234
37	54	54	81	7.4	3.36	9.7	72.8	3,172	31.4	45.5	24.2	20	17	1	62	1,113.156
38	42	33	77	9.7	3.03	10.7	83.5	7,462	11.3	48.7	12.4	41	26	108	58	994.648
39	42	32	76	9.1	3.32	10.5	87.5	6,092	17.5	45.3	13.2	29	32	161	54	1,015.023
40	36	29	72	9.5	3.32	10.6	77.6	3,437	8.1	45.5	13.8	45	59	263	56	991.290
41	37	38	67	11.3	2.99	12.0	81.5	3,387	3.6	50.3	13.5	56	21	44	73	893.991
42	42	29	72	10.7	3.19	10.1	79.5	3,508	2.2	38.8	15.7	6	4	18	56	938.500
43	41	33	77	11.2	3.08	9.6	79.9	4,843	2.7	38.6	14.1	11	11	89	54	946.185
44	44	39	78	8.2	3.32	11.0	79.9	3,768	28.6	49.5	17.5	12	9	48	53	1,025.502
45	32	25	72	10.9	3.21	11.1	82.5	4,355	5.0	46.4	10.8	7	4	18	60	874.281
46	34	32	79	9.3	3.23	9.7	76.8	5,160	17.2	45.1	15.3	31	15	68	57	953.560
47	10	55	70	7.3	3.11	12.1	88.9	3,033	5.9	51.0	14.0	144	66	20	61	839.709
48	18	48	63	9.2	2.92	12.2	87.7	4,253	13.7	51.2	12.0	311	171	86	71	911.701
49	13	49	68	7.0	3.36	12.2	90.7	2,702	3.0	51.9	9.7	105	32	3	71	790.733
50	35	40	64	9.6	3.02	12.2	82.5	3,626	5.7	54.3	10.1	20	7	20	72	899.264
51	45	28	74	10.6	3.21	11.1	82.6	1,883	3.4	41.9	12.3	5	4	20	56	904.155
52	38	24	72	9.8	3.34	11.4	78.0	4,923	3.8	50.5	11.1	8	5	25	61	950.672
53	31	26	73	9.3	3.22	10.7	81.3	3,249	9.5	43.9	13.6	11	7	25	59	972.464
54	40	23	71	11.3	3.28	10.3	73.8	1,671	2.5	47.4	13.5	5	2	11	60	912.202
55	41	37	78	6.2	3.25	12.3	89.5	5,308	25.9	59.7	10.3	65	28	102	52	967.803
56	28	32	81	7.0	3.27	12.1	81.0	3,665	7.5	51.6	13.2	4	2	1	54	823.764
57	45	33	76	7.7	3.39	11.3	82.2	3,152	12.1	47.3	10.9	14	11	42	56	1,003.502
58	45	24	70	11.8	3.25	11.1	79.8	3,678	1.0	44.8	14.0	7	3	8	56	895.696
59	42	33	76	9.7	3.22	9.0	76.2	9,699	4.8	42.2	14.5	8	8	49	54	911.817
60	38	28	72	8.9	3.48	10.7	79.8	3,451	11.7	37.5	13.0	14	13	39	58	954.442

Source: McDonald and Ayers (1978). Reprinted with permission of Dr. Gary C. McDonald.

Data Set B.2 Solid Waste Data

No.	WASTE	INDUS	METAL	WHOLE	RETAIL	REST	FINAN	MISC	HOME
1	.3574	102	69	133	125	36	53	106	5,326
2	1.9673	1,220	723	2,612	953	132	252	574	36,138
3	.1862	139	138	46	35	6	9	27	2,594
4	.3816	221	637	153	115	16	41	83	10,346
5	.1512	12	0	1	9	1	2	7	381
6	.1449	1	50	3	25	2	7	7	3,440
7	.4711	1,046	127	313	392	56	62	212	14,559
8	.6512	2,032	44	409	540	98	106	289	40,182
9	.6624	895	54	168	117	32	20	76	4,600
10	.3457	0	0	2	0	1	0	7	1,540
11	.3355	25	2	24	78	15	25	50	6,516
12	.3982	97	12	91	135	24	28	59	6,011
13	.2044	1	0	15	46	11	35	40	1,922
14	.2969	4	1	18	23	8	13	28	931
15	1.1515	42	4	78	41	61	85	49	538
16	.5609	87	162	599	11	3	3	23	199
17	.1104	2	0	26	24	6	11	50	419
18	.0863	2	9	29	11	2	8	25	1,093
19	.1952	48	18	101	25	4	24	8	328
20	.1688	131	126	387	6	0	2	4	797
21	.0786	4	0	103	49	9	18	59	1,855
22	.0955	1	4	46	16	2	16	24	1,572
23	.0486	0	0	468	56	2	11	21	998
24	.0867	7	0	52	37	5	24	43	2,804
25	.1403	5	1	6	95	11	39	29	2,879
26	.3786	174	113	685	69	18	11	43	6,810
27	.0761	0	0	6	35	4	9	23	3,242
28	.8927	233	153	682	404	85	133	270	26,013
29	.3621	155	56	94	75	17	32	66	4,469
30	.1758	120	74	55	120	8	12	49	2,436
31	.2699	8,983	37	236	77	38	18	49	4,400
32	.2762	59	54	138	55	11	13	18	3,351
33	.3240	72	112	169	228	39	38	99	13,979
34	.3737	571	78	254	162	43	46	147	20,138
35	.9114	853	1,002	1,017	418	57	116	182	19,356
36	.2594	5	0	17	14	13	3	8	1,050
37	.4284	11	34	3	20	4	5	14	2,207
38	.1905	258	1	33	48	13	18	28	810
39	.2341	69	14	126	108	20	95	41	1,017
40	.7759	4,790	2,046	3,719	31	7	6	18	962

Source: Golueke and McGauhey (1970).

APPENDIX C

STATISTICAL TABLES

The distributions tabulated in this appendix include the cumulative standard normal, cumulative Student t, cumulative chi-square, cumulative F, cumulative runs tables, and tables for the Durbin-Watson test. Each has application in different areas of regression methodology. For more extensive tables the reader should consult standard statistical references [e.g., Pearson, E. S. and Hartley, H. O. (1969)].

C.1 CUMULATIVE STANDARD NORMAL DISTRIBUTION

Contained in this table are cumulative probabilities for the standard ($\mu = 0$, $\sigma = 1$) normal distribution. The row and column headings record values of the normal random variable to two decimal places and the body of the table lists the corresponding probability. Thus, for example $P(Z \leqslant 1.83) = 0.9664$ and because of the symmetry of the standard normal distribution $P(Z \leqslant -1.83) = 1 - .9664 = 0.0336$.

Suppose now that one wishes to test $H_0:\beta_j = 0$ vs. $H_a:\beta_j \neq 0$ with a significance level of $\gamma = 0.05$. From Table 6.4, one rejects H_0 if $Z < -Z_{.975} = -1.96$ or $Z > Z_{.975} = 1.96$, where $Z = \hat{\beta}_j/\sigma C_{jj}^{1/2}$ and the standard deviation σ is assumed known. To test $H_0:\beta_j \leqslant 0$ vs. $H_a:\beta_j > 0$ one would reject H_0 if $Z > Z_{.95} = 1.645$. When σ is known, these percentage points can also be used to construct confidence intervals on β_j as is shown in Table 6.10.

C.2. CUMULATIVE STUDENT t DISTRIBUTION

Since the student t distribution is a function of the degrees of freedom, comprehensive tables of percentage points for each degree of freedom are not presented. Instead, values of the t distribution corresponding to some selected probabilities are displayed. For example, if t is a Student t random variable with 8 degrees of freedom, $P(t \leqslant 2.306) = 1 - .025 = 0.975$. Because of the symmetry of the t distribution, $P(t \leqslant -2.306) = 0.025$. Confidence intervals and tests of hypothesis for regression parameters when the error variance is unknown can be performed using these tabulated t values and the statistics in Tables 6.10 and 6.4, respectively. A one-sided test is performed by choosing either the upper or the lower tail value of the t distribution corresponding to a cumulative probability of $1 - \gamma$ or γ, respectively, as with the above example for the normal distribution.

C.3 CUMULATIVE CHI-SQUARE DISTRIBUTION

As with the t distribution, the chi-square distribution depends on the number of degrees of freedom associated with the statistic. Since the chi-square distribution is not symmetric, this table contains chi-square values corresponding to cumulative probabilities of $1 - \gamma$ for selected values of γ. Thus, if u is a chi-square random variable with 15 degrees of freedom, $P(u \leqslant 8.55) = 1 - .900 = 0.100$ and $P(u \leqslant 25.00) = 1 - .050 = 0.950$. This distribution can be used with the statistics in Table 6.4 and 6.10 to test hypotheses and find confidence intervals on the error variance, σ^2.

C.4 CUMULATIVE F DISTRIBUTION

The F distribution depends on two degrees of freedom. If F is written as $F = u/v$, where u is a chi-square random variable divided by its degrees of freedom v_1 (i.e. $u = \chi^2(v_1)/(v_1)$) and v is a chi-square random variable (independently distributed of u) divided by its degrees of freedom v_2 (i.e., $v = \chi^2(v_2)/v_2$), then u/v is an F random variable with v_1 and v_2 degrees of freedom. For example, to test $H_0 : \underline{\beta} = \underline{0}$ vs. $H_a : \underline{\beta} \neq \underline{0}$, eqn. (6.3.1) points out that MSR/MSE is an F random variable with $v_1 = p$ and $v_2 = (n - 1)$ degrees of freedom. This follows since SSR$/\sigma^2 \sim \chi^2(p)$ if H_0 is true and SSE$/\sigma^2 \sim \chi^2(n - p - 1)$ and is independent of SSR. Table C.4 a, b, and c contain F values corresponding to cumulative probabilities $1 - \gamma = 0.75$, 0.90, and 0.95, respectively.

C.5   CRITICAL VALUES OF r IN THE RUNS TEST

This table presents values for the number of runs that would be regarded as significant at the .05 (one-tail) level. For example, if an analysis consists of a set of data for which large positive correlations among the errors are expected, one would anticipate few runs since the large positive correlations suggest that a low value of ε will be followed by several more low values and similarly for large values of ε. One would then consult Table C.5a to determine whether too few runs occur in the residuals. If large negative correlations are suspected, many runs would be anticipated since small errors should be followed by large ones and vice-versa. Table C.5b should be consulted in such cases. If one does not know which type of correlation to expect, both tables can be consulted but the overall significance level of the test doubles to $\gamma = 0.10$.

C.6   CRITICAL VALUES FOR THE DURBIN-WATSON TEST

Contained in this table are critical values for the Durbin-Watson test for first order serial correlations among the errors in a regression model. The table gives lower bounds (d_L) and upper bounds (d_U) for the test $H_0 : \varrho \leqslant 0$ vs. $H_a : \varrho > 0$ as a function of the sample size n and the number of predictor variables p. If d in eqn. (7.2.3) is less than d_L, H_0 is rejected while if $d > d_U$ it is not. A test of $H_0 : \varrho \leqslant 0$ vs. $H_a : \varrho > 0$ can be performed in exactly the same fashion using $d' = 4 - d$.

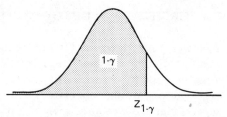

Table C.1. Cumulative Standard Normal Distribution

Z	.00	.01	.02	.03	.04	.05	.06	.07	.08	.09
0.0	.5000	.5040	.5080	.5120	.5160	.5199	.5239	.5279	.5319	.5359
0.1	.5398	.5438	.5478	.5517	.5557	.5596	.5636	.5675	.5714	.5753
0.2	.5793	.5832	.5871	.5910	.5948	.5987	.6026	.6064	.6103	.6141
0.3	.6179	.6217	.6255	.6293	.6331	.6368	.6406	.6443	.6480	.6517
0.4	.6554	.6591	.6628	.6664	.6700	.6736	.6772	.6808	.6844	.6879
0.5	.6915	.6950	.6985	.7019	.7054	.7088	.7123	.7157	.7190	.7224
0.6	.7257	.7291	.7324	.7357	.7389	.7422	.7454	.7486	.7517	.7549
0.7	.7580	.7611	.7642	.7673	.7704	.7734	.7764	.7794	.7823	.7852
0.8	.7881	.7910	.7939	.7967	.7995	.8023	.8051	.8078	.8106	.8133
0.9	.8159	.8186	.8212	.8238	.8264	.8289	.8315	.8340	.8365	.8389
1.0	.8413	.8438	.8461	.8485	.8508	.8531	.8554	.8577	.8599	.8621
1.1	.8643	.8665	.8686	.8708	.8729	.8749	.8770	.8790	.8810	.8830
1.2	.8849	.8869	.8888	.8907	.8925	.8944	.8962	.8980	.8997	.9015
1.3	.9032	.9049	.9066	.9082	.9099	.9115	.9131	.9147	.9162	.9177
1.4	.9182	.9207	.9222	.9236	.9251	.9265	.9279	.9292	.9306	.9319
1.5	.9332	.9345	.9357	.9370	.9382	.9394	.9406	.9418	.9429	.9441
1.6	.9452	.9463	.9474	.9484	.9495	.9505	.9515	.9525	.9535	.9545
1.7	.9554	.9564	.9573	.9582	.9591	.9599	.9608	.9616	.9625	.9633
1.8	.9641	.9649	.9656	.9664	.9671	.9678	.9686	.9693	.9699	.9706
1.9	.9713	.9719	.9726	.9732	.9738	.9744	.9750	.9756	.9761	.9767
2.0	.9772	.9778	.9783	.9788	.9793	.9798	.9803	.9808	.9812	.9817
2.1	.9821	.9826	.9830	.9834	.9838	.9842	.9846	.9850	.9854	.9857
2.2	.9861	.9864	.9868	.9871	.9875	.9878	.9881	.9884	.9887	.9890
2.3	.9893	.9896	.9898	.9901	.9904	.9906	.9909	.9911	.9913	.9916
2.4	.9918	.9920	.9922	.9925	.9927	.9929	.9931	.9932	.9934	.9936
2.5	.9938	.9940	.9941	.9943	.9945	.9946	.9948	.9949	.9951	.9952
2.6	.9953	.9955	.9956	.9957	.9959	.9960	.9961	.9962	.9963	.9964
2.7	.9965	.9966	.9967	.9968	.9969	.9970	.9971	.9972	.9973	.9974
2.8	.9974	.9975	.9976	.9977	.9977	.9978	.9979	.9979	.9980	.9981
2.9	.9981	.9982	.9982	.9983	.9984	.9984	.9985	.9985	.9986	.9986
3.0	.9987	.9987	.9987	.9988	.9988	.9989	.9989	.9989	.9990	.9990
3.1	.9990	.9991	.9991	.9991	.9992	.9992	.9992	.9992	.9993	.9993
3.2	.9993	.9993	.9994	.9994	.9994	.9994	.9994	.9995	.9995	.9995
3.3	.9995	.9995	.9995	.9996	.9996	.9996	.9996	.9996	.9996	.9997
3.4	.9997	.9997	.9997	.9997	.9997	.9997	.9997	.9997	.9997	.9998
3.5	.9998	.9998	.9998	.9998	.9998	.9998	.9998	.9998	.9998	.9998
3.6	.9998	.9998	.9999	.9999	.9999	.9999	.9999	.9999	.9999	.9999
3.7	.9999	.9999	.9999	.9999	.9999	.9999	.9999	.9999	.9999	.9999
3.8	.9999	.9999	.9999	.9999	.9999	.9999	.9999	.9999	.9999	.9999
3.9	1.0000	1.0000	1.0000	1.0000	1.0000	1.0000	1.0000	1.0000	1.0000	1.0000

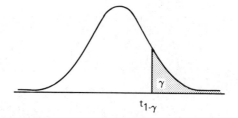

Table C.2. Cumulative Student t Distribution

d.f.	$\gamma = .25$	$\gamma = .10$	$\gamma = .05$	$\gamma = .025$	$\gamma = .01$	$\gamma = .005$
1	1.000	3.078	6.314	12.706	31.821	63.657
2	0.816	1.886	2.920	4.303	6.965	9.925
3	.765	1.638	2.353	3.182	4.541	5.841
4	.741	1.533	2.132	2.776	3.747	4.604
5	0.727	1.476	2.015	2.571	3.365	4.032
6	.718	1.440	1.943	2.447	3.143	3.707
7	.711	1.415	1.895	2.365	2.998	3.499
8	.706	1.397	1.860	2.306	2.896	3.355
9	.703	1.383	1.833	2.262	2.821	3.250
10	0.700	1.372	1.812	2.228	2.764	3.169
11	.697	1.363	1.796	2.201	2.718	3.106
12	.695	1.356	1.782	2.179	2.681	3.055
13	.694	1.350	1.771	2.160	2.650	3.012
14	.692	1.345	1.761	2.145	2.624	2.977
15	0.691	1.341	1.753	2.131	2.602	2.947
16	.690	1.337	1.746	2.120	2.583	2.921
17	.689	1.333	1.740	2.110	2.567	2.898
18	.688	1.330	1.734	2.101	2.552	2.878
19	.688	1.328	1.729	2.093	2.539	2.861
20	0.687	1.325	1.725	2.086	2.528	2.845
21	.686	1.323	1.721	2.080	2.518	2.831
22	.686	1.321	1.717	2.074	2.508	2.819
23	.685	1.319	1.714	2.069	2.500	2.807
24	.685	1.318	1.711	2.064	2.492	2.797
25	0.684	1.316	1.708	2.060	2.485	2.787
26	.684	1.315	1.706	2.056	2.479	2.779
27	.684	1.314	1.703	2.052	2.473	2.771
28	.683	1.313	1.701	2.048	2.467	2.763
29	.683	1.311	1.699	2.045	2.462	2.756
30	0.683	1.310	1.697	2.042	2.457	2.750
60	.679	1.296	1.671	2.000	2.390	2.660
90	.678	1.291	1.662	1.987	2.368	2.632
120	.677	1.289	1.658	1.980	2.358	2.617
∞	.674	1.282	1.645	1.960	2.326	2.576

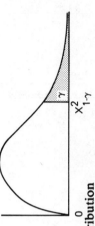

Table C.3. Cumulative Chi-Square Distribution

γ = Probability of a Greater Value

d.f.	0.995	0.990	0.975	0.950	0.900	0.750	0.500	0.250	0.100	0.050	0.025	0.010	0.005	0.001
1					0.02	0.10	0.45	1.32	2.71	3.84	5.02	6.63	7.88	10.83
2	0.01	0.02	0.05	0.10	0.21	0.58	1.39	2.77	4.61	5.99	7.38	9.21	10.60	13.82
3	0.07	0.11	0.22	0.35	0.58	1.21	2.37	4.11	6.25	7.81	9.35	11.34	12.84	16.27
4	0.21	0.30	0.48	0.71	1.06	1.92	3.36	5.39	7.78	9.49	11.14	13.28	14.86	18.47
5	0.41	0.55	0.83	1.15	1.61	2.67	4.35	6.63	9.24	11.07	12.83	15.09	16.75	20.52
6	0.68	0.87	1.24	1.64	2.20	3.45	5.35	7.84	10.64	12.59	14.45	16.81	18.55	22.46
7	0.99	1.24	1.69	2.17	2.83	4.25	6.35	9.04	12.02	14.07	16.01	18.48	20.28	24.32
8	1.34	1.65	2.18	2.73	3.49	5.07	7.34	10.22	13.36	15.51	17.53	20.09	21.96	26.12
9	1.73	2.09	2.70	3.33	4.17	5.90	8.34	11.39	14.68	16.92	19.02	21.67	23.59	27.88
10	2.16	2.56	3.25	3.94	4.87	6.74	9.34	12.55	15.99	18.31	20.48	23.21	25.19	29.59
11	2.60	3.05	3.82	4.57	5.58	7.58	10.34	13.70	17.28	19.68	21.92	24.72	26.76	31.26
12	3.07	3.57	4.40	5.23	6.30	8.44	11.34	14.85	18.55	21.03	23.34	26.22	28.30	32.91
13	3.57	4.11	5.01	5.89	7.04	9.30	12.34	15.98	19.81	22.36	24.74	27.69	29.82	34.53
14	4.07	4.66	5.63	6.57	7.79	10.17	13.34	17.12	21.06	23.68	26.12	29.14	31.32	36.12
15	4.60	5.23	6.26	7.26	8.55	11.04	14.34	18.25	22.31	25.00	27.49	30.58	32.80	37.70
16	5.14	5.81	6.91	7.96	9.31	11.91	15.34	19.37	23.54	26.30	28.85	32.00	34.27	39.25
17	5.70	6.41	7.56	8.67	10.09	12.79	16.34	20.49	24.77	27.59	30.19	33.41	35.73	40.79
18	6.26	7.01	8.23	9.39	10.86	13.68	17.34	21.60	25.99	28.87	31.53	34.81	37.16	42.31
19	6.84	7.63	8.91	10.12	11.65	14.56	18.34	22.72	27.20	30.14	32.85	36.19	38.58	43.82
20	7.43	8.26	9.59	10.85	12.44	15.45	19.34	23.83	28.41	31.41	34.17	37.57	40.00	45.32

Table C.3. Cumulative Chi-Square Distribution, Cont'd

γ = Probability of a Greater Value

d.f.	0.995	0.990	0.975	0.950	0.900	0.750	0.500	0.250	0.100	0.050	0.025	0.010	0.005	0.001
21	8.03	8.90	10.28	11.59	13.24	16.34	20.34	24.93	29.62	32.67	35.48	38.93	41.40	46.80
22	8.64	9.54	10.98	12.34	14.04	17.24	21.34	26.04	30.81	33.92	36.78	40.29	42.80	48.27
23	9.26	10.20	11.69	13.09	14.85	18.14	22.34	27.14	32.01	35.17	38.08	41.46	44.18	49.73
24	9.89	10.86	12.40	13.85	15.66	19.04	23.34	28.24	33.20	36.42	39.36	42.98	45.56	51.18
25	10.52	11.52	13.12	14.61	16.47	19.94	24.34	29.34	34.38	37.65	40.65	44.31	46.93	52.62
26	11.16	12.20	13.84	15.38	17.29	20.84	25.34	30.43	35.56	38.89	41.92	45.64	48.29	54.05
27	11.81	12.88	14.57	16.15	18.11	21.75	26.34	31.53	36.74	40.11	43.19	46.96	49.64	55.48
28	12.46	13.56	15.31	16.93	18.94	22.66	27.34	32.62	37.92	41.34	44.64	48.28	50.99	56.89
29	13.12	14.26	16.05	17.71	19.77	23.57	28.34	33.71	39.09	42.56	45.72	49.59	52.34	58.30
30	13.79	14.95	16.79	18.49	20.60	24.48	29.34	34.80	40.26	43.77	46.98	50.89	53.67	59.70
40	20.71	22.16	24.43	26.51	29.05	33.66	39.34	45.62	51.80	55.76	59.34	63.69	66.77	73.40
50	27.99	29.71	32.36	34.76	37.69	42.94	49.33	56.33	63.17	67.50	71.42	76.15	79.49	86.66
60	35.53	37.48	40.48	43.19	46.46	52.29	59.33	66.98	74.40	79.08	83.30	88.38	91.95	99.61
70	43.28	45.44	48.76	51.74	55.33	61.70	69.33	77.58	85.53	90.53	95.02	100.42	104.22	112.32
80	51.17	53.54	57.15	60.39	64.28	71.14	79.33	88.13	96.58	101.88	106.63	112.33	116.32	124.84
90	59.20	61.75	65.65	69.13	73.29	80.62	89.33	98.64	107.56	113.14	118.14	124.12	128.30	137.21
100	67.33	70.06	74.22	77.93	82.36	90.13	99.33	109.14	118.50	124.34	129.56	135.81	140.17	149.45

Source: Pearson and Hartley (1970). © 1970, Biometrika Trustees, University College London. Reprinted with permission.

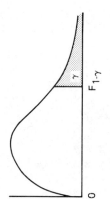

Table C.4a. Cumulative F Distribution

$(\gamma = 0.25)$

ν_2 \ ν_1	1	2	3	4	5	6	7	8	9	10	12	15	20	24	30	40	60	120	∞
1	5.83	7.50	8.20	8.58	8.82	8.98	9.10	9.19	9.26	9.32	9.41	9.49	9.58	9.63	9.67	9.71	9.76	9.80	9.85
2	2.57	3.00	3.15	3.23	3.28	3.31	3.34	3.35	3.37	3.38	3.39	3.41	3.43	3.43	3.44	3.45	3.46	3.47	3.48
3	2.02	2.28	2.36	2.39	2.41	2.42	2.43	2.44	2.44	2.44	2.45	2.46	2.46	2.46	2.47	2.47	2.47	2.47	2.47
4	1.81	2.00	2.05	2.06	2.07	2.08	2.08	2.08	2.08	2.08	2.08	2.08	2.08	2.08	2.08	2.08	2.08	2.08	2.08
5	1.69	1.85	1.88	1.89	1.89	1.89	1.89	1.89	1.89	1.89	1.89	1.89	1.88	1.88	1.88	1.88	1.87	1.87	1.87
6	1.62	1.76	1.78	1.79	1.79	1.78	1.78	1.78	1.77	1.77	1.77	1.76	1.76	1.75	1.75	1.75	1.74	1.74	1.74
7	1.57	1.70	1.72	1.72	1.71	1.71	1.70	1.70	1.69	1.69	1.68	1.68	1.67	1.67	1.66	1.66	1.65	1.65	1.65
8	1.54	1.66	1.67	1.66	1.66	1.65	1.64	1.64	1.63	1.63	1.62	1.62	1.61	1.60	1.60	1.59	1.59	1.58	1.58
9	1.51	1.62	1.63	1.63	1.62	1.61	1.60	1.60	1.59	1.59	1.58	1.57	1.56	1.56	1.55	1.54	1.54	1.53	1.53
10	1.49	1.60	1.60	1.59	1.59	1.58	1.57	1.56	1.56	1.55	1.54	1.53	1.52	1.52	1.51	1.51	1.50	1.49	1.48
11	1.47	1.58	1.58	1.57	1.56	1.55	1.54	1.53	1.53	1.52	1.51	1.50	1.49	1.49	1.48	1.47	1.47	1.46	1.45
12	1.46	1.56	1.56	1.55	1.54	1.53	1.52	1.51	1.51	1.50	1.49	1.48	1.47	1.46	1.45	1.45	1.44	1.43	1.42
13	1.45	1.55	1.55	1.53	1.52	1.51	1.50	1.49	1.49	1.48	1.47	1.46	1.45	1.44	1.43	1.42	1.42	1.41	1.40
14	1.44	1.53	1.53	1.52	1.51	1.50	1.49	1.48	1.47	1.46	1.45	1.44	1.43	1.42	1.41	1.41	1.40	1.39	1.38
15	1.43	1.52	1.52	1.51	1.49	1.48	1.47	1.46	1.46	1.45	1.44	1.43	1.41	1.41	1.40	1.39	1.38	1.37	1.36
16	1.42	1.51	1.51	1.50	1.48	1.47	1.46	1.45	1.44	1.44	1.43	1.41	1.40	1.39	1.38	1.37	1.36	1.35	1.34
17	1.42	1.51	1.50	1.49	1.47	1.46	1.45	1.44	1.43	1.43	1.41	1.40	1.39	1.38	1.37	1.36	1.35	1.34	1.33
18	1.41	1.50	1.49	1.48	1.46	1.45	1.44	1.43	1.42	1.42	1.40	1.39	1.38	1.37	1.36	1.35	1.34	1.33	1.32
19	1.41	1.49	1.49	1.47	1.46	1.44	1.43	1.42	1.41	1.41	1.40	1.38	1.37	1.36	1.35	1.34	1.33	1.32	1.30

Table C.4a. Cumulative F Distribution, Cont'd

(γ = 0.25)

ν_2 \ ν_1	1	2	3	4	5	6	7	8	9	10	12	15	20	24	30	40	60	120	∞
20	1.40	1.49	1.48	1.47	1.45	1.44	1.43	1.42	1.41	1.40	1.39	1.37	1.36	1.35	1.34	1.33	1.32	1.31	1.29
21	1.40	1.48	1.48	1.46	1.44	1.43	1.42	1.41	1.40	1.39	1.38	1.37	1.35	1.34	1.33	1.32	1.31	1.30	1.28
22	1.40	1.48	1.47	1.45	1.44	1.42	1.41	1.40	1.39	1.39	1.37	1.36	1.34	1.33	1.32	1.31	1.30	1.29	1.28
23	1.39	1.47	1.47	1.45	1.43	1.42	1.41	1.40	1.39	1.38	1.37	1.35	1.34	1.33	1.32	1.31	1.30	1.28	1.27
24	1.39	1.47	1.46	1.44	1.43	1.41	1.40	1.39	1.38	1.38	1.36	1.35	1.33	1.32	1.31	1.30	1.29	1.28	1.26
25	1.39	1.47	1.46	1.44	1.42	1.41	1.40	1.39	1.38	1.37	1.36	1.34	1.33	1.32	1.31	1.29	1.28	1.27	1.25
26	1.38	1.46	1.45	1.44	1.42	1.41	1.39	1.38	1.37	1.37	1.35	1.34	1.32	1.31	1.30	1.29	1.28	1.26	1.25
27	1.38	1.46	1.45	1.43	1.42	1.40	1.39	1.38	1.37	1.36	1.35	1.33	1.32	1.31	1.30	1.28	1.27	1.26	1.24
28	1.38	1.46	1.45	1.43	1.41	1.40	1.39	1.38	1.37	1.36	1.34	1.33	1.31	1.30	1.29	1.28	1.27	1.25	1.24
29	1.38	1.45	1.45	1.43	1.41	1.40	1.38	1.37	1.36	1.35	1.34	1.32	1.31	1.30	1.29	1.27	1.26	1.25	1.23
30	1.38	1.45	1.44	1.42	1.41	1.39	1.38	1.37	1.36	1.35	1.34	1.32	1.30	1.29	1.28	1.27	1.26	1.24	1.23
40	1.36	1.44	1.42	1.40	1.39	1.37	1.36	1.35	1.34	1.33	1.31	1.30	1.28	1.26	1.25	1.24	1.22	1.21	1.19
60	1.35	1.42	1.41	1.38	1.37	1.35	1.33	1.32	1.31	1.30	1.29	1.27	1.25	1.24	1.22	1.21	1.19	1.17	1.15
120	1.34	1.40	1.39	1.37	1.35	1.33	1.31	1.30	1.29	1.28	1.26	1.24	1.22	1.21	1.19	1.18	1.16	1.13	1.10
∞	1.32	1.39	1.37	1.35	1.33	1.31	1.29	1.28	1.27	1.25	1.24	1.22	1.19	1.18	1.16	1.14	1.12	1.08	1.00

Table C.4b. Cumulative F Distribution
$(\gamma = 0.10)$

$\nu_2 \backslash \nu_1$	1	2	3	4	5	6	7	8	9	10	12	15	20	24	30	40	60	120	∞
1	39.86	49.50	53.59	55.83	57.24	58.20	58.91	59.44	59.86	60.19	60.17	61.22	61.74	62.00	62.26	62.53	62.79	63.06	63.33
2	8.53	9.00	9.16	9.24	9.29	9.33	9.35	9.37	9.38	9.39	9.41	9.42	9.44	9.45	9.46	9.47	9.47	9.48	9.48
3	5.54	5.46	5.39	5.34	5.31	5.28	5.27	5.25	5.24	5.23	5.22	5.20	5.18	5.18	5.17	5.16	5.15	5.14	5.14
4	4.54	4.32	4.19	4.11	4.05	4.01	3.98	3.95	3.94	3.92	3.90	3.87	3.84	3.83	3.82	3.80	3.79	3.78	3.76
5	4.06	3.78	3.62	3.52	3.45	3.40	3.37	3.34	3.32	3.30	3.27	3.24	3.21	3.19	3.17	3.16	3.14	3.12	3.11
6	3.78	3.46	3.29	3.18	3.11	3.05	3.01	2.98	2.96	2.94	2.90	2.87	2.84	2.82	2.80	2.78	2.76	2.74	2.72
7	3.59	3.26	3.07	2.96	2.88	2.83	2.78	2.75	2.72	2.70	2.67	2.63	2.59	2.58	2.56	2.54	2.51	2.49	2.47
8	3.46	3.11	2.92	2.81	2.73	2.67	2.62	2.59	2.56	2.54	2.50	2.46	2.42	2.40	2.38	2.36	2.34	2.32	2.29
9	3.36	3.01	2.81	2.69	2.61	2.55	2.51	2.47	2.44	2.42	2.38	2.34	2.30	2.28	2.25	2.23	2.21	2.18	2.16
10	3.29	2.92	2.73	2.61	2.52	2.46	2.41	2.38	2.35	2.32	2.28	2.24	2.20	2.18	2.16	2.13	2.11	2.08	2.06
11	3.23	2.86	2.66	2.54	2.45	2.39	2.34	2.30	2.27	2.25	2.21	2.17	2.12	2.10	2.08	2.05	2.03	2.00	1.97
12	3.18	2.81	2.61	2.48	2.39	2.33	2.28	2.24	2.21	2.19	2.15	2.10	2.06	2.04	2.01	1.99	1.96	1.93	1.90
13	3.14	2.76	2.56	2.43	2.35	2.28	2.23	2.20	2.16	2.14	2.10	2.05	2.01	1.98	1.96	1.93	1.90	1.88	1.85
14	3.10	2.73	2.52	2.39	2.31	2.24	2.19	2.15	2.12	2.10	2.05	2.01	1.96	1.94	1.91	1.89	1.86	1.83	1.80
15	3.07	2.70	2.49	2.36	2.27	2.21	2.16	2.12	2.09	2.06	2.02	1.97	1.92	1.90	1.87	1.85	1.82	1.79	1.76
16	3.05	2.67	2.46	2.33	2.24	2.18	2.13	2.09	2.06	2.03	1.99	1.94	1.89	1.87	1.84	1.81	1.78	1.75	1.72
17	3.03	2.64	2.44	2.31	2.22	2.15	2.10	2.06	2.03	2.00	1.96	1.91	1.86	1.84	1.81	1.78	1.75	1.72	1.69
18	3.01	2.62	2.42	2.29	2.20	2.13	2.08	2.04	2.00	1.98	1.93	1.89	1.84	1.81	1.78	1.75	1.72	1.69	1.66
19	2.99	2.61	2.40	2.27	2.18	2.11	2.06	2.02	1.98	1.96	1.91	1.86	1.81	1.79	1.76	1.73	1.70	1.67	1.63

Table C.4b. Cumulative F Distribution, Cont'd

($\gamma = 0.10$)

ν_2 \ ν_1	1	2	3	4	5	6	7	8	9	10	12	15	20	24	30	40	60	120	∞
20	2.97	2.59	2.38	2.25	2.16	2.09	2.04	2.00	1.96	1.94	1.89	1.84	1.79	1.77	1.74	1.71	1.68	1.64	1.61
21	2.96	2.57	2.36	2.23	2.14	2.08	2.02	1.98	1.95	1.92	1.87	1.83	1.78	1.75	1.72	1.69	1.66	1.62	1.59
22	2.95	2.56	2.35	2.22	2.13	2.06	2.01	1.97	1.93	1.90	1.86	1.81	1.76	1.73	1.70	1.67	1.64	1.60	1.57
23	2.94	2.55	2.34	2.21	2.11	2.05	1.99	1.95	1.92	1.89	1.84	1.80	1.74	1.72	1.69	1.66	1.62	1.59	1.55
24	2.93	2.54	2.33	2.19	2.10	2.04	1.98	1.94	1.91	1.88	1.83	1.78	1.73	1.70	1.67	1.64	1.61	1.57	1.53
25	2.92	2.53	2.32	2.18	2.09	2.02	1.97	1.93	1.89	1.87	1.82	1.77	1.72	1.69	1.66	1.63	1.59	1.56	1.52
26	2.91	2.52	2.31	2.17	2.08	2.01	1.96	1.92	1.88	1.86	1.81	1.76	1.71	1.68	1.65	1.61	1.58	1.54	1.50
27	2.90	2.51	2.30	2.17	2.07	2.00	1.95	1.91	1.87	1.85	1.80	1.75	1.70	1.67	1.64	1.60	1.57	1.53	1.49
28	2.89	2.50	2.29	2.16	2.06	2.00	1.94	1.90	1.87	1.84	1.79	1.74	1.69	1.66	1.63	1.59	1.56	1.52	1.48
29	2.89	2.50	2.28	2.15	2.06	1.99	1.93	1.89	1.86	1.83	1.78	1.73	1.68	1.65	1.62	1.58	1.55	1.51	1.47
30	2.88	2.49	2.28	2.14	2.05	1.98	1.93	1.88	1.85	1.82	1.77	1.72	1.67	1.64	1.61	1.57	1.54	1.50	1.46
40	2.84	2.44	2.23	2.09	2.00	1.93	1.87	1.83	1.79	1.76	1.71	1.66	1.61	1.57	1.54	1.51	1.47	1.42	1.38
60	2.79	2.39	2.18	2.04	1.95	1.87	1.82	1.77	1.74	1.71	1.66	1.60	1.54	1.51	1.48	1.44	1.40	1.35	1.29
120	2.75	2.35	2.13	1.99	1.90	1.82	1.77	1.72	1.68	1.65	1.60	1.55	1.48	1.45	1.41	1.37	1.32	1.26	1.19
∞	2.71	2.30	2.08	1.94	1.85	1.77	1.72	1.67	1.63	1.60	1.55	1.49	1.42	1.38	1.34	1.30	1.24	1.17	1.00

Source: Pearson and Hartley (1970). © 1970, Biometrika Trustees, University College London. Reprinted with permission.

Table C.4c. Cumulative F Distribution

$(\gamma = 0.05)$

ν_2 \ ν_1	1	2	3	4	5	6	7	8	9	10	12	15	20	24	30	40	60	120	∞
1	161.00	200.00	216.00	225.00	230.00	234.00	237.00	239.00	241.00	242.00	244.00	246.00	248.00	249.00	250.00	251.00	252.00	253.00	254.00
2	18.51	19.00	19.16	19.25	19.30	19.33	19.35	19.37	19.38	19.40	19.41	19.43	19.45	19.45	19.46	19.47	19.48	19.49	19.50
3	10.13	9.55	9.28	9.12	9.01	8.94	8.89	8.85	8.81	8.79	8.74	8.70	8.66	8.64	8.62	8.59	8.57	8.55	8.53
4	7.71	6.94	6.59	6.39	6.26	6.16	6.09	6.04	6.00	5.96	5.91	5.86	5.80	5.77	5.75	5.72	5.69	5.66	5.63
5	6.61	5.79	5.41	5.19	5.05	4.95	4.88	4.82	4.77	4.74	4.68	4.62	4.56	4.53	4.50	4.46	4.43	4.40	4.36
6	5.99	5.14	4.76	4.53	4.39	4.28	4.21	4.15	4.10	4.06	4.00	3.94	3.87	3.84	3.81	3.77	3.74	3.70	3.67
7	5.59	4.74	4.35	4.12	3.97	3.87	3.79	3.73	3.68	3.64	3.57	3.51	3.44	3.41	3.38	3.34	3.30	3.27	3.23
8	5.32	4.46	4.07	3.84	3.69	3.58	3.50	3.44	3.39	3.35	3.28	3.22	3.15	3.12	3.08	3.04	3.01	2.97	2.93
9	5.12	4.26	3.86	3.63	3.48	3.37	3.29	3.23	3.18	3.14	3.07	3.01	2.94	2.90	2.86	2.83	2.79	2.75	2.71
10	4.96	4.10	3.71	3.48	3.33	3.22	3.14	3.07	3.02	2.98	2.91	2.85	2.77	2.74	2.70	2.66	2.62	2.58	2.54
11	4.84	3.98	3.59	3.36	3.20	3.09	3.01	2.95	2.90	2.85	2.79	2.72	2.65	2.61	2.57	2.53	2.49	2.45	2.40
12	4.75	3.89	3.49	3.26	3.11	3.00	2.91	2.85	2.80	2.75	2.69	2.62	2.54	2.51	2.47	2.43	2.38	2.34	2.30
13	4.67	3.81	3.41	3.18	3.03	2.92	2.83	2.77	2.71	2.67	2.60	2.53	2.46	2.42	2.38	2.34	2.30	2.25	2.21
14	4.60	3.74	3.34	3.11	2.96	2.85	2.76	2.70	2.65	2.60	2.53	2.46	2.39	2.35	2.31	2.27	2.22	2.18	2.13
15	4.54	3.68	3.29	3.06	2.90	2.79	2.71	2.64	2.59	2.54	2.48	2.40	2.33	2.29	2.25	2.20	2.16	2.11	2.07
16	4.49	3.63	3.24	3.01	2.85	2.74	2.66	2.59	2.54	2.49	2.42	2.35	2.28	2.24	2.19	2.15	2.11	2.06	2.01
17	4.45	3.59	3.20	2.96	2.81	2.70	2.61	2.55	2.49	2.45	2.38	2.31	2.23	2.19	2.15	2.10	2.06	2.01	1.96
18	4.41	3.55	3.16	2.93	2.77	2.66	2.58	2.51	2.46	2.41	2.34	2.27	2.19	2.15	2.11	2.06	2.02	1.97	1.92
19	4.38	3.52	3.13	2.90	2.74	2.63	2.54	2.48	2.42	2.38	2.31	2.23	2.16	2.11	2.07	2.03	1.98	1.93	1.88

Table C.4c. Cumulative F Distribution, Cont'd

$(\gamma = 0.05)$

ν_2 \ ν_1	1	2	3	4	5	6	7	8	9	10	12	15	20	24	30	40	60	120	∞
20	4.35	3.49	3.10	2.87	2.71	2.60	2.51	2.45	2.39	2.35	2.28	2.20	2.12	2.08	2.04	1.99	1.95	1.90	1.84
21	4.32	3.47	3.07	2.84	2.68	2.57	2.49	2.42	2.37	2.32	2.25	2.18	2.10	2.05	2.01	1.96	1.92	1.87	1.81
22	4.30	3.44	3.05	2.82	2.66	2.55	2.46	2.40	2.34	2.30	2.23	2.15	2.07	2.03	1.98	1.94	1.89	1.84	1.78
23	4.28	3.42	3.03	2.80	2.64	2.53	2.44	2.37	2.32	2.27	2.20	2.13	2.05	2.01	1.96	1.91	1.86	1.81	1.76
24	4.26	3.40	3.01	2.78	2.62	2.51	2.42	2.36	2.30	2.25	2.18	2.11	2.03	1.98	1.94	1.89	1.84	1.79	1.73
25	4.24	3.39	2.99	2.76	2.60	2.49	2.40	2.34	2.28	2.24	2.16	2.09	2.01	1.96	1.92	1.87	1.82	1.77	1.71
26	4.23	3.37	2.98	2.74	2.59	2.47	2.39	2.32	2.27	2.22	2.15	2.07	1.99	1.95	1.90	1.85	1.80	1.75	1.69
27	4.21	3.35	2.96	2.73	2.57	2.46	2.37	2.31	2.25	2.20	2.13	2.06	1.97	1.93	1.88	1.84	1.79	1.73	1.67
28	4.20	3.34	2.95	2.71	2.56	2.45	2.36	2.29	2.24	2.18	2.12	2.04	1.96	1.91	1.87	1.82	1.77	1.71	1.65
29	4.18	3.33	2.93	2.70	2.55	2.43	2.35	2.28	2.22	2.18	2.10	2.03	1.94	1.90	1.85	1.81	1.75	1.70	1.64
30	4.17	3.32	2.92	2.69	2.53	2.42	2.33	2.27	2.21	2.16	2.09	2.01	1.93	1.89	1.84	1.79	1.74	1.68	1.62
40	4.08	3.23	2.84	2.61	2.45	2.34	2.25	2.18	2.12	2.08	2.00	1.92	1.84	1.79	1.74	1.69	1.64	1.58	1.51
60	4.00	3.15	2.76	2.53	2.37	2.25	2.17	2.10	2.04	1.99	1.92	1.84	1.75	1.70	1.65	1.59	1.53	1.47	1.39
120	3.92	3.07	2.68	2.45	2.29	2.17	2.09	2.02	1.96	1.91	1.83	1.75	1.66	1.61	1.55	1.50	1.43	1.35	1.25
∞	3.84	3.00	2.60	2.37	2.21	2.10	2.01	1.94	1.88	1.83	1.75	1.67	1.57	1.52	1.46	1.39	1.32	1.22	1.00

Source: Pearson and Hartley (1970). © 1970, Biometrika Trustees, University College London. Reprinted with permission.

TABLE C.5a Lower Critical Values of r in the Runs Test*

$(\gamma = .05)$

n_1 \ n_2	2	3	4	5	6	7	8	9	10	11	12	13	14	15	16	17	18	19	20
2											2	2	2	2	2	2	2	2	2
3					2	2	2	2	2	2	2	2	2	3	3	3	3	3	3
4				2	2	2	3	3	3	3	3	3	3	3	4	4	4	4	4
5			2	2	3	3	3	3	3	4	4	4	4	4	4	4	5	5	5
6		2	2	3	3	3	3	4	4	4	4	5	5	5	5	5	5	6	6
7		2	2	3	3	3	4	4	5	5	5	5	5	6	6	6	6	6	6
8		2	3	3	3	4	4	5	5	5	6	6	6	6	6	7	7	7	7
9		2	3	3	4	4	5	5	5	6	6	6	7	7	7	7	8	8	8
10		2	3	3	4	5	5	5	6	6	7	7	7	7	8	8	8	8	9
11		2	3	4	4	5	5	6	6	7	7	7	8	8	8	9	9	9	9
12	2	2	3	4	4	5	6	6	7	8	8	9	9	9	9	9	9	10	10
13	2	2	3	4	5	5	6	6	7	7	8	8	9	9	9	10	10	10	10
14	2	2	3	4	5	5	6	7	7	8	8	8	8	8	10	10	10	11	11
15	2	3	3	4	5	6	6	7	7	8	8	9	9	10	10	11	11	11	12
16	2	3	4	4	5	6	6	7	8	8	9	9	10	10	11	11	11	12	12
17	2	3	4	4	5	6	7	7	8	9	9	10	10	11	11	11	12	12	12
18	2	3	4	5	5	6	7	8	8	9	9	10	10	11	11	12	12	13	13
19	2	3	4	5	6	6	7	8	8	9	10	10	11	11	12	12	13	13	13
20	2	3	4	5	6	6	7	8	9	9	10	10	11	12	12	13	13	13	14

*Any value of r that is equal to or smaller than that shown in the body of this table for given value of n_1 and n_2 is significant at the 0.05 level.

Source: Swed, F. S. and Eisenhart, C. (1943). ©1943 by Institute of Mathematical Statistics. Reprinted with permission.

Table C.5b Upper Critical Values of r in the Runs Test*

$(\gamma = .05)$

n_1 \ n_2	2	3	4	5	6	7	8	9	10	11	12	13	14	15	16	17	18	19	20
2																			
3																			
4				9	9														
5			9	10	10	11	11												
6			9	10	11	12	12	13	13	13	13								
7				11	12	13	13	14	14	14	14	15	15	15					
8				11	12	13	14	14	15	15	16	16	16	16	17	17	17	17	17
9					13	14	14	15	16	16	16	17	17	18	18	18	18	18	18
10					13	14	15	16	16	17	17	18	18	18	19	19	19	20	20
11					13	14	15	16	17	17	18	19	19	19	20	20	20	21	21
12					13	14	16	16	17	18	19	19	20	20	21	21	21	22	22
13						15	16	17	18	19	19	20	20	21	21	22	22	23	23
14						15	16	17	18	19	20	20	21	22	22	23	23	23	24
15						15	16	18	18	19	20	21	22	22	23	23	24	24	25
16							17	18	19	20	21	21	22	23	23	24	25	25	25
17							17	18	19	20	21	22	23	23	24	25	25	26	26
18							17	18	19	20	21	22	23	24	25	25	26	26	27
19							17	18	20	21	22	23	23	24	25	26	26	27	27
20							17	18	20	21	22	23	24	25	25	26	27	27	28

*Any value of r that is equal to or smaller than that shown in the body of this table for given values of n_1 and n_2 is significant at the 0.05 level.

Source: Swed, F. S. and Eisenhart, C. (1943). © 1943 by Institute of Mathematical Statistics. Reprinted with permission.

Table C.6 Critical Values for Durbin-Watson Test Statistic
($\gamma = 0.05$)

n	p = 1		p = 2		p = 3		p = 4		p = 5	
n	d_L	d_U	d_L	d_U	d_L	d_U	d_L	d_U	d_L	d_U
15	1.08	1.36	0.95	1.54	0.82	1.75	0.69	1.97	0.56	2.21
16	1.10	1.37	0.98	1.54	0.86	1.73	0.74	1.93	0.62	2.15
17	1.13	1.38	1.02	1.54	0.90	1.71	0.78	1.90	0.67	2.10
18	1.16	1.39	1.05	1.53	0.93	1.69	0.82	1.87	0.71	2.06
19	1.18	1.40	1.08	1.53	0.97	1.68	0.86	1.85	0.75	2.02
20	1.20	1.41	1.10	1.54	1.00	1.68	0.90	1.83	0.79	1.99
21	1.22	1.42	1.13	1.54	1.03	1.67	0.93	1.81	0.83	1.96
22	1.24	1.43	1.15	1.54	1.05	1.66	0.96	1.80	0.86	1.94
23	1.26	1.44	1.17	1.54	1.08	1.66	0.99	1.79	0.90	1.92
24	1.27	1.45	1.19	1.55	1.10	1.66	1.01	1.78	0.93	1.90
25	1.29	1.45	1.21	1.55	1.12	1.66	1.04	1.77	0.95	1.89
26	1.30	1.46	1.22	1.55	1.14	1.65	1.06	1.76	0.98	1.88
27	1.32	1.47	1.24	1.56	1.16	1.65	1.08	1.76	1.01	1.86
28	1.33	1.48	1.26	1.56	1.18	1.65	1.10	1.75	1.03	1.85
29	1.34	1.48	1.27	1.56	1.20	1.65	1.12	1.74	1.05	1.84
30	1.35	1.49	1.28	1.57	1.21	1.65	1.14	1.74	1.07	1.83
31	1.36	1.50	1.30	1.57	1.23	1.65	1.16	1.74	1.09	1.83
32	1.37	1.50	1.31	1.57	1.24	1.65	1.18	1.73	1.11	1.82
33	1.38	1.51	1.32	1.58	1.26	1.65	1.19	1.73	1.13	1.81
34	1.39	1.51	1.33	1.58	1.27	1.65	1.21	1.73	1.15	1.81
35	1.40	1.52	1.34	1.58	1.28	1.65	1.22	1.73	1.16	1.80
36	1.41	1.52	1.35	1.59	1.29	1.65	1.24	1.73	1.18	1.80
37	1.42	1.53	1.36	1.59	1.31	1.66	1.25	1.72	1.19	1.80
38	1.43	1.54	1.37	1.59	1.32	1.66	1.26	1.72	1.21	1.79
39	1.43	1.54	1.38	1.60	1.33	1.66	1.27	1.72	1.22	1.79
40	1.44	1.54	1.39	1.60	1.34	1.66	1.29	1.72	1.23	1.79
45	1.48	1.57	1.43	1.62	1.38	1.67	1.34	1.72	1.29	1.78
50	1.50	1.59	1.46	1.63	1.42	1.67	1.38	1.72	1.34	1.77
55	1.53	1.60	1.49	1.64	1.45	1.68	1.41	1.72	1.38	1.77
60	1.55	1.62	1.51	1.65	1.48	1.69	1.44	1.73	1.41	1.77
65	1.57	1.63	1.54	1.66	1.50	1.70	1.47	1.73	1.44	1.77
70	1.58	1.64	1.55	1.67	1.52	1.70	1.49	1.74	1.46	1.77
75	1.60	1.65	1.57	1.68	1.54	1.71	1.51	1.74	1.49	1.77
80	1.61	1.66	1.59	1.69	1.56	1.72	1.53	1.74	1.51	1.77
85	1.62	1.67	1.60	1.70	1.57	1.72	1.55	1.75	1.52	1.77
90	1.63	1.68	1.61	1.70	1.59	1.73	1.57	1.75	1.54	1.78
95	1.64	1.69	1.62	1.71	1.60	1.73	1.58	1.75	1.56	1.78
100	1.65	1.69	1.63	1.73	1.61	1.74	1.59	1.76	1.57	1.78

BIBLIOGRAPHY

Afifi, A. A. and Elashoff, R. M. (1967). "Missing Observations in Multivariate Statistics II. Point Estimation in Simple Linear Regression," *Journal of the American Statistical Association*, **62**, 10-29.

Anderson, T. W. (1958). **An Introduction to Multivariate Statistical Analysis**. New York: John Wiley and Sons, Inc.

Anscombe, F. J. (1960). "Rejection of Outliers," *Technometrics*, **5**, 123-47.

Anscombe, F. J. (1973). "Graphs in Statistical Analysis," *The American Statistician*, **27**, 17-22.

Anscombe, F. J. and Tukey, J. W. (1963). "The Examination and Analysis of Residuals," *Technometrics*, **2**, 141-60.

Behnken, D. W., and Draper, N. R. (1972). "Residuals and Their Variance Patterns," *Technometrics*, **14**, 101-11.

Bendel, R. B. and Afifi, A. A. (1977). "Comparison of Stopping Rules in Forward Stepwise Regression," *Journal of the American Statistical Association*, **72**, 46-53.

Blalock, H. M. (1963). "Correlated Independent Variables: The Problem of Multicollinearity," *Social Forces*, **42**, 233-7.

Boardman, T. J. and Bryson, M. C. (1978). "A Review of Some Smoothing and Forecasting Techniques," *Journal of Quality Technology*, **10**, 1-11.

Bock, M. E., Yancey, T. A., and Judge, G. G. (1973). "Statistical Consequences of Preliminary Test Estimators in Regression," *Journal of the American Statistical Association*, **68**, 109-16.

Bohrnstedt, G. W., and Carter, T. M. (1971). "Robustness in Regression Analysis," in H. L. Costner (ed.) **Sociological Methodology 1971**. San Francisco: Jossey-Bass, Inc.

Box, G. E. P. (1953). "Non-Normality and Tests on Variances," *Biometrika*, **40**, 318-335.

Chatterjee, S. and Price, B. (1977). **Regression Analysis By Example**. New York: John Wiley & Sons, Inc.

Coale, A. J. and Zelnik, M. (1963). **New Estimates of Fertility and Population in the United States**. Princeton, New Jersey: Princeton University Press.

Cochran, W. G. (1947). "Some Consequences When the Assumptions for the Analysis of Variance are not Satisfied," *Biometrics*, **3**, 22-38.

Cochran, W. G., Mosteller, F., and Tukey, J. (1954). **Statistical Problems of the Kinsey Report**. Washington, D. C.: The American Statistical Association.

Cohen, J. and Cohen, P. (1975). **Applied Multiple Regression/Correlations Analysis for the Behavioral Sciences**. New York: John Wiley and Sons, Inc. (Halstead Press Division).

Coleman, J. S., Campbell, E. Q., Hobson, C. J., McPartland, J., Mood, A. M., Weinfield, F. D., and York, R. L. (1966). **Equality of Educational Opportunity**. Washington, D. C.: U. S. Department of Health, Education and Welfare, Office of Education.

Cook, R. D. (1977). "Detection of Influential Observations in Linear Regression," *Technometrics*, **19**, 15-8.

Cox, D. R. and Snell, E. J. (1974). "The Choice of Variables in Observational Studies," *Applied Statistics*, **23**, 51-9.

Crane, D. (1965). "Scientists at Major and Minor Universities: A Study of Productivity and Recognition," *American Sociological Review*, **30**, 699-714.

Daniel, C. (1959). "Use of Half-Normal Plots in Interpreting Factorial Two-Level Experiments," *Technometrics*, **1**, 311-41.

Daniel, C. and Wood, F. S. (1971). **Fitting Equations to Data**. New York: John Wiley and Sons, Inc.

Davies, O. L. (1967). **Design and Analysis of Industrial Experiments**. New York: Hafner Publishing Co.

Dempster, A. P., Schatzoff, M., and Wermuth, N. (1977). "A Simulation Study of Alternatives to Ordinary Least Squares," *Journal of the American Association*, **72**, 77-91.

Draper, N. R. and Smith, H. (1966). **Applied Regression Analysis**. New York: John Wiley & Sons, Inc.

Durbin, J. and Watson, G. C. (1950). "Testing For Serial Correlation in Least Squares Regression. I." *Biometrika*, **37**, 409-28.

Durbin, J. and Watson, G. S. (1951). "Testing For Serial Correlation in Least Squares Regression. II." *Biometrika*, **38**, 159-78.

Durbin, J. and Watson, G. S. (1971). "Testing For Serial Correlation in Least Squares Regression. III." *Biometrika*, **58**, 1-19.

Ehrenberg, A. S. C. (1968). "The Elements of Lawlike Relationships," *Journal of the Royal Statistical Society*, Series A, **131**, 280-302.

Ehrenberg, A. S. C. (1975). "Children Heights and Weights in 1905," *Journal of the Royal Statistical Society*, Series A, **138**, 239-41.

Elashoff, R. M. and Afifi, A. A. (1966). "Missing Values in Multivariate Statistics I. Review of the Literature," *Journal of the American Statistical Association*, **61**, 595-604.

Elashoff, R. M. and Afifi, A. A. (1969a). "Missing Values in Multivariate Statistics III," *Journal of the American Statistical Association*, **64**, 337-58.

Elashoff, R. M. and Afifi, A. A. (1969b). "Missing Values in Multivariate Statistics IV," *Journal of the American Statistical Association*, **64**, 359-65.

Farrar, D. E. and Glauber, R. R. (1967). "Multicollinearity in Regression Analysis: The Problem Revisited," *Review of Economics and Statistics*, **49**, 92-107.

Fisher, J. C. (1976). "Homicide In Detroit: The Role of Firearms," *Criminology*, **14**, 387-400.

Furnival, G. M. and Wilson, R. W., Jr. (1974). "Regression By Leaps and Bounds," *Technometrics*, **16**, 499-512.

Gibbs, J. P. (1969). "Martial Status and Suicide in the United States, A Special Test of the Status Integration Theory," *American Journal of Sociology*, **74**, 521-33.

Glejser, H. (1969). "A New Test for Heteroscedasticity," *Journal of the American Statistical Association*, **64**, 316-23.

Gnanadesikan, R. (1977). **Methods for Statistical Data Analysis of Multivariate Observations**. New York: John Wiley and Sons, Inc.

Goldfeld, S. M. and Quandt, R. E. (1965). "Some Tests for Homoscedasticity," *Journal of the American Statistical Association*, **60**, 539-47.

Golueke, C. G. and McGauhey, P. H. (1970). **Comprehensive Studies of Solid Waste Management**, U. S. Department of Health Education, and Welfare: Public Health Services Publication No. 2039.

Gorman, J. W. and Toman, R. J. (1966). "Selection of Variables For Fitting Equations to Data," *Technometrics*, **8**, 27-51.

Graybill, F. A. (1976). **Theory and Application of the Linear Model**. North Scituate, Mass.: Duxbury Press.

Gunst, R. F., Webster, J. T., and Mason, R. L. (1976). "A Comparison of Least Squares and Latent Root Regression Estimators," *Technometrics*, **18**, 75-83.

Gunst, R. F. and Mason, R. L. (1977a). "Advantages of Examining Multicollinearities in Regression Analysis," *Biometrics*, **33**, 249-260.

Gunst, R. F. and Mason, R. L. (1977b). "Biased Estimation in Regression: An Evaluation Using Mean Squared Error," *Journal of the American Statistical Association*, **72**, 616-28.

Hadley, G. (1961). **Linear Algebra**. Reading, Mass.: Addision-Wesley Publishing Co.

Hamaker, H. C. (1962). "On Multiple Regression Analysis," *atistica Neerlandica*, **16**, 31-56.

Hare, C. T. and Bradow, R. L. (1977). "Light Duty Diesel Emission Correction Factors for Ambient Conditions," Paper No. 770717, Society of Automotive Engineers Off-Highway Vehicle Meeting, MECCA Milwaukee, Sept. 12-15, 1977.

Hare, C. T. (1977). "Light Duty Diesel Emission Correction Factors For Ambient Conditions." Final Report to the Environmental Protection Agency under Contract No. 68-02-1777, April, 1977.

Hawkins, D. M. (1973). "On the Investigation of Alternative Regression by Principal Component Analysis," *Applied Statistics*, **22**, 275-86.

Hoaglin, D. C. and Welsch, R. F. (1978). "The Hat Matrix in Regression and ANOVA," *The American Statistician*, **32**, 17-22.

Hocking, R. R. and Leslie, R. N. (1967). "Selection of the Best Subset in Regression Analysis," *Technometrics*, **9**, 531-40.

Hocking, R. R. (1972). "Criteria For Selection of a Subset Regression: Which One Should Be Used," *Technometrics*, **14**, 967-70.

Hocking, R. R. (1976). "The Analysis and Selection of Variables in Linear Regression," *Biometrics*, **32**, 1-49.

Hocking, R. R., Speed, F. M., and Lynn, M. T. (1976). "A Class of Biased Estimators in Linear Regression," *Technometrics*, **18**, 425-88.

Hoerl, A. E. (1954). "Fitting Curves to Data," **Chemical Business Handbook** (J. H. Perry, ed). New York: McGraw-Hill Book Company, Inc.

Hoerl, A. E. and Kennard, R. W. (1970a). "Ridge Regression Biased Estimation for Nonorthogonal Problems," *Technometrics*, **12**, 55-67.

Hoerl, A. E. and Kennard, R. W. (1970b). "Ridge Regression: Applications to Nonorthogonal Problems," *Technometrics*, **12**, 69-82.

Kennedy, W. J. and Bancroft, T. A. (1971). 'Model-Building for Prediction in Regression Based on Repeated Significance Tests," *Annals of Mathematical Statistics*, **42**, 1273-84.

Kinsey, A. C., Pomeroy, W. B., and Martin, C. E. (1948). **Sexual Behavior in the Human Male**. Philadelphia: Saunders.

LaMotte, L. R. and Hocking, R. R. (1970). "Computational Efficiency in the Selection of Regression Variables," *Technometrics*, **12**, 83-93.

LaMotte, L. R. (1972). "The SELECT Routines: A Program For Identifying Best Subset Regression," *Applied Statistics*, **21**.

Larsen, W. A. and McCleary, S. J. (1972). "The Use of Partial Residual Plots in Regression Analysis," *Technometrics*, **14**, 781-9.

Lave, L. B. and Seskin, E. P. (1970). "Epidemiology, Causality, and Public Policy," *American Scientist*, **67**, 178-86.

Loether, H. J., McTavish, D. G., and Voxland, P. M. (1974). **Statistical Analysis for Sociologists: A Student Manual**. Boston: Allyn and Bacon, Inc.

Mallows, C. L. (1973). "Some Comments on C_p," *Technometrics*, **15**, 661-75.

Mantel, N. (1970). "Why Stepdown Procedures in Variable Selection," *Technometrics*, **12**, 591-612.

Mansfield, E. R., Webster, J. T., and Gunst, R. F. (1977). "An Analytic Variable Selection Technique for Principal Component Regression," *Applied Statistics*, **26**, 34-40.

Marquardt, D. W. (1970). "Generalized Inverses, Ridge Regression, Biased Linear Estimation and Nonlinear Estimation," *Technometrics*, **12**, 591-612.

Marquardt, D. W. and Snee, R. D. (1975). "Ridge Regression in Practice," *American Statistician*, **29**, 3-20.

Mason, R. L., Gunst, R. F., and Webster, J. T. (1975). "Regression Analysis and Problems of Multicollinearity," *Communications in Statistics*, **4**, 277-292.

Massey, W. F. (1965). "Principal Component Regression in Exploratory Statistical Research," *Journal of the American Statistical Association*, **60**, 234-56.

McDill, E. L. (1961). "Anomie, Authoritarianism, Prejudice, and Socioeconomic Status: An Attempt at Clarification," *Social Forces*, **39**, 239-45.

McDonald, G. C. and Schwing, R. C. (1973). "Instabilities of Regression Estimates Relating Air Pollution to Mortality," *Technometrics*, **15**, 463-82.

McDonald, G. C., and Galarneau, D. I. (1975). "A Monte Carlo Evaluation of Some Ridge-Type Estimators," *Journal of the American Statistical Association*, **70**, 407-16.

McDonald, G. C. and Ayers, J. A. (1978). "Some Applications of the 'Chernoff Faces': A Technique for Graphically Representing Multivariate Data." in **Graphical Representation of Multivariate Data** (Wang, ed.). New York: Academic Press.

Meier, P. (1972). "The 1954 Field Trial of the Salk Poliomyelitis Vaccine," in J. M. Tanur et al., ets., **Statistics: A Guide to the Unknown**, San Francisco: Holden-Day, Inc. 2-13.

Merrill, S. (1977). **Draft Report on Housing Expenditures and Quality, Part III: Hedonic Indices as a Measure of Housing Quality**. Cambridge, Mass.: Abt Associates, Inc.

Mood, A. M., Graybill, F. A., and Boes, D. C. (1974). **Introduction to the Theory of Statistics**. New York: McGraw-Hill Book Co.

Mosteller, F. and Moynahan, D. (1972). **On Equality of Educational Opportunity**. New York: Random House.

Mosteller, F. and Tukey, J. W. (1977). **Data Analysis and Regression**. Reading, Massachusetts: Addison-Wesley Publishing Co.

National Center for Health Statistics (1970). "Natality Statistics Analysis, 1965-1967," *Vital and Health Statistics*, Series 20, No. 19, Washington, D.C.: U.S. Government Printing Office.

Pearson, E. S. and Hartley, H. O. (1969). **Biometrika Tables for Statisticians**. Cambridge: Cambridge University Press.

Pope, P. T. and Webster, J.T. (1972). "The Use of an F-Statistic in Stepwise Regression Procedures," *Technometrics*, **14**, 327-40.

Porter, J. (1974). "Race, Socialization, and Mobility in Educational and Early Occupational Attainment," *American Sociological Review*, **39**, 303-16.

Roberts, A. and Rokeach, M. (1956). "Anomie, Authoritarianism, and Prejudice: A Replication," *American Journal of Sociology*, **61**, 355-8.

Ruben, D. B. and Stroud, T. W. F. (1977). "Comparing High Schools With Respect to Student Performance in University," *Journal of Educational Statistics*, **2**, 139-55.

Schmidt, P. (1976). **Econometrics**. New York: Marcel Dekker, Inc.

Searle, S. R. (1971). **Linear Models**. New York: John Wiley and Sons, Inc.

Seber, G. A. F. (1977). **Linear Regression Analysis**. New York: John Wiley & Sons, Inc.

Sewell, H. W., Haller, A. D. and Ohlendorf, G. W. (1970). "The Educational and Early Occupational Status Attainment Process: Replication and Revision," *American Sociological Review*, **70**, 1014-27.

Srole, L. (1956). "Social Integration and Some Corollaries: An Exploratory Study," *American Sociological Review*, **21**, 709-16.

Swed, F. S. and Eisenhart, C. (1943). "Tables for Testing Randomness of Grouping in a Sequence of Alternatives," *Annals of Mathematical Statistics*, **14**, 66-87.

Tatsuoka, M. M. (1971). **Multivariate Analysis: Techniques for Educational and Psychological Research**. New York: John Wiley and Sons, Inc.

Theil, H. (1971). **Principles of Econometrics**. New York: John Wiley & Sons, Inc.

Toro-Vizcarrondo, C. and Wallace, T. D. (1968). "A Test of the Mean Square Error Criterion for Restrictions in Linear Regression," *Journal of the American Statistical Association*, **63**, 558-72.

Tukey, J. (1977). **Exploratory Data Analysis**. Reading, Mass.: Addison-Wesley Publishing Co.

U.S. Bureau of the Census (1961). **Historical Statistics of the United States, Colonial Times to 1957**. Washington, D.C.: U.S. Government Printing Office.

U.S. Bureau of the Census (1975a). "Estimates of the Population of the United States and Components of Change: 1974." *Current Population Reports*, Series P-25, No. 545, Washington, D.C.: U.S. Government Printing Office.

U.S. Bureau of the Census (1975b). "Projections of the Population of the United States: 1975 to 2050." *Current Population Reports*, Series P-25, No. 601, Washington D.C.: U.S. Government Printing Office.

U.S. Bureau of the Census (1976). "Population Profiles of the United States: 1975," *Current Population Reports*, Series P-20, No. 292, Washington, D.C.: U.S. Government Printing Office.

U.S. Department of Health, Education and Welfare (1976). "Projections of Education Statistics to 1984-85." National Center for Education Statistics, Washington, D.C.: U.S. Government Printing Office.

Webster, J. T., Gunst, R. F., and Mason, R. L. (1974). "Latent Root Regression Analysis," *Technometrics*, **16**, 513-22.

Wichern, D. A. and Churchill, G. A. (1978). "A Comparison of Ridge Estimators," *Technometrics*, **20**, 301-11.

Wood, F. S. (1973). "The Use of Individual Effects and Residuals in Fitting Equations to Data," *Technometrics*, **15**, 677-95.

INDEX

All possible regression procedure, 268-270, 275, 278, 306
Analysis of variance table, 154-158, 266-267, 322-326, 335-338, 346-348
Assumptions, 169-174
    verification, 231-241

Backward elimination procedure, 264, 278, 282-284, 291
Best possible regression procedure, 268, 270-272, 274, 291, 306
Beta weights, 75-77, 112, 130, 145-148, 159, 179, 194
Bias, 178-182, 201-202, 268
Biased estimator
    coefficient, 122
    σ^2, 200-201
    types, 317
Biased regression, 315-317

Categorical variable
    definition, 33
    Interpretation, 36-37
    linear dependent, 120
    specification, 34-36, 108
Causation, 17-18, 61-63, 92, 129
Chi-square
    distribution, 189, 191, 193, 195-196
    statistic, 195-196, 217-218, 374
    table, 378-379

C_k plot, 268, 272-275
Coefficient of determination, 82-83, 89-90, 157-158, 197
Computer usage, 128-131
Confidence interval
    coefficient, 203-204
    definition, 202-203
    elipsoids, 205
    mean response, 206-210
    regions, 205, 255
    σ^2, 204
Correlated errors, 231, 234-236
Correlation
    definition, 8-9
    estimator, 82-83, 185
    matrix, 114-115, 184
    time, 234-236
    two variable, 82, 93

Covariation, 150-151, 183-186
Covariance matrix, 182-183, 214-216
Crosstabs, 27

Data
    aberrations, 23-26, 32
    analysis, 1
    collection, 2-3
    deficiencies, 3-4
    editing, 23
    exploration, 19, 23
    inferences, 5-6
    missing, 25

Data sets
  A.1, 23-33, 45-47, 50-51, 72-74, 76-77, 97-
    100, 109-110, 107, 240-241, 243-247,
    252,258, 353, 358
  A.2, 34-41, 48-50, 107-108, 120, 146-148,
    153,154, 159, 180-181, 198-201, 353, 359
  A.3, 58-61, 84-85, 107-108, 115-120, 134-
    140, 150, 153, 232-236, 332-335, 353-
    354, 360
  A.4, 129-130, 145, 354, 361
  A.5, 167-169, 292-293, 299, 309, 355-362
  A.6, 174-177, 192-195, 197-198, 205,206,
    237-238, 241, 296-299, 301-304, 308,
    310-312, 355-356, 363-364
  A.7, 262-264, 266-267, 269-278, 280-286,
    290-291, 306-308, 356-357, 365-366
  A.8, 316-317, 321-322, 325-326, 328-329,
    332-333, 336-337, 340, 345-356, 349-350,
    367
  B.1, 368, 370-371
  B.2, 369,372
Degrees of freedom, 155-158
Durbin-Watson
  table, 388
  test, 234-236, 375

Error variance, 158-160
Estimator
  centered, 133-135, 142-243, 247-248, 156-
    157, 185, 204, 208, 212
  coefficient, 10, 52, 66, 85-88
  coefficient vector, 96, 132
  derivation, 85-87, 160-163
  distribution, 187-189, 216-218
  interpretation, 149-152
  moments, 178-186, 210-213
  normal deviate, 87-88, 144-145, 147-148,
    204, 208, 212
  properties, 174-189
  σ^2, 83-85, 158-160, 189
  unit length, 144-148, 156-159, 204, 208,
    212-213, 318,320
Examples
  average annual salary, 20-23
  Coleman report, 5-6
  faculty survey, 3, 5
  height and weight, 16-18
  Kinsey report, 4
  mortality and pollution, 6-7

occupational attainment, 11-12
population projections, 13-16
prejudice, 92-94
Salk vaccine, 1-2
suicide rate, 61-66, 70-72
tuition data, 53-55, 68-70, 79-82, 209-
  210
Expectation, 178-183, 210-213
Experimental design, 142
Exponential fit, 221-223
  function, 43-44
Extrapolation, 12-16, 33, 62-63, 109, 310
Eyeballing fit, 56

Fit
  by stages, 135-140, 161-163
  measures of, 77-85, 154-160
Forward selection procedure, 263-265, 278-
  282, 284
F statistic, 195-198, 202-206, 255-256, 266-
  267, 270, 278-286, 324-326, 274
  distribution, 196
  table, 380-385

Generalization, 15-17, 93, 109

Heteroscedasticity, 236-239
Hypothesis testing
  coefficients, 189-206
  mean response, 196
  σ^2, 196

Inference, 167
Interaction
  definition, 37-39
  multicollinearity, 292-293
  plots, 39, 242-243
  specification, 39
Intercept
  centered, 143
  normal deviate, 144
  raw, 142-143
  unit length, 144
Interrelationships, 92-94, 104, 152, 194

Lack of fit test, 198-202
Latent roots & vectors, 104-107, 118-120,
  124-126, 299-301, 318-321, 329-332

Latent root regression
  analysis of variance, 335-338
  diagnostic, 330-335
  distributional properties, 332, 338-340
  estimator, 332
  F statistic, 336-339
  hypothesis test, 338-340
  motivation, 330-332
Lawlike relationships, 16
Least squares
  geometry, 174-178
  principle, 66-71, 131
Leverage values, 226-227, 256-257
Linear
  dependences, 106-108, 115-120, 151
  independences, 102
Logarithmic transformation, 43-44, 222, 246

Matrix
  addition, 97
  algebra, 94-108, 132
  alias, 179
  definition, 95
  determinant, 102, 106, 122-126
  diagonal, 100
  identity, 101
  inverse, 100, 106, 122-126
  multiplication, 97-100
  multiplication, 97-100
  nonsingular, 101
  null, 100-101
  operations, 96-102
  orthogonal, 105, 140
  rank, 106-108
  singular, 101
  square, 100
  symmetric, 104
  transpose, 96
Misspecification bias, 108-110
Model
  assumptions, 71-72, 169-174
  building, 48-49
  centered, 142-143
  interpretation, 150-152, 315
  misspecification, 33, 179, 206, 224-225,
    241-242, 255, 265-266
  multiple variable, 94-96, 103, 131-132, 142,
    170-171, 188, 202, 265-266
  no-intercept, 63-66, 148-149
  single variable, 56
  specification, 7-8, 10, 33, 131, 198, 265

Multicollinearity
  coefficients, 293, 299-301
  definition, 39, 115-116
  detection, 118-120
  effects, 152-153, 263, 290, 293, 299-301,
    315-317
  estimator correlation, 296, 298
  estimator covariance, 296, 298
  estimator variance, 294-295, 298-299
  exact, 120
  nonpredictive, 319, 332-335
  population, 120-122, 308, 315
  prediction, 310-312
  predictive, 329-330, 332-335
  remedial steps, 122, 303-304, 315
  sample, 120-122, 308, 315
  test statistics, 302-308
  variable selection, 304-308
Multivariate normal distribution, 188-189,
  216-217

Near-neighbor estimate, 202
No-intercept model, 63-66, 70-71, 148-149
Normal probability plot, 239-241, 247, 263

Orthogonal predictors, 140-142
Outlier
  definition, 25, 252
  deletion, 28-33
  detection, 220, 228-231, 252-258
  test, 255-258
Overspecification, 110-111

Parameter estimation, 11-12, 67, 131
Partial residual plot, 247-252
Polynomial regression, 42-44, 281
Prediction, 9-10, 48-49
Prediction equation
  single variable, 66
  multiple variable, 10, 103-104, 131-132
  no-intercept, 70, 148-149
Prediction interval, 207-208
Principal component regression
  analysis of variance, 322-326
  deleted components, 323
  deletion strategy, 324-328
  estimator, 319, 323
  F statistic, 324-326
  hypothesis tests, 323-329
  moments, 320-321
  motivation, 318-322
  procedure, 324-325
  variable elimination, 328-329

Pure error, 200

Randomness check, 232-234
Random predictor, 172-173, 181-183
Redundancy, 110-111, 115, 290
Regression
    abuses, 12-18
    analysis, 6-9
    approximation, 61-63
    empirical verification, 61
    linearity, 8
    multiple variable, 94-95, 103-104, 128-131
    partitioned, 266-268
    single variable, 52-53, 56
    theoretical model, 57-61
    uses, 9-12
Repeated predictor values, 198-202
Residual
    adjusted, 136-140, 152-153
    analysis, 220
    average, 67-69
    definition, 20, 66, 225-226
    distribution, 228
    expectation, 226, 228
    histogram, 252-253
    variability, 83-85
    variance, 226, 228
    vector, 132
Residual plot
    definition, 220-221
    categorical variable, 242-243
    cross-product term, 242-243
    deleted residual, 254-255
    observed response, 237
    predicted response, 236-238, 247, 252
    predictor variable, 234-235, 242-247, 252
    squared residual, 237-239
    time, 221-222, 234-236
Residual type
    deleted, 225-231, 258-259
    partial, 247-252
    raw, 225-226, 242
    standardized, 228
    studentized, 228, 230
Response interval, 206-210
Response variable distribution, 207
Ridge regression
    analysis of variance, 346-348
    estimator, 341
    F statistic, 347
    moments, 342-343

motivation, 341-343
parameter, 341-342
parameter selection, 343-346
procedure, 340-341
trace, 343-346
t statistic, 348
Runs
    table, 386-387
    test, 232-234, 375

Sample variance, 83-84
Scatter plot, 21, 28-32, 39, 118, 173
Serial correlation test, 234-236
Significance level, 197, 205-206, 256-257,
    280, 303, 327-328
Simultaneous prediction interval, 204-205
Smoothing, 39-42
Specification
    categorical variables, 33-37
    initial, 225, 241-243
    interaction terms, 37-39
    predictor variable, 45-47
    problems, 93-94
    response variable, 47-48
Standardization
    benefits, 112-113, 227-228
    centered, 133, 142-143, 148, 156-157,
        185, 192, 204-205, 212, 293
    definition, 74-76
    effects, 77
    normal deviate, 74-75, 114-115, 144-146,
        148-149, 185, 192, 205, 212, 293
    rationale, 72-73
    raw, 142-143, 145, 147, 293
    response variable, 75-76, 112, 145
    unit length, 75-76, 114-118, 144-149,
        156-157, 185, 192, 205, 212, 293-301
Stepwise selection procedure, 278, 284-287
Straight line fit, 52
Sum of squares
    adjusted regression, 266-267, 286
    error, 79-81, 155-157, 194, 196, 267
    mean, 78-81, 155-157
    regression, 79-81, 155-157, 164, 195-196,
        266-267
    total, 78-81, 88-89, 155-157

t-directed search procedure, 275-278, 305
t distribution, 189, 218
    statistic, 190-195, 207, 209, 218, 224-231,
        275-278, 327, 330, 339,340, 348, 374
    table, 377

Transformations
  predictor, 45-47
  response, 41, 47-48, 237
  types of, 41-44

Unbiasedness, 178-180, 315
Unequal variances, 236-239

Variability, 78, 154-160
Variable
  deletion, 120-122
  selection, 262-264, 268, 278
Variance inflation factor, 295-299, 321, 345-
    346
Variable selection criteria
  C_k, 268, 270, 272-277, 304
  F, 267, 279-286
  MSE, 267, 270, 272, 279, 283, 304
  R^2, 263-264, 267, 269-272, 304

Vector
  addition, 97
  definition, 95
  expectation, 178
  linear combination of, 97-98
  linear independence, 102
  multiplication, 97-101
  normalization, 105
  orthogonal, 105
  orthonormal, 105
  subtraction, 97

$X'X$, 103-104

Z distribution, 171, 187-189, 203, 373
  statistic, 190-191, 203, 207-208, 228, 233,
    376
  table, 373